GAMMA-RAY ASTRONOMY

GEOPHYSICS AND ASTROPHYSICS MONOGRAPHS

AN INTERNATIONAL SERIES OF FUNDAMENTAL TEXTBOOKS

VOLUME 14

GAMMA-RAY ASTRONOMY

Nuclear Transition Region

by

E. L. CHUPP

University of New Hampshire, Dept. of Physics,
Durham, N.H. 03824, U.S.A.

D. REIDEL PUBLISHING COMPANY

DORDRECHT-HOLLAND/BOSTON-U.S.A.

Library of Congress Cataloging in Publication Data

Chupp, E L 1927–
Gamma-ray astronomy: nuclear transition region.

(Geophysics and astrophysics monographs; v. 14)
Bibliography: p.
Includes index.
1. Gamma ray astronomy. 2. Nuclear astrophysics.
I. Title. II. Series.
QB471.C47 522′.6 76–21711
ISBN 90–277–0695–6
ISBN 90–277–0696–4 pbk.

Published by D. Reidel Publishing Company,
P.O. Box 17, Dordrecht, Holland

Sold and distributed in the U.S.A., Canada and Mexico
by D. Reidel Publishing Company, Inc.
Lincoln Building, 160 Old Derby Street, Hingham,
Mass. 02043, U.S.A.

Printed in The Netherlands

Dedicated to my family

Mary, Timothy, Christine

and Geoffrey

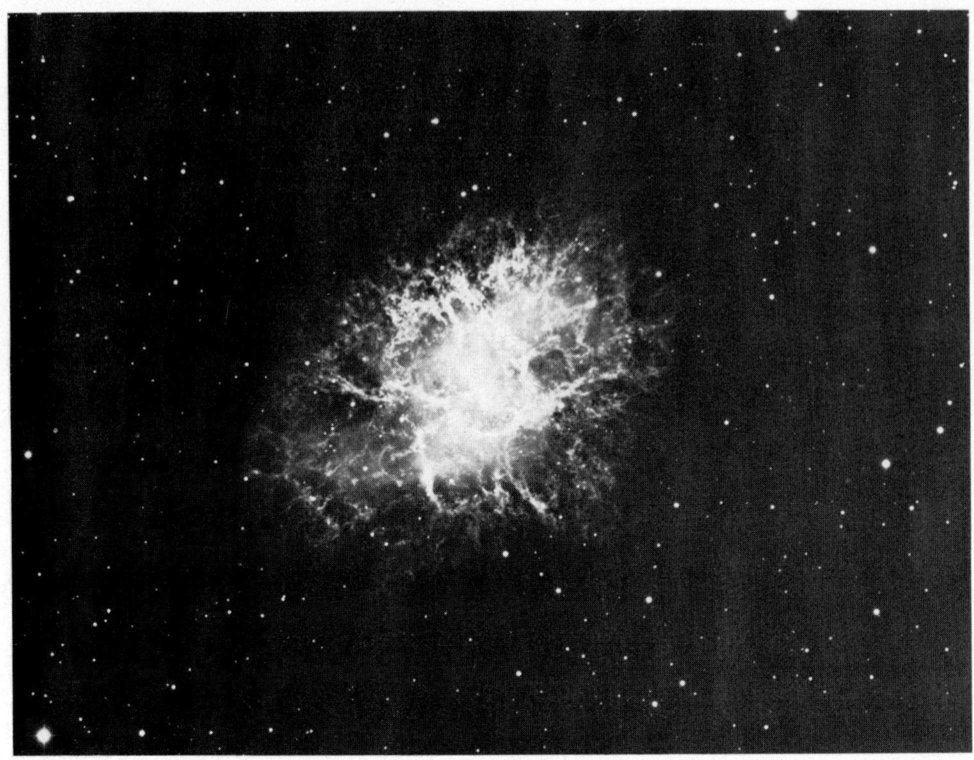

This object in our galaxy is said to embody all of astronomy (and possibly physics). The emerging field of gamma ray line astronomy could provide direct experimental evidence that nucleosynthesis had occurred here in 1054 A.D. and that here cosmic rays are currently being produced. (*NGC 1952 'Crab' Nebula in Taurus. Messier 1. Taken in red light. Remains of supernova of AD 1054. 200-inch photograph.* Used by permission Hale Observatories.)

TABLE OF CONTENTS

PREFACE

Observation of discrete energy electromagnetic emissions from celestial objects in the radio, IR, optical, UV, and X-ray spectral regions has dramatically advanced our knowledge in the field of astrophysics. It is expected that identification of nuclear γ-ray line emissions from any cosmic source would also prove to be a powerful new tool for probing the Universe.

Since the publication of Morrison's work in 1958, many experiments were carried out searching for evidence of γ-ray lines from cosmic sources, however with little success. Only a few positive experimental results have been reported, in spite of an expenditure of considerable effort by many people: in particular, the possible Galactic Center emission line (473 to 530 keV) and γ-ray lines at several energies (e.g., 0.5 MeV and 2.2 MeV) associated with large solar flares. Both of these observations are unconfirmed by independent observations (ca. 1975). The high energy γ-rays ($>$30 MeV) from the Galactic Center are at least partly due to the decay of π^0 mesons, which are of unique energy (67.5 MeV) in the π^0 rest frame only. The reasons for the limited amount of data available in this field, even though early theoretical predictions were very optimistic regarding fluxes of nuclear lines, are that experimental efforts are plagued with high backgrounds and low fluxes, and that development of instruments with telescopic properties in the energy range of interest is difficult. Also, detection of cosmic γ-ray line fluxes from many sources may be beyond the capabilities of most instruments available today.

This monograph then is to provide a source of information in the field of γ-ray astronomy in the energy region 10 keV to $\sim$100 MeV, where nuclear lines are expected. Specifically, γ-ray lines from ^{57}Fe at 14.1 keV to π^0 decay γ-rays at 70 MeV bracket the major energy range of direct interest. Consideration must also be given to the sources of continuous spectra that radiate in this energy region.

The author believes that a review of the major theoretical and experimental efforts made in the past twelve years is essential. It is hoped that the collection of references and basic data contained in this monograph will be a helpful starting point for experimentalists and new workers in the field.

Durham, New Hampshire, July 1975

ACKNOWLEDGMENTS

Acknowledgments are usually an inadequate measure of the debt incurred. This author feels strongly the need to express his gratitude to several persons and organizations, knowing full well that his words will be insufficient and that it is impossible to acknowledge everyone.

First, there would be no necessity at all for an Acknowledgment page if it were not for my wife, Mary, who provided that essential encouragement whenever I was about to falter and who also gave of her wisdom, time, and energy in the many editorial tasks required.

The work on this monograph began while the author was a guest at the Max-Planck Institute for Physics and Astrophysics, Institute for Extraterrestrial Physics, Garching, F.R.G., during 1972–1973, supported by an award from the Alexander von Humboldt Stiftung and by a Fulbright Hayes Senior Fellowship. The work was completed with the partial support of NASA through research grant NGL 30-002-021 and the University of New Hampshire at Durham.

I am much indebted to Professors Reimar Lüst and Klaus Pinkau for their hospitality, kindness, and their time, and for providing a stimulating intellectual environment while I was in Germany. Many people have given me their support in this effort but I want to thank, particularly, Drs Reuven Ramaty, Philip Dunphy, Gottfried Kanbach and Amar Suri for their assistance and valuable suggestions in reading and editing various drafts of the manuscript. Dr David Forrest has also been a most appreciated critic on many experimental questions. The author, of course, is solely responsible for any errors that may exist in this monograph and for any inaccuracies or infelicities in describing the work of colleagues.

LIST OF SYMBOLS

a_0	1st Bohr radius $= \hbar^2/m_0 e^2 = 0.53 \times 10^{-8}$ cm.
$B_\perp$	component of magnetic field strength (G) perpendicular to particle motion.
E, E_γ	energy of particle or γ-ray.
K_e	coefficient of differential electron spectrum.
k	Boltzmann's constant.
m_0	electron's proper mass.
m, M	proper mass of particle other than electron.
n_{ph}	photon number density.
r_0	classical radius of the electron $= e^2/m_0 c^2 = 2.82 \times 10^{-13}$ cm.
r, R	distance to source volume element.
T	absolute temperature in degrees Kelvin.
v	velocity of a particle.
α	exponent of a differential spectrum of electrons of the form $K_e E^{-\alpha}$, $\eta_0 E^{-\alpha}$, or $\eta_0 \gamma^{-\alpha}$ or fine structure constant $= 1/137 = 2\pi e^2/hc$.
γ	Lorentz factor or symbol for a γ-ray.
ϵ	photon energy $\epsilon = h\nu$ or $E_\gamma = h\nu$.
η_0	coefficient of differential electron spectrum.
$\kappa \equiv {}_a\sigma_p$	pair production cross-section.
μ	electron rest energy $= m_0 c^2$.
ν	frequency of a photon.
σ_T	Thomson cross section $= 8\pi/3 \cdot r_0^2 = 6.65 \times 10^{-25}$ cm^2.
σ_0	$1/137 (r_0)^2 \rightarrow$ bremsstrahlung cross section unit, or $\pi r_0^2 = 2.5 \times 10^{-25}$ cm^2.
τ	optical depth for radiation or photoelectric absorption coefficient.
${}_a\tau_K$	photoelectric absorption coefficient (K shell).

INTRODUCTION

The full expanse of the cosmic γ-ray energy spectrum covers roughly sixteen orders of magnitude of energy from $\sim 10^5$ eV to 10^{21} eV, though in this monograph we discuss only the three decades of energy from $\sim 10^5$ eV to 10^8 eV, where nuclear γ-ray lines are expected. The most recent comprehensive review of the general fields of X-ray and γ-ray astronomy has been given by Greisen (1971) and this should be consulted for a broad perspective of the astrophysical questions. Topics in the field of greatest current experimental and theoretical interest were discussed recently at a Goddard Space Flight Conference (Stecker and Trombka, 1973). Stecker (1975) has reviewed the theoretical and experimental work concerned with energetic γ-rays ($>$ 30 MeV). A recent textbook (Harwit, 1973) is recommended for reference on fundamental astrophysical questions at the advanced introductory level.

It is clear that the ultimate task in this field is the experimental determination of the characteristics of the cosmic and solar γ-ray spectra in the nuclear transition region, since this is the only means by which theoretical predictions can be tested.

1.1. Brief History of Attempts to Detect Celestial γ-Rays $<$ 50 MeV

In 1895 when Roentgen, by accident, discovered X-rays while investigating the UV light given off by electrical discharge tubes, atomic spectroscopy was already sufficiently developed to be applied to the study of the spectra of the stars. Becquerel's (1896) later discovery of radioactivity while studying the phosphorescence of U and K salts exposed to light was identified by Rutherford (1899) as being composed of charged α and β particles. Villard (1900) showed that there was also emitted a penetrating radiation, not deflected by a magnetic field, which was later named γ-radiation. It was not until 1914 that Rutherford and Andrade measured the wavelength of these γ-rays with crystalline diffraction techniques establishing that some had discrete energies. The theoretical models describing the γ-ray line spectra that result from transitions between excited nuclear states have been long established. However, in contrast to atomic spectroscopy, nuclear spectroscopy is yet to be proven a viable tool in the study of cosmic phenomena.

In a sense, the conception of γ-ray line astronomy can be marked by the publication of theoretical predictions by Morrison (1958) of strong γ-ray line fluxes from several cosmic sources. This paper stimulated considerable experimental activity; however, several exploratory experiments were conducted before 1958. Even though designed to search for an extraterrestrial γ-ray flux, the experiments did not have precise energy determination capability and were mostly designed for photons with energies greater than 50 MeV; however, they should be noted because of their historical significance.* Probably the first cosmic γ-ray experiment was carried out in 1948 by Hulsizer and Rossi (1949) using a balloon-borne ionization chamber. They demonstrated that γ-rays (and electrons) with energies >1 GeV comprised less than 1% of the incoming cosmic ray flux. Critchfield *et al.* (1952) arrived at the same conclusion after carrying out a cloud chamber experiment. In the low energy range of <50 MeV, the first experiments by Rest *et al.* (1951) and Perlow and Kissinger (1951) used Geiger-Mueller tubes in short rocket flights.

Early experiments with instruments capable of providing spectral information were carried out by Anderson (1961), Jones (1961), and Vette (1962), all with alkali-halide scintillation counters. Only the detector of Jones (1961), in which CsI(Tl) crystals were used in a phoswich arrangement, thus eliminating charged particle events, could be considered as having the capability of detecting spectral features. The energy loss spectrum in this detector was measured in the energy range between 0.1 and 2.4 MeV with a resolution of ~ 70 KeV/channel; however, this was not sufficient to resolve a spectral line at 0.511 MeV. In addition, the presence of a very large local background γ-ray spectrum produced in a passive Pb collimator limited the sensitivity of this experiment.

The first experiments which gave encouraging evidence for an extraterrestrial γ-ray flux were those carried out on two Ranger spacecraft in *cis*-lunar space and reported by Arnold *et al.* (1962) and Metzger *et al.* (1964). The detectors used in these observations were CsI(Tl) spectrometers in 4π phoswhich charged particle shields. Since a 32-channel pulse height analyzer was used to study the energy loss spectrum in the range 70 keV to 4.4 MeV in two gain modes, the instrument could resolve individual nuclear γ-ray lines if any were present. Inflight calibration sources of ^{57}Co and ^{203}Hg were also provided, and the detector could be extended from the spacecraft on a 6-ft boom, so the effect of spacecraft background could be evaluated. The most significant result of this experiment was the evidence for a diffuse celestial γ-ray spectral flux which can be expressed as (Greisen, 1966a)

$$\mathrm{d}n(E) = 0.012\,E^{-2.2}\,\mathrm{d}E\ (\text{photons cm}^{-2}\ \text{s}^{-1}\ \text{sr}^{-1})$$

where E is the γ-ray energy in MeV.

The Ranger observations were also able to provide the first significant upper limits to the important spectral lines at 0.51 MeV and 2.23 MeV due to positron-electron annihilation and neutron-proton capture, respectively. The upper limit results were 0.014 photons cm^{-2} s^{-1} and 0.005 photons cm^{-2} s^{-1} for the 0.51 MeV and 2.23 MeV lines, respectively. These upper limits apply to all discrete (point) sources and any diffuse source contribution to the lines. Further deep space measurements on Apollo 15 (Trombka *et al.*, 1973) have extended the Ranger γ-ray spectrum above 1 MeV. However, at the present time (ca. 1975) the true extraterrestrial nature of this radiation can be questioned.

* For a review of the history of γ-ray astronomy before 1966, see Greisen (1966a).

While our primary interest is in photon energies <20 MeV, it should be mentioned that prior to 1966 the most sophisticated experiment carried out to investigate the higher energy photons (>120 MeV) was that of Kraushaar *et al.* (1965) on the Explorer 11 satellite. The instrument consisted of scintillation and Cerenkov detectors having an angular resolution of $\pm 20^\circ$. These properties made it possible to determine that the atmospheric flux upwards from the Earth was about a factor of 10 greater than the downward flux from the sky and that the flux from the Earth's horizon was as much as a factor of 20 higher than the apparent flux from the sky. A subsequent experiment on the OSO-3 satellite demonstrated convincingly that photons with energies greater than 50 MeV were emanating from the galactic disk in proportion to the H gas concentration (Clark *et al.*, 1968). A portion, or all, of this radiation may be a result of the decay of π^0 mesons produced by charged cosmic ray interactions in the interstellar gas and, therefore, may represent the first observation of celestial γ-rays from nuclear reactions. This observation has now been confirmed by the SAS-2 and TD-1 satellite experiments, as well as several balloon experiments (see also Fazio, 1970; Clark, 1971 for reviews of the OSO-3 observations and also Section V-5.2.2).

In the lower energy region (<10 MeV), a large number of experiments were conducted since 1961 using high resolution NaI(Tl) scintillation spectrometers carried aloft by high altitude balloons with the capability of detecting γ-ray lines. In the first of these experiments by Peterson (1963) there was no evidence of cosmic γ-rays; however, a strong line at 0.51 MeV was measured at an intensity of 0.62 ± 0.06 photons cm^{-2} s^{-1} over Minnesota at a geomagnetic cutoff of 3.0 GV and at an atmospheric depth of 6.0 g cm^{-2}. This line, thought to come from the atmosphere, can always be seen in γ-ray experiments when instruments of sufficient energy resolution are used. Several subsequent experiments have recorded this line, but the search for variation in intensity of this line indicating a flux from the Sun or other celestial sources had met with no success (cf. Frost *et al.*, 1966; Rocchia *et al.*, 1965; Chupp *et al.*, 1970; Nakagawa *et al.*, 1971; Kasturirangan *et al.*, 1972). The 0.511 MeV line flux is lowest near the geomagnetic equator where it has a value of 0.08 ± 0.01 photons cm^{-2} s^{-1} at 7.6° N at an atmospheric depth of 6.0 g cm^{-2}. This value is 5 to 6 times higher than the upper limit celestial flux at the same energy recorded by the Ranger 3 and 5 flights. This fact clearly illustrates the difficulty of searching for celestial γ-ray lines in experiments conducted near the Earth, since one must shield against this high background.

The first evidence for an extraterrestrial γ-ray line (from the direction of the Galactic Center at ~0.5 MeV) was reported by Johnson *et al.* (1972) and reportedly confirmed in later flights by Johnson and Haymes (1973). The flux of this apparent line is $\sim 1 \times 10^{-3}$ photons cm^{-2} s^{-1} or less (Haymes *et al.*, 1975).

Possible evidence for solar γ-ray lines, but not at any specific energy, was reported by Hirasima *et al.* (1969). In August 1972 the first direct evidence for specific γ-ray lines associated with solar flares was reported by Chupp *et al.* (1973a, b). Strong emissions were seen at 0.51 and 2.23 MeV with weaker fluxes at 4.4 and 6.1 MeV. The average flux values recorded during the rising phase of an intense solar flare were, respectively: 0.065, 0.28, 0.03, and 0.03 photons cm^{-2} s^{-1}.

The latest and most dramatic development in γ-ray astronomy has been the announcement in June 1973 of the discovery on Vela satellites of intense bursts ($\sim 10^{-4}$ erg cm^{-2} s^{-1}) of extraterrestrial γ-rays (Klebesadel *et al.*, 1973). The average photon energy is typically ~150 keV, but the spectra are known to extend to higher energies ($\gtrsim$ 500 keV) in some cases. These bursts were observed as early as 1967 and, as yet, their origin is completely unexplained. None of the Vela instruments observing these bursts was capable of determining whether any γ-ray line structure was present (see Section V-5.4).

Further experiments are in progress by several experimental groups to search for γ-ray lines from any celestial source up to energies of 10 MeV. Perhaps by the time this volume is published there will be further evidence of nuclear lines from specific celestial sources.

1.2. Astrophysical Significance of γ-Ray Line Astronomy

Though γ-ray lines from celestial sources were predicted by Morrison as early as 1958, unexpectedly, cosmic X-rays were *first* discovered in 1962 by Giaconni *et al.* Morrison (1966, 1967) has reviewed the arguments, in retrospect, why X-ray sources might have been expected to be more intense than γ-ray sources. The essential conclusion is that all conceivable sources give a differential intensity spectrum falling off strongly with increasing photon energy, so the stronger X-ray fluxes are more easily detectable. X-ray astronomy is now a flourishing field, and the association of variable X-ray sources with binary systems may have led to one of the most profound recent discoveries in astronomy, the existence of Black Holes. Advances in γ-ray astronomy have, however, been limited by a formidable trio of experimental problems: (1) high instrument background, (2) extremely low fluxes, and (3) lack of a true γ-ray telescope which functions in the energy region 50 keV to 10 MeV.

Why then the continued interest in γ-ray line astronomy which has progressed at a snail's pace in terms of accomplishments and which presents such experimental difficulties? Certainly there is the experimental challenge to build more elegant experiments, but there is much more. Gamma-ray lines by themselves have two basic characteristics which give them a unique power in probing the Universe. The first, their discrete energy can identify specific nuclear processes at work in the cosmos, and the second, their penetrability of matter allows us to probe back to the earliest epoch of the Universe, possibly to $z \sim 100$ depending on the density of matter in the Universe. To be more specific, we can list several important astrophysical questions that can be directly probed using γ-ray line astronomy as a tool:

1. The site of nucleosynthesis;
2. The location of any discrete sources of cosmic rays;
3. The nature of exotic sources such as QSO's and Radio Galaxies;
4. The conditions in the Universe at an early cosmological epoch;
5. The possible existence of antimatter in the Universe;
6. The properties of the solar flare particle accelerating mechanism; and
7. The nature of the Vela bursts.

Certainly these are among the most fundamental questions in astrophysics. We must therefore consider the prospects for studying these problems with γ-ray line astronomy.

In the following text we begin with a summary (Chapter II) of the important mechanisms which produce γ-rays in the energy range of interest, followed in Chapter III by a discussion of some of the more significant theoretical predictions in the field. The status of the observational work in the field in Chapter V is prefaced by a review of the interactions of γ-rays with matter in Chapter IV. In the last chapter (VI), experimental problems and some future experimental directions are summarized.

CHAPTER II

MECHANISMS FOR γ–RAY LINE AND CONTINUUM PRODUCTION

There are a large number of physical mechanisms which give rise to photons in the nuclear line energy region. These basic mechanisms must then be applied in the vast number of possible astrophysical settings in order to estimate the expected photon flux and spectral characteristics at the observer. The emphasis here is on identifiable nuclear lines; however, the actual astronomical spectra observed will be a mixture of photons from both continuum and line (or discrete) spectra. We shall use throughout this discussion the term 'line' to refer to γ-rays of discrete energy unless otherwise noted as, for example, in the case of a one dimensional source distribution such as the galactic disk line source (Section III-3.2.4)

The general problem in interpreting any cosmic γ-ray measurement is to relate the measured γ-ray spectrum at the Earth to the physical conditions in the cosmic source region. Consider a unit volume in the cosmic γ-ray production region at a location relative to the Earth specified by the radius vector **r** to the volume element. The directional intensity of γ-rays of energy E_γ to $E_\gamma + \mathrm{d}E_\gamma$ from this volume element may be expressed as:

$$I_\gamma(E_\gamma)\mathrm{d}E_\gamma = \frac{\mathrm{d}E_\gamma}{4\pi} \int \mathrm{d}r\, q(E_\gamma, \mathbf{r}) \exp\left[-\int_0^{\mathbf{r}} \mathrm{d}r'\, \kappa(E_\gamma, \mathbf{r}')\right]$$

$$(\text{photons cm}^{-2}\,\text{s}^{-1}\,\text{sr}^{-1}) \qquad \text{(II.1)}$$

where $q(E_\gamma, \mathbf{r})$ is the number of γ-rays produced with an energy E_γ to $E_\gamma + \mathrm{d}E_\gamma$ per unit volume and time at **r**, and therefore has the units (photons cm^{-3} s^{-1} (unit energy)$^{-1}$). The argument of the exponential term accounts for any absorption of the γ-rays in the intervening space between the source element and the Earth, where $\kappa(E_\gamma, \mathbf{r})$ is the absorption coefficient of the medium per unit path length (see Sections II-2.5.3 and IV-4.2). The γ-rays are assumed to be emitted isotropically.

In this chapter, the production mechanisms which can contribute to $q(E_\gamma, \mathbf{r})$ are considered. For the case of two interacting nuclei, j and k, the volume production rate may be written as

$$q(E_\gamma, \mathbf{r}) = 4\pi \sum_{\mathrm{j,k}} n_\mathrm{j}(\mathbf{r}) \int \mathrm{d}E_\mathrm{k}\, \sigma_{\mathrm{jk}}(E_\gamma \,|\, E_\mathrm{k}) I(E_\mathrm{k}, \mathbf{r})$$

$$(\text{photons cm}^{-3}\ \text{s}^{-1}\ \Delta E_\gamma^{-1}) \qquad \text{(II.2)}$$

so the total source function is found by summing over all species of the ambient medium, j, and energetic cosmic ray nuclei of type k. $n_\mathrm{j}(\mathbf{r})$ (cm^{-3}) is the number density of the target nuclei at location **r**, and $I(E_k, \mathbf{r})$ is the directional cosmic ray intensity (cm^{-2} s^{-1} sr^{-1}) at location **r** for species k with an energy E_k. The cross section $\sigma_{jk}(E_\gamma | E_k)$ (cm^2/unit energy) gives the probability that the energetic particle, k, interacting with ambient species, j, will produce a γ-ray of energy E_γ to $E_\gamma + dE_\gamma$. This cross section is, in turn, a sum over all the secondary processes (or particles) which are involved in the primary interaction between particles j and k.

It is implied in the above relations that the charged particle flux is isotropic in the source volume and that the resulting γ-ray spectrum given by the interactions is also emitted isotropically. Within the constraints implied by these assumptions, Equations (II.1) and II.2) are the two basic relations that are needed to relate any experimental observation to physical parameters in the source region. On the other hand, the fundamental theoretical problem is the calculation of $q(E_\gamma, \mathbf{r})$. Also, the equations must be modified if cosmological effects are important.

2.1. Basic Mechanisms and Sources

There are many ways to classify the processes that give rise to γ-ray photons. For example, Fazio (1967) has grouped the mechanisms into two classes:

(a) Quantum state transitions involving nuclei, baryons, and leptons, where the transition is caused by the electromagnetic interaction or the strong interaction.

TABLE II-1

Physical mechanisms for γ-ray production and modification

Black body radiation field

Particle-field interaction:

- Compton effect
- Magnetobremsstrahlung
- Photomeson production ($\pi^0 \to 2\gamma$)
- "Curvature radiation"

Particle-matter interaction:

- Bremsstrahlung
- π^0 production by proton-proton interactions[a]
- π^0 production by proton-antiproton annihilation[a]
- Nuclear transitions[b]
- Fission[b]
- $e^+ - e^-$ annihilation[b]
- Capture γ-rays (e.g. $H^1(n, \gamma)D^2$)[b]

Energy modifying mechanisms:

- Direct Compton scattering of all the above radiations in the source
- Doppler, gravitational and cosmological shifts

a Broad line structure.

b Sharp line structure possible.

(b) Matter-antimatter annihilation of nucleons, baryons, and leptons, which produces γ-rays directly or indirectly through π mesons giving γ-rays through $\pi^0 \to 2\gamma$ or positron-electron annihilation.

We have chosen here for convenience, a method used by several authors (cf. Ginzburg, 1969; Lüst and Pinkau, 1967), which essentially classifies the γ-ray production mechanisms as particle-field or particle-matter interactions. This scheme is summarized in Table II-1.

The standard treatment for the basic physical mechanisms involved in producing γ-ray radiation through electro-magnetic interactions of electrons and positrons is that of Heitler (1954). This authoritative work does not, however, treat the sources of γ-rays originating in nuclear interactions. Our summary of source mechanisms is, therefore, taken basically from many recent references, particularly Fazio (1967), Lüst and Pinkau (1967), Boldt (1969), Ginzburg (1969), Stecker (1971), and Greisen (1971). However, in the discussion that follows reference will be given where appropriate to more fundamental work.

Figure II-1 shows a schematic representation of the photon energy region where the different source mechanisms listed in Table II-1 can contribute photons. Each source mechanism indicated in the figure is described in the following sections of this chapter. The reference to the basic equation giving the characteristic γ-ray energy produced by each mechanism is shown by the equation number at the right side of the figure. From this figure it can readily be seen that the mechanisms which produce a continuous spectrum of photons can, in principle, overlap the energy region where the γ-rays of discrete

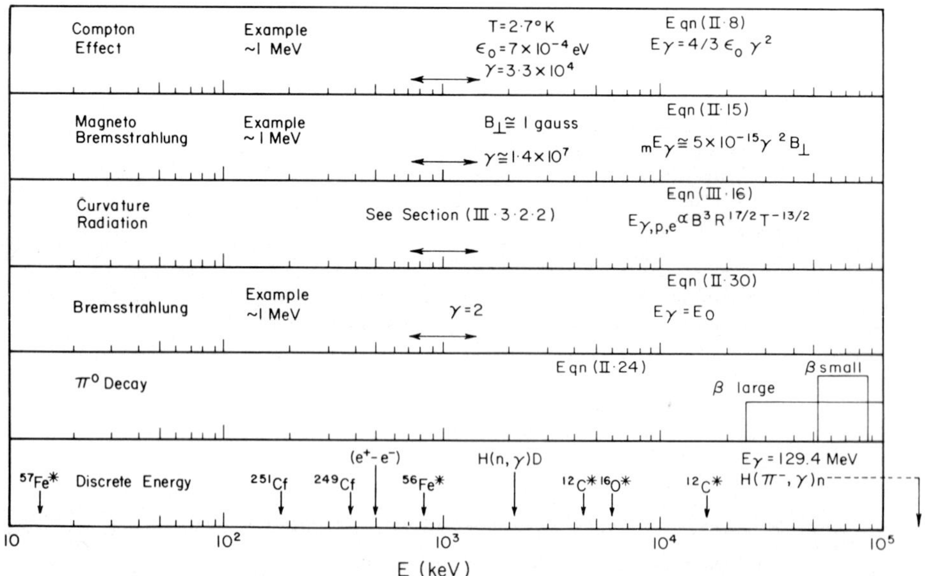

Fig. II-1. The photon energy regions where different γ-ray source mechanisms can contribute are shown on a logarithmic scale.

TABLE II-2

Astrophysical sites where γ-ray source mechanisms may operate

Sun (solar activity or quasi-continuous production)

Cosmic (point or localized sources)

- Supernovae and their remnants
- Neutron stars – pulsars
- Flare stars
- Galactic core and disk
- Radio sources – QSO's – black holes

Cosmic (diffuse sources)

- Continuous spectrum
- Cosmological π^0 and H(n, γ)D
- Low energy cosmic rays in galaxy

Planetary atmospheres

energy are present. In Table II-2 are listed the astrophysical sites where, again in principle, the several γ-ray production mechanisms can operate. Predictions that have been made for the γ-ray flux at the Earth from several of these sources are discussed in Chapter III.

2.2. Thermal Radiation Field

An astrophysical object sufficiently hot, behaving as a black body (optically thick source) can give photons in the energy range of interest to nuclear γ-ray line astronomy. The temperature range required for γ-ray energy photons due to this mechanism is $10^8 \mathrm{K} < T < 10^{10}$ K. At the lower extreme, solar flare temperatures as high as this have been suggested to account for some hard solar X-ray bursts (Chubb *et al.*, 1966). At the upper extreme, one is dealing with temperatures appropriate to the early history of a big bang Universe or to temperatures occurring in explosive cosmic sources. In the present epoch a cosmological black body radiation spectrum would be strongly modified by the red shift.

For reference here, we give the unshifted thermal photon number spectrum for a black body source in terms of the photon energy. The number of photons emitted from a unit area of a black body per second at temperature T(K) according to the Planck formula is

$$N(E_\gamma) = 9.9 \times 10^{40} E_\gamma^2 \left[\exp(1.2 \times 10^{10} E_\gamma / T) - 1\right]^{-1} \quad (\text{photons cm}^{-2}\ \text{s}^{-1}\ \text{MeV}^{-1}) \tag{II.3}$$

where E_γ must be expressed in MeV. In units appropriate here, Wien's law gives the position of the maximum in this distribution at $E_{max}(\text{MeV}) = 4.7 \times 10^{-10}\ T(\text{K})$. The average photon energy $E(\text{MeV}) = 2.7\,kT = 2.3 \times 10^{-10}\ T(\text{K})$.

For more details on the fundamental aspects of the Planck thermal radiation see Allen (1973), Unsöld (1969), Boldt (1969), and Hayakawa (1969).

2.3. Particle-Field Interactions

2.3.1. COMPTON EFFECT

The effect of scattering a highly relativistic electron from a photon and thereby raising the energy of the photon has been exhaustively studied. The problem was first treated by Feenberg and Primakoff (1948) in discussing the energy loss of cosmic ray electrons in interstellar space, and is often called the inverse Compton effect.

The basic theory for calculating the photon energies resulting from scattering electrons on ambient (starlight or black body) photons has been further treated or reviewed by many authors, for example: Donahue (1951), Felten and Morrison (1963, 1966), Hayakawa *et al.* (1964), Garmire and Kraushaar (1965), Weekes (1969), and Stecker (1971).

The basic problem of an electron scattering on a photon is shown in Figure II-2a in the laboratory or observer's frame of reference and in Figure II-2b in the electron's rest frame (cf. Felten and Morrison, 1966; Stecker, 1971; Weekes, 1969). In Figure II-2b the kinematical situation is the same as the familiar Compton scattering of a photon by an electron at rest. Therefore, the method of obtaining the desired results is to transform all kinematical quantities after the collision, from the electron's rest frame (Figure II-2b) to the observer's frame of reference, in which the electron has a high velocity and the photons are usually of very low energy (<1 eV). The symbols in Figure II-2 are:

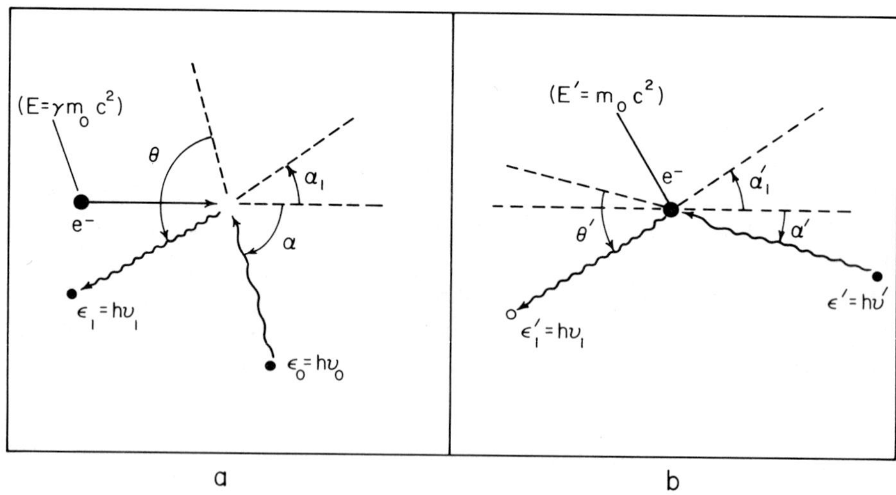

Fig. II-2. a. Electron scattering of a photon in the laboratory frame of reference, "Inverse Compton Effect." b. Scattering of a photon in the electron's rest frame. (From J. E. Felten and P. Morrison, *Astrophys. J.* **146,** 686. Copyright 1966, The American Astronomical Society. Used by permission of the University of Chicago Press. Also from F. W. Stecker: 1971, *NASA SP-249* and Monobook Company, Baltimore, Maryland, 1971.)

$E = \gamma m_0 c^2$ = electron's total energy, where
$\gamma = (1 - v^2/c^2)^{-1/2}$ and $m_0 c^2$ is the electron's rest energy, v is the electron's velocity
α = angle the incident photon makes with electron's velocity vector
α_1 = the angle the recoil photon makes with the electron's trajectory
$\theta = (\alpha + \alpha_1)$ = scattering angle of the photon
$\epsilon_0 = h\nu_0$ = photon's energy before the collision
$\epsilon_1 = h\nu_1$ = photon's energy after the collision.

Quantities in the electron's rest frame are denoted by primes (′).

The exact expression for the energy of the scattered photon as seen by the observer is (Felten and Morrison, 1966)

$$\epsilon_1 = h\nu_1 = \frac{\gamma^2 \epsilon_0 (1 + \beta \cos\alpha)(1 - \beta \cos\alpha_1')}{1 + (\gamma\epsilon_0/m_0c^2)(1 + \beta \cos\alpha)(1 - \cos\theta')} \tag{II.4}$$

where θ' is the angle of scattering of the photon in the electron's rest system, and $\beta = v/c$.

For any practical application in astrophysics, an assumption must be made concerning the distribution of the angles of incidence (α) of the initial photons with the electrons' direction, since this, along with the cross section, will determine the average energy of the scattered photons. The usual assumption made is that the photon distribution is isotropic, which then permits a determination of the mean energy loss of the electron and the mean energy of a scattered photon. There are then two cases to be considered depending on whether:

$$\gamma\epsilon_0 \ll m_0c^2$$

or

$$\gamma\epsilon_0 \gg m_0c^2$$

and in either case $\gamma \gg 1$.

(a) $\gamma\epsilon_0 \ll m_0c^2$. The former case, in which the total Compton scattering cross section is just the classical Thomson cross section, has been treated in detail by Ginzburg and Syrovatskii (1964b), who give the differential cross section for a single electron of energy E scattering on a photon of energy ϵ_0 to give a quantum of energy E_γ as

$$\sigma(E_\gamma, \epsilon_0, E) = \frac{\pi r_0^2 (m_0c^2)^4}{4\epsilon_0^2 E^3}\left(2\frac{E_\gamma}{E} - \frac{(m_0c^2)^2\, E_\gamma^2}{\epsilon_0 E^3} + 4\frac{E_\gamma}{E}\cdot \ln\frac{(m_0c^2)^2\, E_\gamma}{4\epsilon_0 E^2} + \frac{8\epsilon_0 E}{(m_0c^2)^2}\right) \tag{II.5}$$

for $\epsilon_0 \leqslant E_\gamma \leqslant 4\epsilon_0\gamma^2$, where $r_0 = e^2/m_0c^2$.

Integration of this expression over all E_γ gives the Thomson cross section, $\sigma_T = (8\pi/3)\, r_0^2 = 6.65 \times 10^{-25}$ cm^2.

The mean energy loss of an electron in a photon field is then found from the collision rate $cn_{ph}\sigma_T(s^{-1})$ and the mean energy loss per collision giving:

$$-dE/dt = cn_{ph}\sigma_T\left[\tfrac{4}{3}\,\bar{\epsilon}_0\,(E/m_0c^2)^2\right] \quad (\text{erg s}^{-1}) \tag{II.6}$$

where n_{ph} is the total number of photons cm^{-3}, and $\bar{\epsilon}_0$ is the average energy of the incident photons which depends on the distribution of the photons $n_{\mathrm{ph}}(\epsilon)$ (photons/unit energy) according to $n_{\mathrm{ph}} = \int n_{\mathrm{ph}}(\epsilon)\, \mathrm{d}\epsilon$ and

$$\bar{\epsilon}_0 = \int \epsilon_0 n_{\mathrm{ph}}(\epsilon_0) \mathrm{d}\epsilon_0 \Big/ \int n_{\mathrm{ph}}(\epsilon_0)\, \mathrm{d}\epsilon_0. \tag{II.7}$$

From the above expression for the electron's energy loss (Equation (II.6)), the quantity in brackets [ ] is the mean energy of the scattered photon,

$$\bar{E}_\gamma = \tfrac{4}{3}\, \bar{\epsilon}_0 (E/m_0 c^2)^2. \tag{II.8}$$

Figure II-3 shows a plot of this relation vs. the electron's γ for two cases: that is, when

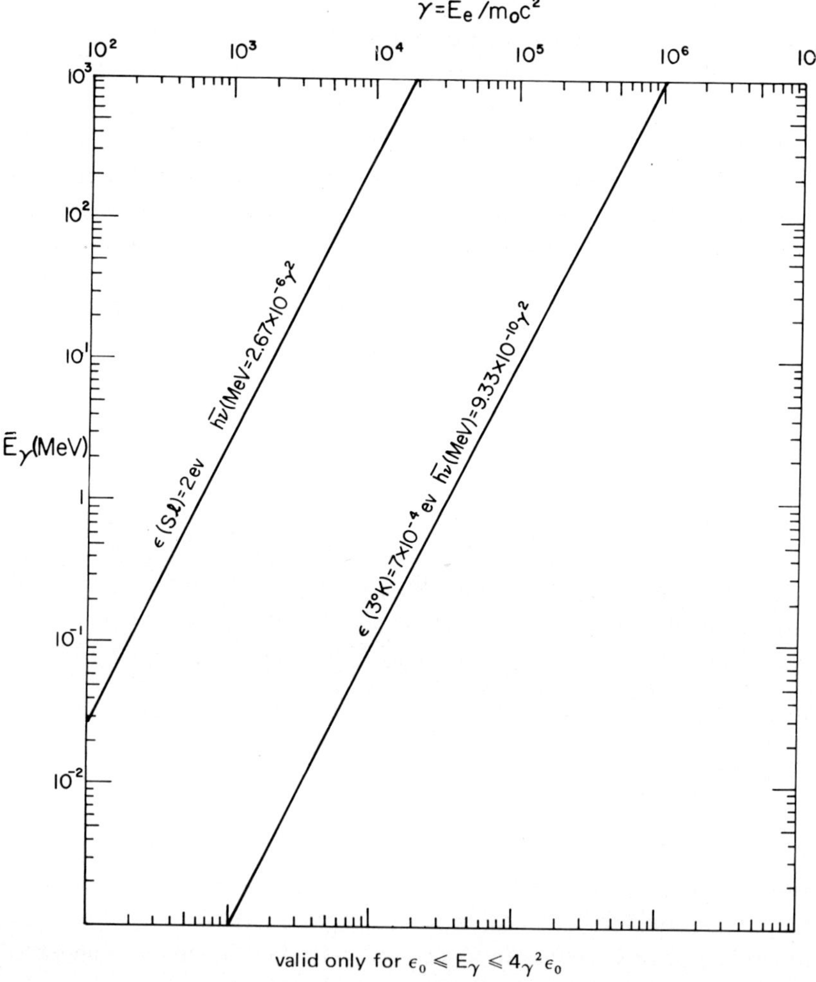

Fig. II-3. The average energy of the scattered photon vs. the energy of the incident electron. The average energies of the cold photons are $\bar{\epsilon}_0 = 7 \times 10^{-4}$ eV (2.7 K BB) and $\bar{\epsilon}_0 = 2$ eV (starlight).

the electrons are incident on, for example, starlight photons of average energy $\bar{\epsilon}_0 = 2\,\text{eV}$ or black body photons at a temperature of 2.7 K where

$$\bar{\epsilon}_0 = 2.7\,kT \cong 6 \times 10^{-4}\,\text{eV}.$$

Thus far only, the mean photon energy expected for monoenergetic electrons in an isotropic photon field has been discussed. Now the more general case involving a distribution of electron energies must be considered. For an electron intensity spectrum $I_e(E)(\text{cm}^{-2}\ \text{s}^{-1}\ \text{sr}^{-1}\ \text{MeV}^{-1})$ isotropically incident on an isotropic photon field the resulting γ-ray spectrum (Ginzburg and Syrovatskii, 1964b) due to scatterings along a path length L is

$$I_\gamma(E_\gamma) = L_{\text{eff}} \int_0^\infty n_{\text{ph}}(\epsilon_0)\, \text{d}\epsilon_0 \int_{E_{\text{min}}}^\infty \sigma(E_\gamma, \epsilon_0, E)\, I_e(E)\text{d}E. \qquad \text{(II.9)}$$

For this approximation, $\epsilon_0 \leqslant E_\gamma \leqslant 4\epsilon_0(E/m_0c^2)^2$, so the lower limit, E_{min}, is determined by the lowest energy electron to produce a given E_γ, or $E_{\text{min}} = m_0c^2(E_\gamma/4\epsilon_0)^{1/2}$.

The most extensive calculations for the differential photon spectrum, $I_\gamma(E_\gamma)$, are based on a power law electron energy spectrum of the form

$$I_e(E) = K_e E^{-\alpha}\ (\text{electrons cm}^{-2}\ \text{s}^{-1}\ \text{sr}^{-1}\ \text{MeV}^{-1}) \qquad \text{(II.10)}$$

interacting with a black body radiation field, where the photon spectral density is

$$n_{\text{ph}}(\epsilon_0) = \frac{(8\pi)}{h^3c^3}\frac{\epsilon_0^2}{e^{\epsilon_0/kT} - 1}\ (\text{cm}^{-3}). \qquad \text{(II.11)}$$

The result in numerical form given by Stecker (1971) is

$$I(E_\gamma) = 6.22 \times 10^{-24} L_{\text{eff}}[10^{-2.962\alpha}]\ f(\alpha)\ K_e T^{(\alpha+5)/2} E_\gamma^{-(\alpha+1)/2}$$
$$(\text{cm}^{-1}\ \text{s}^{-1}\ \text{sr}^{-1}\ \text{MeV}^{-1}) \qquad \text{(II.12)}$$

The quantity $f(\alpha)$, which depends on the exponent of the electron spectrum and is near unity, has for $\alpha = 1$, 2, 3, or 4 the values 0.84, 0.86, 0.99, and 1.4, respectively. The above differential intensity spectrum for the γ-rays depends only on the effective path length, L_{eff} in cm, for γ-ray production and the electron spectral exponent α.

Another useful approximate form for the γ-ray spectrum in terms of the photon energy density (given by Felten and Morrison, 1966) is

$$I(E_\gamma) \cong 10^9(56.9)^{3-\alpha}\ n_0 L_{\text{eff}} \rho T^{(\alpha-3)/2} E_\gamma^{-(\alpha+1)/2}$$
$$(\text{cm}^{-2}\ \text{s}^{-1}\ \text{sr}^{-1}\ \text{MeV}^{-1}) \qquad \text{(II.13)}$$

where L_{eff} is the distance in light years along the line of sight through the interaction region, $\rho = \epsilon_0 n_{\text{ph}}$ (eV cm^{-3}) the photon energy density and, in this case, the electron spectrum is given in the form $n(\gamma)\,\text{d}(\gamma) = n_0\gamma^{-\alpha}\,\text{d}\gamma$ where n_0 has the units (cm^{-3}).

(b) $\gamma\epsilon_0 \gg m_0c^2$. In this extreme, the energy of the scattered photon in the laboratory or observation frame is $\epsilon_1 \sim \gamma m_0c^2 \sim E$, which is the electron's energy, so the scattered photon energy is independent of ϵ_0. This case requires use of the Klein-Nishina formula for the Compton scattering and is not usually treated in astrophysics. However, this approximation must be used when dealing with very hot photons ($\epsilon_0 \gg 1$ eV) and electrons with a small γ or, inversely, electrons with $\gamma \gg 1$ and cool photons.

2.3.2. MAGNETOBREMSSTRAHLUNG

A charged particle moving in a magnetic field accelerates due to the Lorentz force $\mathbf{F} = q(\mathbf{v} \times \mathbf{B})$, which causes the electron to spiral around the magnetic field direction. According to classical theory, the electron must radiate electromagnetic energy, continuously producing what is usually called synchrotron radiation. This mechanism apparently is responsible for the general X-ray emissions (<30 keV) from the Crab Nebula as well as the radio emission from Jupiter. In the case when the pitch angle of an electron with respect to a magnetic field is zero, so the electron's motion is along the field line, γ-radiation can be produced as discussed by Sturrock (1971) if the field line is curved. This is termed curvature radiation and will be discussed briefly in Section III-3.2.2 in considering the predicted γ-ray spectrum from pulsars.

The theory of emission of radiation by relativistic electrons in magnetic fields was discussed in detail by Schwinger (1949), and a basic treatment is given by Jackson (1962) The most useful treatment for astrophysical application is given by Ginzburg (1969). Consider an electron moving in a magnetic field as shown in Figure II-4. As can be seen,

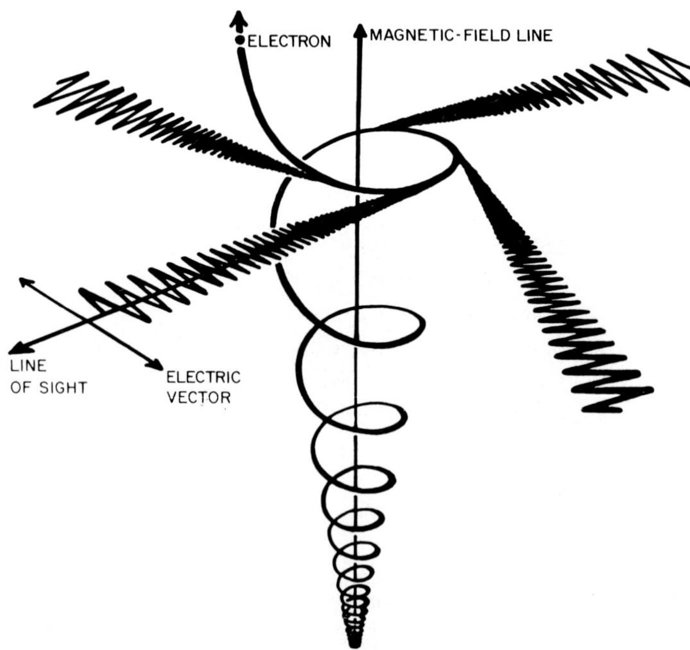

Fig. II-4. Idealized geometry, in perspective, for an electron moving in a magnetic field, emitting magnetobremsstrahlung (synchrotron radiation).

the radiation emitted is normal to the instantaneous acceleration vector of the electron with a polarization parallel to the plane of the electron's motion as in the case of classical dipole radiation. This is only approximately correct (Jackson, 1962), since some component (~12%) is *polarized* normal to the instantaneous acceleration vector for relativisitic electrons. The important point is that the radiation is beamed in a narrow cone along the instantaneous velocity vector of the electron and sweeps across a point of observation as a searchlight. As is shown approximately by Ginzburg and Syrovatskii (1964b) and Ginzburg (1969), the characteristic frequency emitted by a single electron above which the emitted radiation becomes negligible is given by

$$\nu_s = \frac{3e}{4\pi m_0 c^2}\gamma^2 B_\perp \quad \text{(Hz)} \tag{II.14}$$

where $\gamma = E/m_0c^2$ is the Lorentz factor for the electron, $B_\perp$ is the magnetic field (in G) normal to the electron's motion and the other quantities are standard notation. The frequency distribution of the emitted radiation is actually a discrete asymmetric distribution with a maximum of the envelope at $\nu_m \sim 0.29\,\nu_s$. We can then express the energy of the photon at the maximum of the distribution as

$${}_mE_\gamma = 0.29\,h\nu_s \cong 5 \times 10^{-15}\gamma^2 B_\perp \quad \text{(MeV)} \tag{II.15}$$

where $B_\perp$ is the perpendicular magnetic field (in G).

Clearly, tremendously energetic electrons or strong magnetic fields are needed to produce γ-ray photons in the MeV energy range.

For completeness, we will give here the total rate of energy emission by the synchrotron process from a spectrum of electrons, taken from Ginzburg (1969, p. 9). First, the total power emitted over all frequencies for a single charged particle of mass M and charge Z is given, in the ultra relativistic case, by

$$P_s = -\mathrm{d}E/\mathrm{d}t = 0.98 \times 10^{-3} B_\perp^2 \frac{(Z^2 m_0)^2}{M^2}\left(\frac{E}{Mc^2}\right)^2 \quad (\text{eV s}^{-1}) \tag{II.16}$$

where $B_\perp$ is in G and m_0 is the electron rest mass.

For electrons, $M = m_0$, so

$${}_eP_s \simeq 10^{-3} B_\perp^2 \gamma^2 \quad (\text{eV s}^{-1})$$

and for protons of the same energy

$${}_pP_s \simeq (m_0/M_p)^4\ {}_eP_s \simeq 10^{-13}\ {}_eP_s.$$

Based upon the above expression for the power loss of the electron by magnetobremsstrahlung, Ginzburg (1969, p. 43), deduces that the time to reduce the energy of an electron by 1/2 is

$$T_M \cong 5 \times 10^8/\gamma B_\perp^2 \quad \text{(s)} \tag{II.17}$$

where $B_\perp$ is in units of G.

Therefore, for electrons of energy 3×10^8 MeV in a field with $B_\perp = 3 \times 10^{-6}$ G, the lifetime is only 3×10^3 yr, which is very short on a cosmic scale. Thus, it is generally considered that magnetobremsstrahlung at optical, X-ray, and γ-ray energies will be naturally weak. The situation is vastly different though in supernovae remnants or on condensed objects where a continuous energy supply may be available. Again using the Crab Nebula as an example, presumably here the rotating neutron star pulsar continuously supplies energy to electrons at a rate fast enough to overcome the extremely high loss rate.

It appears, therefore, that the usual astrophysical sites where magnetobremsstrahlung could give γ-ray photons in the MeV energy range are near condensed objects where extremely strong magnetic fields and ultrarelativistic electrons may be present. Therefore, we give here the pertinent formula for estimating the γ-ray energy spectrum for the case of a power law spectrum of electrons trapped in a random magnetic field of dimension L and volume V at a distance R from the observer as illustrated in Figure II-5. Taking the trapped electron density as

$$N(E) = K_e E^{-\alpha} \mathrm{d}E \quad (\text{electrons cm}^{-3}) \tag{II.18}$$

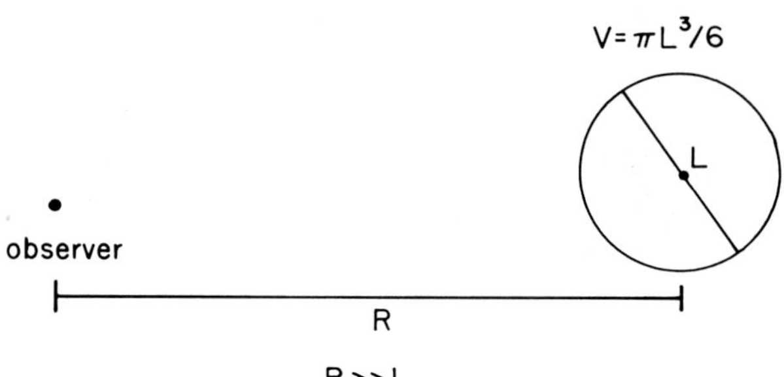

Fig. II-5. An idealized picture showing a cloud of electrons trapped in a magnetic field B contained in volume V.

Ginzburg (1969, p. 42), gives the γ-ray spectral intensity as

$$I_\nu = 1.35 \times 10^{-22} a(\alpha) \frac{K_e V B^{(\alpha+1)/2}}{R^2} \left(\frac{6.26 \times 10^{18}}{\nu} \right)^{(\alpha-1)/2}$$

$$(\text{erg cm}^{-2}\ \text{s}^{-1}\ \text{Hz}^{-1}). \tag{II.19}$$

Here the trapped electron cloud is assumed to be spherical, so $V = \pi L^3/6$ and all other quantities have been previously defined. The coefficient $a(\alpha)$ has been given by Ginzburg (1969), and a list of values is shown in Table II-3.

TABLE II-3

The coefficients $a(\alpha)$ for Equation (II-19). (From Ginzburg, V.L., *Elementary Processes for Cosmic Ray Astrophysics,* in *Topics in Astrophysics and Space Physics,* A.G.W. Cameron and G.B. Field (eds.), Gordon and Breach Science Publishers, Inc., New York, 1969. Used by permission.)

α	1	1.5	2	2.5	3	4	5
$a(\alpha)$	0.283	0.147	0.103	0.0852	0.0742	0.0725	0.0922

For our purposes it is convenient to transform to a γ-ray number spectrum by making the appropriate change in units, which gives

$$F_\gamma = 3.27 \times 10^{-2}\, a(\alpha) K_e V \frac{B^{(\alpha+1)/2}}{R^2} \left(\frac{2.59 \times 10^{-2}}{E_{\text{MeV}}}\right)^{(\alpha+1)/2}$$

$$(\text{photons cm}^{-2}\ \text{s}^{-1}\ \text{MeV}^{-1}). \qquad \text{(II.20)}$$

It is interesting to observe that the γ-ray spectral shape for a power law electron spectrum is the same as for Compton scattering. As discussed by Jones (1965), Boldt (1969) and Stecker (1974), this is due to the fact that the synchroton process can be viewed as the interaction of an electron with 'virtual' photons of the magnetic field with energy density $\rho_H = B^2/8\pi$.

Finally, it is important to note that these theoretical formulas for synchrotron radiation apply only if certain physical conditions are met. Some of these are listed by Ginzburg (1969, p. 11), and, pertaining to total power loss, are as follows: (1) ultra-relativistic electrons (or positrons) are moving in a constant and homogeneous magnetic field with no other fields present; (2) the influence of the ambient medium is negligible; (3) the direct or indirect influence of other relativistic electrons is neglected; and (4) the following condition on the electron energy is met,

$$\gamma = E/m_0c^2 \ll 10^8 B^{-1/2}$$

where B is in G.

For the case of sunspot magnetic fields of 10^4 G these conditions require that $\gamma \ll 10^6$ or electron energies $E \ll 10^6$ MeV. For a neutron star with $B \sim 10^{12}$ G, then $E \ll 10^2$ MeV.

Clearly, the brief discussion of magnetobremsstrahlung we have given only serves the purpose of giving the reader an appreciation of the physical quantities involved in the process. In Chapter III, we present some examples of the application of these ideas to specific astrophysical sources with due regard to the hazards of oversimplification.

Hayakawa (1969) has summarized the important features of magnetobremsstrahlung as shown in Figure II-6. Here the half-life of an electron (Equation (II-17)) against synchrotron radiation loss is shown as a family of parametric curves for a range of electron kinetic energies and magnetic field strengths. The characteristic photon energies (Equa-

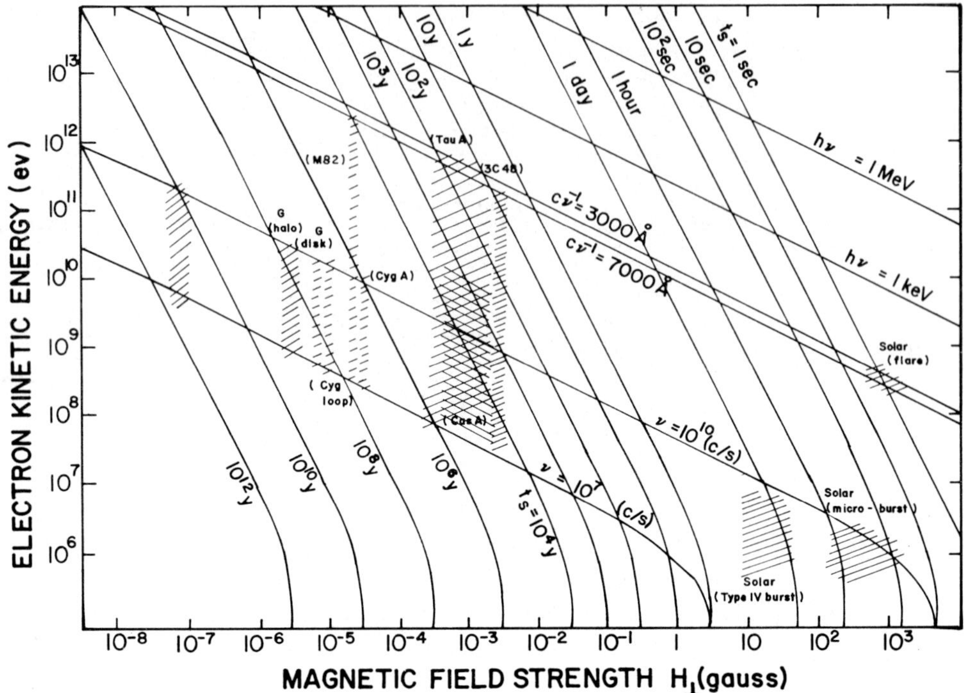

Fig. II-6. The characteristic magnetobremsstrahlung photon energies and electron lifetimes are given for different electron energies vs. the magnetic field strength. (From S. Hayakawa, *Cosmic Ray Physics, Nuclear and Astrophysical Aspects,* p. 641. Copyright 1969, John Wiley & Sons, Inc. Used by permission.)

tion (II.15)) are also shown as another set of parametric curves. The spectral regions where synchrotron emissions are observable from various sources are also shown as hatched lines.

2.3.3. PHOTOMESON PRODUCTION

The generation of γ-rays by production of π^0 mesons from photons and the subsequent decay of π^0 mesons can be an important process in the astrophysical case. In the terrestrial laboratory or the rest frame of the proton (the target), this process can be initiated only by γ-rays. However, in the astrophysical 'laboratory' system this process can be initiated by energetic protons colliding with low energy 'cold' optical or radio photons. The basic process is

$$\gamma + \mathrm{p} = \mathrm{p} + \pi^0 \tag{II.21}$$

and is shown schematically in Figure II-7, although multiple meson production is also possible. In the rest frame of the proton the 'cold' photon of initial lab energy $\epsilon_0 = h\nu$ has an energy:

$$\epsilon^* = \epsilon_0(1 + \cos\theta)\,(E_\mathrm{p}/m_\mathrm{p}c^2); \qquad \epsilon_0 \ll m_\mathrm{p}c^2 \ll E_\mathrm{p} \tag{II.22}$$

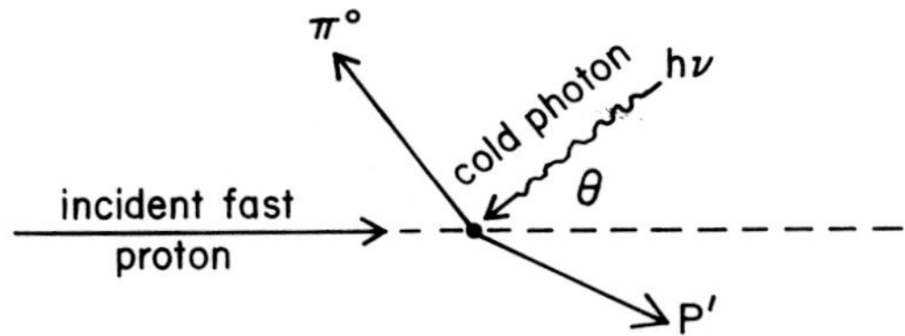

Fig. II-7. Photomeson production by a fast proton colliding with a cold photon.

where E_p is the laboratory energy of the proton, $m_p c^2$ is its rest energy, and θ is the angle between the initial photon direction and the proton's direction.

This then gives the proton threshold energy for creating a single π^0 meson in the laboratory system as

$$E_p^{th} = \frac{m_\pi c^2 (m_\pi c^2 + 2m_p c^2)}{2\epsilon_0(1+\cos\theta)}; \qquad m_\pi c^2 \sim 132\,\text{MeV}. \tag{II.23}$$

Jackson (1962) and Fazio (1967) present a thorough discussion of the necessary kinematics. For thermal radiation, the average photon energy is $\bar{\epsilon}_0 = 2.7\,kT = 2.3 \times 10^{-4}\,T$ (K) eV.

Consider then the case of a photon with $\epsilon_0 \sim 1$ eV, corresponding to a temperature of 4300 K, colliding head-on ($\theta = 0$) with a fast proton. From the above equation, the threshold proton kinetic energy is 7×10^{16} eV!

The proton and π° resulting from the interaction of the primary proton and the cold photon have, therefore, extremely high energy in the laboratory system. In the rest system of the π_0 the two decay γ-rays have the same energy $E'_\gamma = m_{\pi^\circ} c^2/2 \sim 68$ MeV. In the laboratory system the maximum and minimum energies of the γ-rays $E_{max,\gamma}$ and $E_{min,\gamma}$ must add up to the total energy of the pion, E_π. Stecker (1971) shows that, in the extreme case when the two γ-rays are emitted in the direction of motion of the π_0, the maximum and minimum photon energies in the laboratory system are given by

$$E_{max,\gamma} = \frac{E_\pi}{2}(1+\beta_\pi)$$

$$E_{min,\gamma} = \frac{E_\pi}{2}(1-\beta_\pi) \tag{II.24}$$

where $\beta_\pi = v_\pi/c$.

In Figure II-8 the π^0 decay γ-ray spectra are shown for various simple pion energy distributions. In the extreme relativistic case the pions have $\beta \sim 1$, and the π^0 γ-ray energies can go all the way from zero to $E_{max,\gamma} = E_\pi$, which is the case here.

Hayakawa *et al.* (1964) have calculated the γ-ray spectrum expected from this process by metagalactic cosmic ray protons on starlight and give the integral γ-ray spectrum down to $\sim 10^{13}$ eV, and at that energy the integral γ-ray flux at the Earth is only 10^{-16} cm^2 s^{-1} sr^{-1}. Greisen (1966b) has also considered this process for interactions on the 3 K background radiation, which gives a mean photon energy $\bar{\epsilon} = 2.7\,kT \sim 6 \times 10^{-4}$ eV,

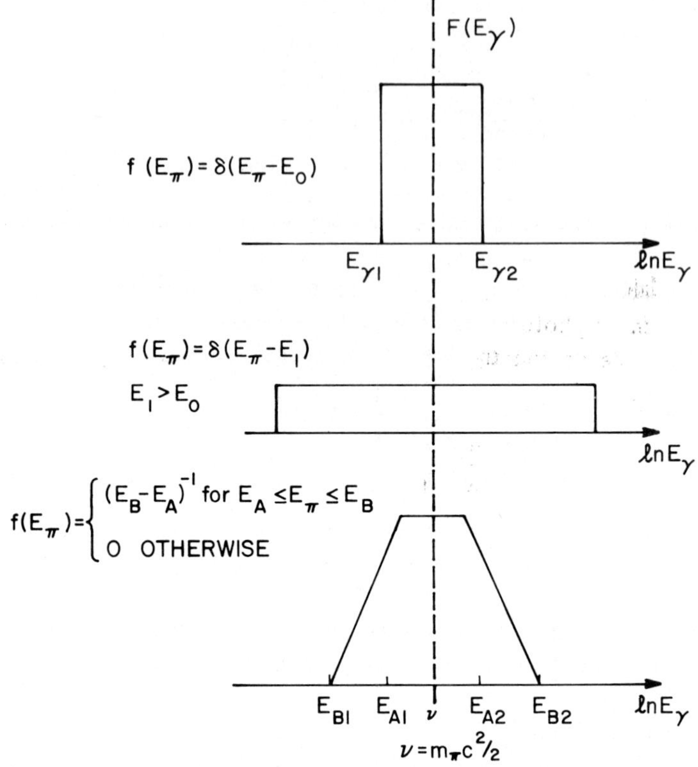

Fig. II-8. Some ideal γ-ray spectra emitted from the decay of ideal spectra of neutral pions. (From F.W. Stecker: 1971, *NASA SP-249* and Monobook Company, Baltimore, Maryland, 1971.)

so the proton threshold energy is $\sim 10^{20}$ eV. On the other hand, the number density of the isotropic radiation is higher by a factor of 5×10^4 than the average number density of the more energetic starlight, so the high energy cosmic ray proton spectrum could be cut off by this mechanism.

2.4. Particle-Matter Interactions

2.4.1. BREMSSTRAHLUNG

In this process a charged particle interacting with matter will be deflected by interactions with the Coulomb fields of the electrons and protons; the resulting acceleration gives rise to radiation. Consider an incident electron of energy E_0 and momentum $\mathbf{P}_0$ interacting with a nuclear field giving rise to a bremsstrahlung photon of energy E_γ. The process can be thought of in quantum mechanical terms as the transition of the electron from an energy-momentum state $(E_0, \mathbf{P}_0)$ to a final state $(E, \mathbf{P})$ with the emission of a photon with energy E_γ such that $E_\gamma = E_0 - E$, neglecting the recoil energy of the nucleus. Since the interaction is with the nuclear field, the nucleus, which is heavy compared to the electron, can take any amount of momentum and this must be taken into account to determine the momentum of the emitted photon. We will summarize some basic conclusions for the nuclear case following Fazio (1967), which is based on the cross section derived by Heitler (1954).

The differential cross section for emission of a photon of energy E_γ in $\mathrm{d}E_\gamma$ is

$$\sigma_B(E_0, E_\gamma)\ \mathrm{d}E_\gamma = 4\sigma_0 Z^2 \frac{\mathrm{d}E_\gamma}{E_\gamma} f(E_\gamma, E_0) \qquad \text{(II.25)}$$

where $f(E_\gamma, E_0)$ is a slowly varying function of energy and

$$\sigma_0 = \frac{1}{137}\left(\frac{e^2}{m_0 c^2}\right)^2 = 5 \times 10^{-28}\ (\mathrm{cm}^2 \text{ per nucleus}).$$

In the extreme relativistic case $E_0 \gg m_0 c^2$, for complete screening by atomic electrons (distant collisons), the cross section can be approximated by

$$\sigma_B(E_0, E_\gamma)\mathrm{d}E_\gamma \cong \left(\frac{m}{X_0}\right)\frac{\mathrm{d}E_\gamma}{E_\gamma}(\mathrm{cm}^2); \quad E_0 \gg 137\ m_0 c^2 Z^{-1/3} \qquad \text{(II.26)}$$

where m is the mass of the target atom in grams, Z is the atomic number, and the quantity X_0 (g cm^{-2}) is known as the radiation length. From Rossi (1952), $X_0 = [4\alpha N_0 Z^2 A^{-1} r_0^2(\ln 183\ Z^{-1/3})]^{-1}$ (g cm^{-2}), where N_0 is Avogadro's number and A is the target mass number. Physically this quantity may be thought of as the path length in g cm^{-2}

TABLE II-4

Values of the radiation length, X_0 (g cm^{-2}) and the critical energy, ϵ_0, in several materials. (From B. Rossi, *High Energy Particles,* 1952, p. 55. Reprinted by permission of Prentice-Hall, Inc., Englewood Cliffs, New Jersey.)

Substance	Z	A	X_0 (g cm^{-2})	ϵ_0(MeV) Without density effect	With density effect
Carbon	6	12	44.6	102	76
Nitrogen	7	14	39.4	88.7	
Oxygen	8	16	35.3	77.7	
Aluminum	13	27	24.5	48.8	
Argon	18	39.9	19.8	35.2	
Iron	26	55.84	14.1	24.3	21
Copper	29	63.57	13.1	21.8	
Lead	82	207.2	6.5	7.8	
Air	7.37	14.78	37.7	84.2	
Water	7.23	14.3	37.1	83.8	65
Hydrogen (neutral)	1	1	63	See Stecker (1975)	
Helium (neutral)	2	4	93	See Stecker (1975)	
Interstellar gas H(80); He(20)			65	See Stecker (1975)	

that an electron must pass through in order for its energy to be reduced e^{-1} by the radiation (bremsstrahlung) process. In Table II-4, we show several values of X_0 for different elements of interest from Rossi (1952), using an equation slightly modified from that above to account for electron effects and inaccuracies in the Born approximation. Rossi (1952) also gives an excellent intuitive discussion of the cascade shower process in which radiation loss plays a fundamental role. The values of X_0 for H, He, and interstellar matter are from Dovshenko and Pomanskii (1964) as discussed by Stecker (1975). The critical energy in Table II-4 is the energy for which radiation and ionization losses are equal. In the highly relativistic case, with no screening by atomic electrons (near-collisions), the radiation cross section is

$$\sigma_B(E_0, E_\gamma)\, dE_\gamma \simeq 4\sigma_0 Z^2 \ln\left(\frac{2E_0}{m_0c^2}\right)\frac{dE_\gamma}{E_\gamma}\,;$$
$$m_0c^2 \ll E_0 \ll 137\, m_0c^2 Z^{-1/3}. \tag{II.27}$$

This form of the cross section also applies for an ionized gas. If the interaction includes relativistic electrons with atomic electrons, then the above cross sections apply if Z^2 is replaced by $Z(Z+1)$. In the nonrelativistic region other forms of the cross section must be used, and, in this case, electron-electron collisions are much weaker than electron-proton collisons. (See Bethe and Ashkin, 1953).

Also, in an ionized gas where no atomic electron screening is present, the shielding distance from a point charge is given by the Debye-length $\lambda_D \cong 7(T/n_e)^{1/2}$ (cm), where the electron density is n_e and the plasma temperature is T(K). The radiation length in a plasma is then (Lüst and Pinkau, 1967)

$$X_p = X_0 (\ln a_0/\lambda_c)/(\ln \lambda_D/\lambda_c) \tag{II.28}$$

where the electron Compton wavelength is $\lambda_c = h/m_0c \cong 2.4 \times 10^{-10}$ cm and the first Bohr radius is $a_0 = \hbar^2/m_0c^2 \cong 0.53 \times 10^{-8}$ cm. For example, in ionized hydrogen $X_p \simeq 10$ g cm^{-2} compared to 63 g cm^{-2} for neutral hydrogen.

We wish now to review the nature of the resulting γ-ray spectrum for the case of a nonthermal distribution of high energy electrons (e.g., cosmic rays and solar electrons) and for a Maxwellian distribution of electron energies.

Consider first the rate of energy loss $(dE/dt)_B$ per second by bremsstrahlung, following Fazio (1967), of a relativistic electron traversing atoms, each of mass m(g) and number density n(cm^{-3})

$$-(dE/dt)_B = nc \int_0^{E_0} E_\gamma \sigma_B(E, E_\gamma)\, dE_\gamma \tag{II.29}$$

and for $E \gg 137\ m_0c^2 Z^{-1/3}$

$$-(dE/dt)_B = \frac{nmc}{X_0} E_0 \equiv \frac{E^*}{\tau} \tag{II.30}$$

where τ is a mean loss time to lose an energy E^*. Because of the definition of X_0 given above, the factor before E_0 may be thought of as the probability per second for radiation loss. Therefore, $E^* = E_0$ is the characteristic photon energy radiated by bremsstrahlung. The differential number spectrum for bremsstrahlung by a monoenergetic relativistic electron is $\propto E_\gamma^{-1}$, with a cutoff at $E_{\gamma max} = E_0$ (Evans, 1955).

a. *Non-Thermal Bremsstrahlung*

If an incident electron spectrum is given in the form $I_e(E)\ dE = K_e E_e^{-\alpha}\ dE$, then Fazio (1967) shows that the resulting differential spectrum of gamma rays is given by

$$I_\gamma(E_\gamma)\ dE_\gamma = \frac{mN(L)}{X_0}\left(\frac{dE_\gamma}{E_\gamma}\right)\int_{E_\gamma}^{\infty} I_e(E)\ dE \tag{II.31}$$

or

$$I_\gamma(E_\gamma)\ dE_\gamma = \frac{mN(L)}{X_0}\left(\frac{K_e}{\alpha - 1}\right) E_\gamma^{-\alpha}\ dE_\gamma \tag{II.32}$$

where m (g) is the mass of the target nuclei and $N(L)$ (cm^{-1}) is the integrated number of target nuclei in the line of sight, $N(L) = \int_0^L n\ dl$. The spectral shape given in Equation (II.32) refers to the case of relativistic electrons, although Ramaty *et al.* (1975) have pointed out that this spectrum should be $\propto E_\gamma^{-\alpha + 0.1}$, which is slightly flatter. Refinements must also be made for the case of nonrelativistic electrons (see Brown, 1971 and Table II-11).

Bremsstrahlung by fast protons incident on electrons at rest has been considered as a possible source of cosmic γ-rays by several authors (Boldt and Serlemitsos, 1969; Hayakawa, 1970; Brown 1970a, b). Jones (1971) has extended the calculations and concluded that the resulting γ-ray spectrum above ~5 MeV would have a power law spectrum that is one power steeper than that of the proton spectrum. It is concluded that proton bremsstrahlung most likely does not make a significant contribution to the galactic cosmic γ-ray spectrum above ~5 MeV, since a differential γ-ray spectrum of the form E_γ^{-3} would be expected for a proton energy spectrum that varies as E_p^{-2} above 20 GeV. This appears to be much steeper than the diffuse cosmic ray spectrum above ~1 MeV (see Section V-5.3).

b. *Thermal Bremsstrahlung*

The emission of electromagnetic radiation from a sufficiently hot plasma can extend into the X-ray and γ-ray spectral region. In the case for which the emitting volume is optically thick ($\tau \gg 1$) to its own radiation, the spectrum is given by the Planck law (Section II-2.2). If the source is optically thin ($\tau \ll 1$), then free-free (thermal bremsstrahlung) transitions are possible among the electrons and ions of the plasma. Also, in general, if the plasma is sufficiently cool, then so-called free-bound and bound-bound transitions can occur; but for γ-ray energy photons (>100 keV) only free-free transitions need to be considered.

Hayakawa (1969) has discussed this problem in some detail. The emission spectrum from a unit volume of such a plasma with electron density n_e and ion density n_i is

$$p_{ff}(E_\gamma) = 8.1 \times 10^{-13} n_e T^{-1/2} \left(\sum_Z Z^2 n_i \bar{g}_{ff}(Z, T, E_\gamma) \right) \frac{1}{E_\gamma} e^{-E_\gamma/kT}$$

$$(\text{cm}^{-3}\ \text{s}^{-1}\ \text{sr}^{-1}\ \text{keV}^{-1}) \qquad \text{(II.33)}$$

where Z is the ion atomic number and $\bar{g}_{ff}(Z, T, E_\gamma)$ is the average Gaunt factor, which is ~0.6 for the temperature range such that $1 \lesssim (E_\gamma/kT) \lesssim 3$. The temperature-averaged Gaunt factor is given also by Green (1959). A spectral shape of this form has been fit to the recently observed Vela γ-ray burst spectrum and is briefly discussed in Section V-5.4.

Ginzburg (1969, p. 57) has treated approximately the case of a hot, completely ionized H plasma in which thermal bremsstrahlung predominates. The major contribution to the radiation comes from electron-proton collisions, and the total radiation from a unit volume of plasma at a temperature T with ion density n_i (cm^{-3}) and electron density n_e (cm^{-3}) is

$$W_T = \frac{32\sqrt{2}\, e^6 Z^2 n_i n_e (kT/m_0)^{1/2}}{3\sqrt{\pi}\, m_0 c^3 \hbar} \quad (\text{erg cm}^{-3}\ \text{s}^{-1}) \qquad \text{(II.34)}$$

where n_e is the electron density and all other symbols have been previously defined. For a neutral plasma $n_e = Zn_i$, giving to a good approximation,

$$W_T = 1.57 \times 10^{-27} n_e^2 T^{1/2} \quad (\text{erg cm}^{-3}\ \text{s}^{-1}) \qquad \text{(II.35)}$$

which is valid for $10^5\ \text{K} < T < 10^{10}\ \text{K}$ and $Z = 1$.

This total emissivity is related to the spectral density of the radiation by $W_T = 4\pi \int_0^\infty \epsilon_\nu\, d\nu$ where ν is the photon frequency (Hz). Ginzburg (1969) gives several forms for ϵ_ν valid for different temperature regions, and we will give a form valid for many cases of interest:

$$\epsilon_\nu = \frac{7.7 \times 10^{-38}}{4\pi T^{1/2}} n_e^2 e^{-h\nu/kT} \quad (\text{erg cm}^{-3}\ \text{s}^{-1}\ \text{Hz}^{-1}) \qquad \text{(II.36)}$$

This is valid for $10^5\ \text{K} < T < 10^{10}\ \text{K}$ for H. Refinements including quantum mechanical effects may be found, for example, in the work of Allen (1973) and Green (1959).

2.4.2. π^0 PRODUCTION BY (p-p) AND (p-α) INTERACTIONS AND (p-$\bar{\text{p}}$) INTERACTIONS: ($\pi^0 \rightarrow 2\gamma$)

a. (p-p) *and* (p-α) *Interactions*

Whenever proton energies are sufficiently high, meson production can occur in p-p and p-α interactions. If π^0 mesons are produced, then γ-ray production follows from the decay mode $\pi^0 \xrightarrow{10^{-16}\,\text{s}} 2\gamma$ where each γ-ray has an energy of $m_{\pi^0} c^2/2 \sim 68$ MeV in the rest frame of the meson (see also Section II-2.3.3).

There are numerous modes by which π^0 mesons can be produced, and the problem has been treated quite generally by Stecker (1971). For example:

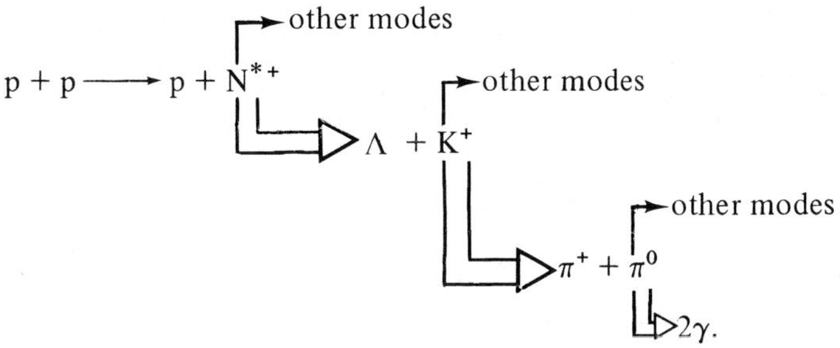

$$
\begin{array}{l}
\qquad\qquad\quad \ulcorner\!\rightarrow \text{other modes} \\
p + p \longrightarrow p + N^{*+} \qquad\quad \ulcorner\!\rightarrow \text{other modes} \\
\qquad\qquad\quad \llcorner\!\Rightarrow \Lambda + K^{+} \qquad\qquad \ulcorner\!\rightarrow \text{other modes} \\
\qquad\qquad\qquad\qquad\quad \llcorner\!\Rightarrow \pi^{+} + \pi^{0} \\
\qquad\qquad\qquad\qquad\qquad\qquad \llcorner\!\Rightarrow 2\gamma.
\end{array}
\tag{II.37}
$$

The decay chain indicated by the double arrows eventually leads to a π^0 meson and is started by the production of an "intermediate nonstrange isobar", N^{*+}, which decays in $\sim 10^{-23}$ s into a lambda hyperon and a kaon (K^+). The K^+ decays in 10^{-8} s into two pions: π^+ and the desired π^0. At each stage in this chain, possible branches to other decay modes are shown, which would have specified branching ratios but also lead to π^0 mesons. In general the energy distribution of the π^0's will be different for each mode and, consequently, the differential energy spectrum of the γ-rays in the observer's frame of reference will be different. The case where the intermediate N^{*+} state decays directly into a proton and several mesons is the most probable and has been studied thoroughly at accelerator energies. Several important examples are:

$$
\begin{aligned}
p + p &\longrightarrow p + p + x_1(\pi^+ + \pi^-) + y_1\pi^0 \\
p + p &\longrightarrow p + n + \pi^+ + x_2(\pi^+ + \pi^-) + y_2(\pi^0) \\
p + p &\longrightarrow n + n + 2\pi^+ + x_3(\pi^+ + \pi^-) + y_3(\pi^0) \\
p + p &\longrightarrow d + \pi^+ + x_4(\pi^+ + \pi^-) + y_4(\pi^0)
\end{aligned}
\tag{II.38}
$$

where the numbers (x_i, y_i) are 0 or positive integers. In the simplest case $x_1 = 0$ and $y_1 = 1$ and a single π^0 meson is produced. The threshold for this case is

$$T_{\text{th}} = m_{\pi^0}c^2\left(2 + \frac{m_{\pi^0}c^2}{2m_p c^2}\right) = 279.7\ \text{MeV}. \tag{II.39}$$

The threshold is higher for producing more than one π^0 or any number of charged mesons (cf. Fazio, 1967).

The two fundamental quantities of interest are the production cross section for π^0's via all possible modes and the resulting γ-ray spectrum which, of course, depends on the energy distribution of the emitted π^0 mesons. Stecker (1973a) has given a complete summary of the experimentally determined cross sections times multiplicity, $\sigma \cdot \zeta$, for π^0 production in p-p interactions. This is shown in Figure II-9 plotted versus the proton kinetic energy T(GeV). The data were fitted fairly closely by two power laws as follows:

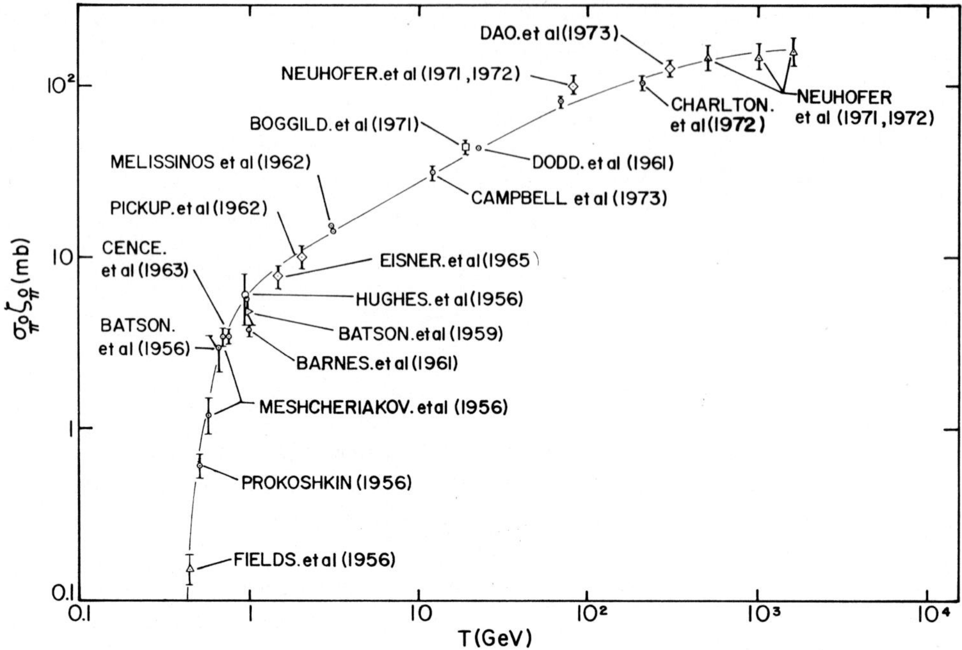

Fig. II-9. The experimental production cross section times multiplicity for π^0 meson production from p-p interactions. (From F.W. Stecker, *Astrophys. J.* **185**, 499. Copyright 1973, The American Astronomical Society. Used by permission of the University of Chicago Press.)

$$\sigma_{\pi^0}(T)\ \zeta_{\pi^0}(T) \simeq \begin{cases} 10^{-25}\ T^{7.64}\ \text{cm}^2,\ 0.4 \leqslant T \leqslant 0.7\ \text{GeV} \\ 8.4 \times 10^{-27}\ T^{0.53}\ \text{cm}^2,\ T \geqslant 0.7\ \text{GeV}. \end{cases} \tag{II.40}$$

The references shown on the figure are to the original experimental data summarized by Stecker (1973a).

In order to find the observable energies of the decay γ-rays, the usual procedure is to carry out a Lorentz transformation to the laboratory system from the center of mass system of the π^0 where the energies of the two γ-rays are equal and the angular distribution of emission is presumed to be isotopic. The normalized γ-ray differential energy spectrum that results for a single π^0 energy is given by (Stecker, 1971, p. 19)

$$f(E_\gamma\,|\,E_\pi) = (E_\pi^2 - m_\pi^2)^{-1/2} \tag{II.41}$$

which is valid for $(E_\pi/2)\ (1-\beta_\pi) \leqslant E_\gamma \leqslant (E_\pi/2)\ (1+\beta_\pi)$ (see Equation II.24).

The γ-ray source function, or the production rate of γ-rays of energy E_γ from a unit volume at distance $\mathbf{r}$ from the Earth, is given by Stecker (1971) as

$$q(E_\gamma,\,\mathbf{r}) = 4\pi n(\mathbf{r}) \int \mathrm{d}E_\mathrm{p} I(E_\mathrm{p},\,\mathbf{r}) \int_{E_{\pi,\min}}^{E_{\pi,\max}} \mathrm{d}E_\pi\ \sigma(E_\pi\,|\,E_\mathrm{p}) \cdot 2 \cdot f(E_\gamma\,|\,E_\pi) \tag{II.42}$$

where $n(\mathbf{r})$ is the hydrogen number density (cm^{-3}) at $\mathbf{r}$, and $I(E_\mathrm{p}, \mathbf{r})$ is the average directional intensity of cosmic rays ($\text{cm}^{-2}\ \text{s}^{-1}\ \text{sr}^{-1}$) of energy E_p in the specified source volume. The limits on the second integral are shown to be generally (Stecker, 1971): $E_{\pi,\mathrm{max}} \to \infty$ and $E_{\pi,\mathrm{min}} \to E_\gamma + m_\pi^2/4E_\gamma$. The predicted cosmic ray π^0 γ-ray spectrum at the Earth from this process is discussed in Section III-3.2.4. The units of Equation (II.42) are the same as those of Equation (II.2) with the cross section in appropriate units.

Production of π^0 mesons by p-α and α-p reactions is also important since the quantity $\zeta_{\pi^0}\sigma_{\pi^0}$ (mb), the π^0 multiplicity times the production cross section, is significantly larger for alpha reactions at all energies than for p-p interactions. Even though the He abundance is about one tenth that of H in most astrophysical media, the contribution of these reactions to the total γ-ray production is of about the same magnitude as for p-p interactions (Stecker, 1971).

b. *Matter-Antimatter* (p-$\bar{\mathrm{p}}$) *Interactions*

If antimatter is distributed in the Universe, as assumed by Alfvén (1965), Harrison (1967) and Omnes (1969), it is likely that the most direct method of detecting its existence would be through the observation of the γ-ray spectrum that results from the decay of the π^0 mesons produced when annihilation takes place. A considerable amount of literature is now available on the expected results of matter-antimatter annihilation. In particular, Stecker (1971) has reviewed in detail the dominant γ-ray producing processes and Stecker *et al.* (1971a, b) have calculated the resulting γ-ray spectrum, in particular, from p-$\bar{\mathrm{p}}$ interactions (see Section II-2.4.4 for positron-electron annihilation). The γ-rays which result are mainly through the intermediary of meson production and not direct annihilation γ-rays. The main processes for p-$\bar{\mathrm{p}}$ annihilation of importance to cosmological problems have been reviewed by Stecker (1971), who concludes that the following three cases should dominate:

1. $(10^{11}\ \mathrm{K} \lesssim T \lesssim 10^{13}\ \mathrm{K})$

$$\mathrm{p} + \bar{\mathrm{p}} \longrightarrow \text{bosons} \longrightarrow \pi^0 \longrightarrow 2\gamma\,.$$

In this case, where the kinetic energy of one of the protons is greater than 10 MeV, the cross section is given by

$$\sigma_\mathrm{A} = \pi r_\mathrm{p}^2/\beta_\mathrm{cms}\ \ (\mathrm{cm}^2)$$

where the proton (antiproton) radius is $r_\mathrm{p} = 0.87 \times 10^{-13}$ cm and $\beta_\mathrm{cms} \sim \frac{1}{2}\beta$ for this case. For this temperature region

$$\sigma_\mathrm{A}(\mathrm{I}) = 4.8 \times 10^{-26}\,\beta^{-1}\ \ (\mathrm{cm}^2). \qquad \text{(II.43)}$$

2. $(10^4\,\mathrm{K} \lesssim T \lesssim 10^{11}\,\mathrm{K})$

For direct annihilation below 10 MeV, the Coulomb attraction of the proton and antiproton causes an increase in the direct annihilation cross section. This modifies the cross section given in Equation (II.43), so that at ~4 MeV the velocity dependence of σ_A

changes from $\propto \beta^{-1}$ to $\propto \beta^{-2}$. For this case

$$\sigma_A(\text{II}) = 2.2 \times 10^{-27}\ \beta^{-2}\ (\text{cm}^2) \tag{II.44}$$

3. $(10\ \text{K} \lesssim T \lesssim 10^4\ \text{K})$

When the annihilation region is cool and consists of neutral atomic gases of H and $\overline{\text{H}}$, then the following so-called H-$\overline{\text{H}}$ rearrangement collision takes place: $\overline{\text{H}} + \text{H} \longrightarrow \Pi_p + (\Pi_e$ or $e^+ + e^-)$ where Π_p and Π_e are bound states of the indicated antipairs. This reaction leads to the dominant mode of annihilation whose cross section has been given by Morgan and Hughes (1970) as

$$\sigma_{\overline{\text{H}}\text{-H, A}} \simeq (0.31\, a_0^2)\ \beta^{-0.64}$$

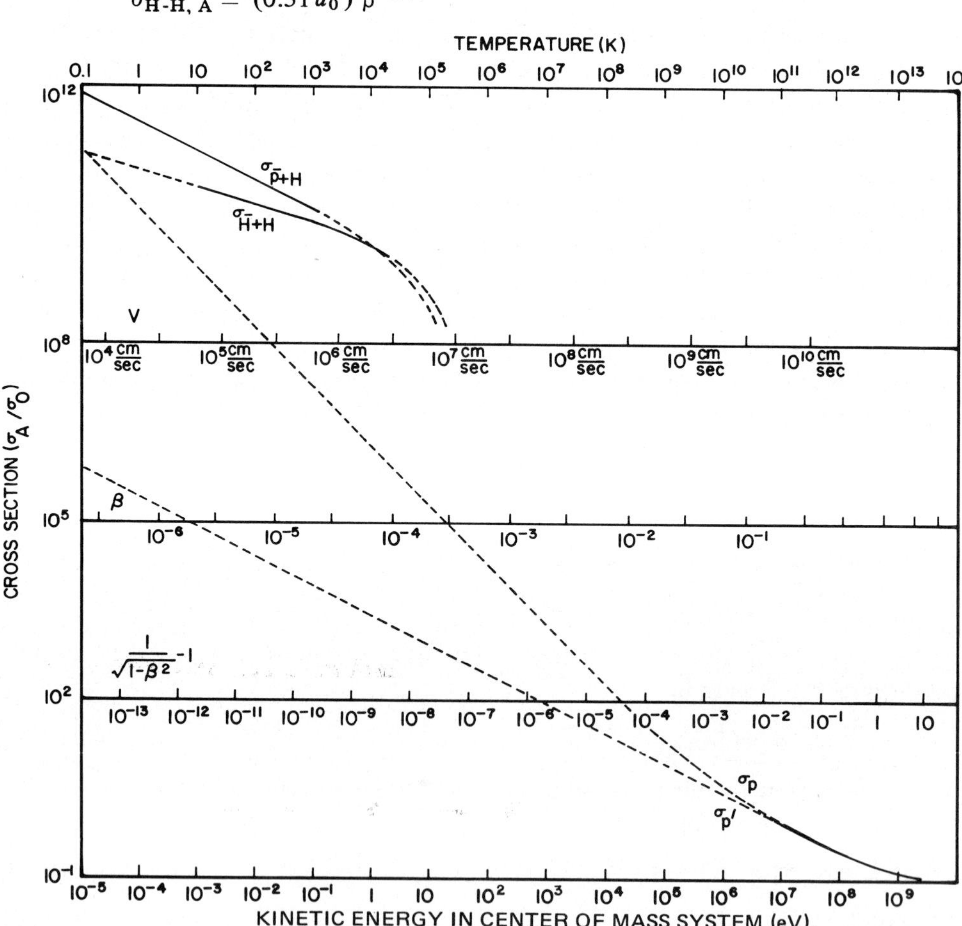

Fig. II-10. The annihilation cross sections for p-$\overline{\text{p}}$ and H-$\overline{\text{H}}$ interactions are given vs. the kinetic energy in the center of mass system. The curve marked σ_p takes into account the mutual Coulomb attraction between the proton and antiproton; curve marked $\sigma_p{}'$ is an extrapolation from the accelerator data that fails to take this effect into account. (From F.W. Stecker: 1971, *NASA SP-249* and Monobook Company, Baltimore, Maryland, 1971.)

where a_0 (first Bohr radius) ~ 0.51 Å. The cross section in this case is:

$$\sigma_A(\text{III}) = 2.6 \times 10^{-18}\ \beta^{-0.64}\ (\text{cm}^2)\,. \tag{II.45}$$

The p-$\bar{\text{p}}$ annihilation cross section appropriate over the full energy range of conceivable interest is shown in Figure II-10 plotted versus the kinetic energy in the center of mass system (see Morgan and Hughes, 1970; Stecker, 1971). Abscissas corresponding to temperature and velocity (v and β) are also shown as well as the kinetic energy in rest mass units $[(1-\beta^2)^{-1/2}-1]$. The cross section of the ordinate is expressed in units of $\sigma_0 = \pi r_0^2$ where $r_0 = e^2/m_0c^2 = 2.8 \times 10^{-13}$cm is the classical radius of the electron so $\sigma_0 = 2.46 \times 10^{-25}\,\text{cm}^2$.

It is of particular interest to know the γ-ray spectrum resulting from p-$\bar{\text{p}}$ annihilation which is due primarily to the decay of the π^0 mesons produced. The cross section for direct annihilation into γ-rays is lower than that involving meson producing by $\alpha^2 = (1/137)^2$.

The γ-ray spectrum resulting will of course depend on the energy distribution of the π^0 mesons and that in turn depends on the astrophysical situation. Stecker *et al.* (1971a, b) and Frye and Smith (1966) have calculated the γ-ray spectrum resulting from p-$\bar{\text{p}}$ annihilations at rest. The normalized spectrum resulting is shown in Figure II-11. For this case the lowest photon energy is ~5 MeV and the highest is ~919 MeV but the region of significant photon flux only is shown in Figure II-11. The results of Frye and Smith (1966) also generally agree with this curve.

In Section III-3.3 the effects of a cosmological red shift on the γ-ray spectral shape are discussed in connection with the interpretation of a possible diffuse cosmic γ-ray

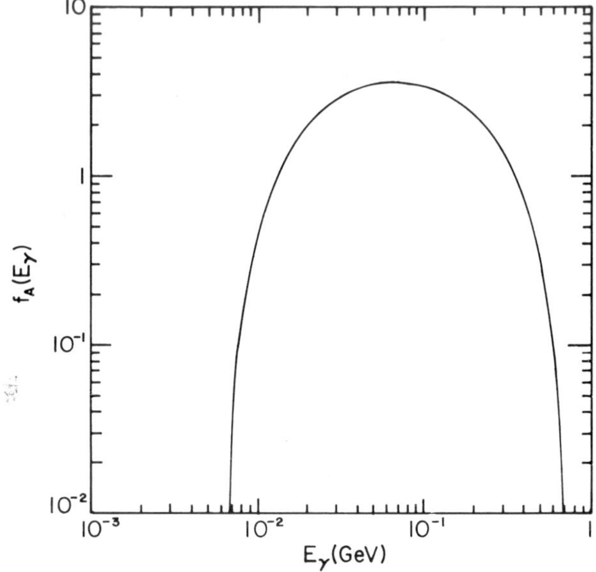

Fig. II-11. The normalized local π^0 γ-ray spectrum resulting from p-$\bar{\text{p}}$ annihilation at rest. (From F.W. Stecker: 1971, *NASA SP-249* and Monobook Company, Baltimore, Maryland, 1971.)

component. In addition, the matter density changes with the red shift, so a complicated photon transport calculation must be performed in order to determine the spectrum. This problem has been treated by Arons (1971a, b) and Stecker *et al.* (1971a).

2.4.3. NUCLEAR EXCITATIONS

Truly nuclear γ-radiation takes place only when a nucleus in an excited state undergoes a transition to a lower energy state. This, in principle, includes the capture γ-radiation emitted when the compound nucleus decays in the (p, γ) and (n, γ) capture reactions. The intensity of γ-ray emission for a specific nuclear transition depends on the total angular momentum of the nucleus in the initial and final states and their respective parities. The detailed theory of nuclear radiative transitions is well developed; a thorough treatment may be found in Blatt and Weisskopf (1952) and a modern summary is presented in Marmier and Sheldon (1969). Also, an intermediate level intuitive approach is given by Cohen (1971). We will summarize the most significant results necessary for a proper interpretation of experimental data.

The ultimate challenge of nuclear line γ-ray astronomy is similar to that of the optical atomic or molecular spectroscopy of the stars, except the physical conditions required for γ-ray line emission are much different, since energetic particles with energies above 10 MeV are required. In the optical case, one observes a spectral line at a certain wavelength or energy and with a particular intensity. The wavelength of the line is often sufficient to determine the atomic or molecular species responsible, and the line shape gives information about the temperature and the density of the stellar medium. The intensity of the spectral line is directly related to the number of atoms or molecules in a particular excited state through the transition probability associated with the particular line. In many cases, these are unknown and theoretical values must be used. The same is true for the nuclear case. In Figure II-12, we show a hypothetical situation where a particular nucleus with atomic number Z and mass number A has two excited states. The total angular momentum, parities, and energies of the states are denoted, respectively, by J_i, π_i, and E_i. Three nuclear decay transitions possible are denoted as γ_1, γ_2,

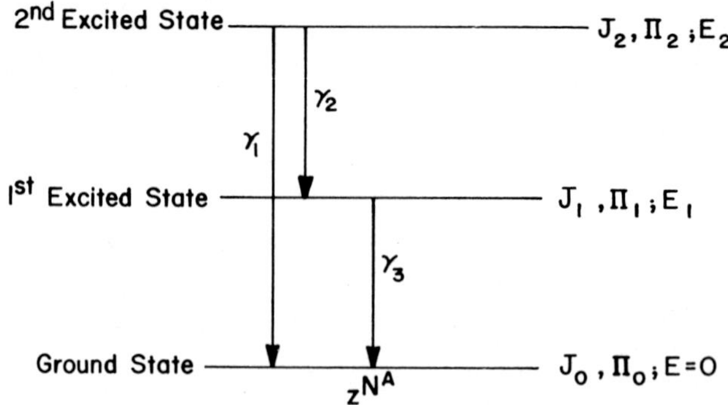

Fig. II-12. The first two excited states of a typical nucleus showing possible γ-ray line transitions.

γ_3 with transition energies E_2, $E_2 - E_1$, and E_1, respectively, if nuclei have been excited to the second level, or both the first and second levels, by some process. The energy of the emitted γ-ray is the transition energy, E, less the energy shift due to the recoil of the nucleus. This correction is $\Delta E_\gamma \sim E^2/2mc^2$ for a recoil nucleus of mass m and is, for example, 1.3 keV for the γ-ray from neutron capture on protons.

The instantaneous intensity of γ-rays for a particular transition is

$$N(\mathrm{s}^{-1}) = N_\mathrm{i}(t) A_{\mathrm{i} \to \mathrm{f}} \tag{II.46}$$

where N_i is the number of nuclei in a particular initial excited state and $A_{\mathrm{i}\to\mathrm{f}}$ (s^{-1}) is the transition probability from the initial state, i, to the final state, f, giving a γ-ray of energy $E_\mathrm{i} - E_\mathrm{f}$.

The population $N_\mathrm{i}(t)$ is determined by the excitation mechanism and is usually discussed in terms of the cross section for the process. We are interested here primarily in the nuclear properties that relate to $A_{\mathrm{i}\to\mathrm{f}}$. The emission of nuclear γ-rays is described (as in atomic radiation) in terms of the classical multipole description of an oscillating charge or current distribution. In the quantum mechanical description, a photon carries off from the excited nucleus an energy $E = E_\mathrm{i} - E_\mathrm{f} = \hbar\omega$ and an angular momentum $L = \hbar\,[l(l+1)]^{1/2}$. In terms of vector conservation of angular momentum for a nuclear transition $\mathbf{J}_\mathrm{i} \xrightarrow[l]{\hbar\omega} \mathbf{J}_\mathrm{f}$ then

$$\mathbf{J}_\mathrm{i} - \mathbf{J}_\mathrm{f} = \mathbf{L}$$

which requires that the quantum number l be bounded by

$$|J_\mathrm{i} - J_\mathrm{f}| \leq l \leq J_\mathrm{i} + J_\mathrm{f}.$$

The J values in normal type face are the total angular momentum quantum numbers describing the states shown, for example, in Figure II-12. There are two possibilities (Cohen, 1971):

(a) The state of the oscillating charge distribution changes, so *Electric Multipole Radiation* known as *$E - l$ radiation* is produced.

(b) The oscillating current distribution changes, so *Magnetic Multipole Radiation* known as *$M - l$ radiation* is produced.

Note that the multipole order for the radiation is determined by the angular momentum quantum number belonging to the photon. The transition probabilities for the two cases are:

$$(E - l) \qquad A_{\mathrm{i}\to\mathrm{f}} \equiv \lambda(E - l) = \frac{2(l+1)}{\hbar l\,[(2l+1)!!]^2} \left(\frac{\omega}{c}\right)^{2l+1} Q_l^2 \ (\mathrm{s}^{-1})$$

$$(M - l) \qquad A_{\mathrm{i}\to\mathrm{f}} \equiv \lambda(M - l) = \frac{2(l+1)}{\hbar l\,[(2l+1)!!]^2 c^2} \left(\frac{\omega}{c}\right)^{2l+1} A_l^2 \ (\mathrm{s}^{-1}) \tag{II.47}$$

The quantities Q_l and A_l are closely related to the electric and magnetic multipole moment oscillations causing the transition. In quantum mechanical terms, they are the

the matrix elements given by time-dependent perturbation theory

$$\begin{matrix} Q_l \\ = \\ A_l \end{matrix} \int \psi_f^* \begin{pmatrix} H(E-l) \\ \text{or} \\ H(M-l) \end{pmatrix} \psi_i \, d\tau \tag{II.48}$$

The operators $H(E-l)$ and $H(M-l)$ of the transition are analogous to the classical quadrupole moments of oscillating charge or current distributions, respectively. As an intuitive example, Cohen (1971) gives the following illustrations for an electric transition: In the single particle picture of a nuclear transition, gamma radiation is emitted when a nucleon changes from one orbit to another. On the other hand, in the above description the electromagnetic radiation emitted arises from the change in the state of oscillation of electric charges. Thus, when a proton changes orbits, its energy and hence its frequency of oscillation is charged, which is what is necessary for a system to emit electromagnetic radiation. Needless to say, the matter is entirely analogous to the atomic case, but unfortunately the problem of the basic nuclear interaction is not solved yet.

The parity change involved in the two types of transitions are

$$E-l \qquad \Delta\pi = (-1)^l$$

$$M-l \qquad \Delta\pi = (-1)^{l-1}$$

so magnetic multipole radiation has opposite parity to electric multipole radiation of the same order. It is best to illustrate the application of these momentum and parity selection rules with an example taken from Marmier and Sheldon (1969), shown in Figure II-13. It is seen that, depending on the initial and final J values, simple multipolarities or more complex multipolarities (or mixtures) are possible. The problem is even more complicated in practice since there are isospin selection rules that must be applied when the isotopic spin is a good quantum number. This is also discussed in Marmier and Sheldon (1969).

For the purpose of avoiding errors in γ-ray astronomy, the only sure approach is to use the experimental data on lifetimes of specific transitions as given in the Lederer *et al.* (1968) tables or the Landolt/Börnstein tables (Hellwege, 1961). Also, one can use experimental cross sections for producing a given γ-ray line which implicitly includes the transition probability. A useful tabulation of excited nuclear state lifetimes is given by Lindskog *et al.* (1966). In general, the lifetime of a nuclear level $\tau_i = (\sum_f A_{i\to f})^{-1}$.

In cases where no experimental lifetimes or transition rates are available, then recourse can be made to theoretical predictions with due regard for caution. As an example, if the actual nuclear changes are due to single nucleon transitions (single particle picture) then the so-called Weisskopf transition rates may be used for a guide. See Blatt and Weisskopf (1952) for a detailed discussion.

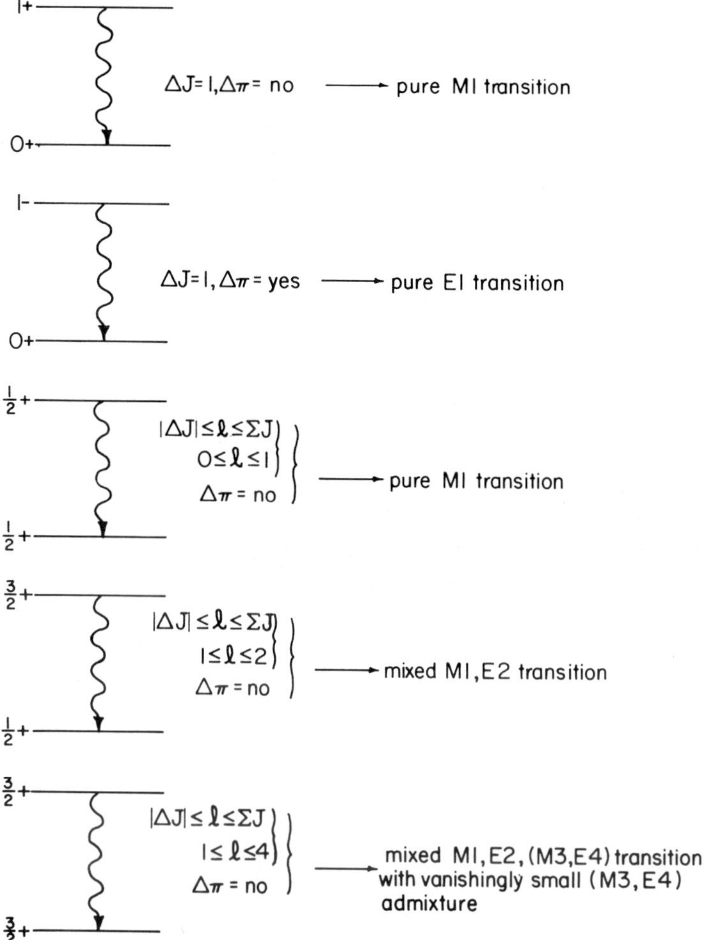

Fig. II-13. Several examples illustrating the angular momentum and parity selection rules for multipole radiation. (From P. Marmier and E. Sheldon, *Physics of Nuclei and Particles, Vol. 1,* Academic Press, Inc., New York, 1969. Used by permission.)

In Figure II-14 we show $\lambda(E\text{-}l)$ versus the energy of the transition for $E\text{-}l$ radiation for $1 \leqslant l \leqslant 5$ and a range of atomic numbers. In Figure II-15 is shown the corresponding values for magnetic multipole radiation, $\lambda(M\text{-}l)$. Notice that at lower energies the decay rates can be increased over γ-ray emission by the internal conversion process which suppresses γ-ray emission. In these cases, one should be careful to use the lower branch of the curve for a given A in order to estimate the γ-ray intensity. Table II-5, taken from Marmier and Sheldon (1969), summarizes the nomenclature, selection rules, and typical half-lives for transitions of energy 1 MeV.

As a final practical example of particular interest in γ-ray astronomy, we consider the actual case of the γ-rays expected from excited states of ^{12}C and ^{16}O which, for

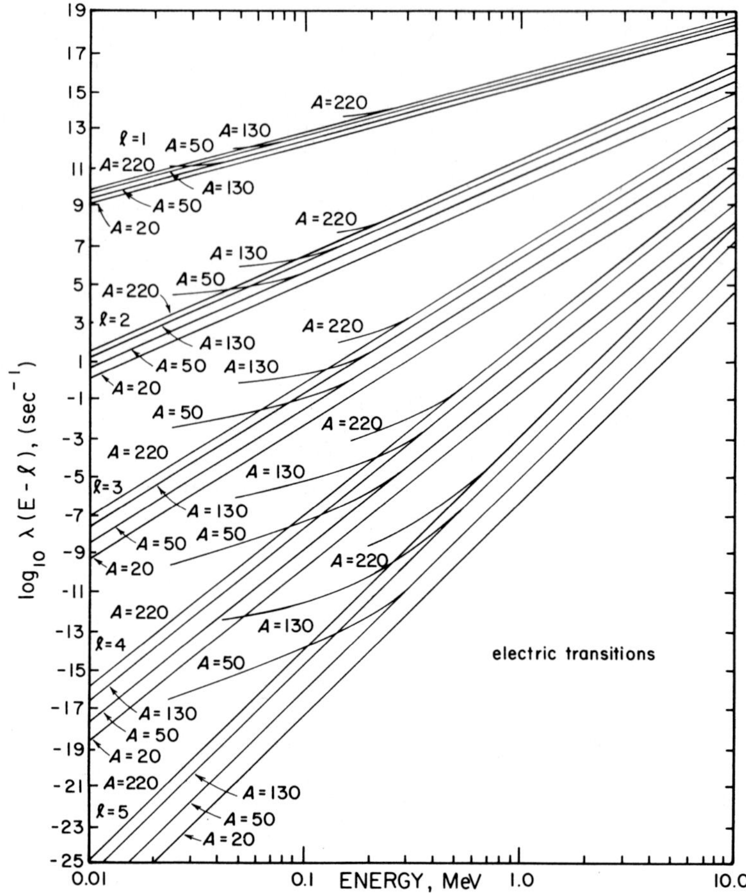

Fig. II-14. The theoretical single-particle transition probabilities for *electric* multipole radiation vs. photon energy. The results are shown for a range of mass numbers A for each value $l = 1 \to 5$. The lower branch is for γ-ray emission only, and the upper branch includes internal conversion. (From R.W. Hayward, in E.U. Condon and H. Odishaw (eds.), *Handbook of Physics.* Copyright 1967, McGraw-Hill Book Company. Used by permission.)

TABLE II-5

Nomenclature, selection rules, and typical half-lives for 1 MeV γ-ray transitions. (From P. Marmier and E. Sheldon, *Physics of Nuclei and Particles, Vol. 1,* Academic Press, Inc., New York, 1969. Used by permission.)

$T_{1/2}$ for $E_\gamma \approx 1$ MeV	10^{-12} s		$10^{-7.5}$ s	10^{-3} s
E-l radiation	Dipole	Quadrupole	Octopole	Hexadecapole
$\left.\begin{array}{l}\lvert\Delta J\rvert \equiv \lvert J_i - J_f\rvert \leqslant \\ \Sigma J \equiv J_i + J_f \geqslant\end{array}\right\} l$	1	2	3	4
$\Delta\pi = (-1)^l$	Yes	No	Yes	No
$M - l$ radiation	Dipole		Quadrupole	Octopole
$\left.\begin{array}{l}\lvert\Delta J\rvert \equiv \lvert J_i - J_f\rvert \leqslant \\ \Sigma J \equiv J_i + J_f \geqslant\end{array}\right\} l$	1		2	3
$\pi = (-1)^{l-1}$	No		Yes	No

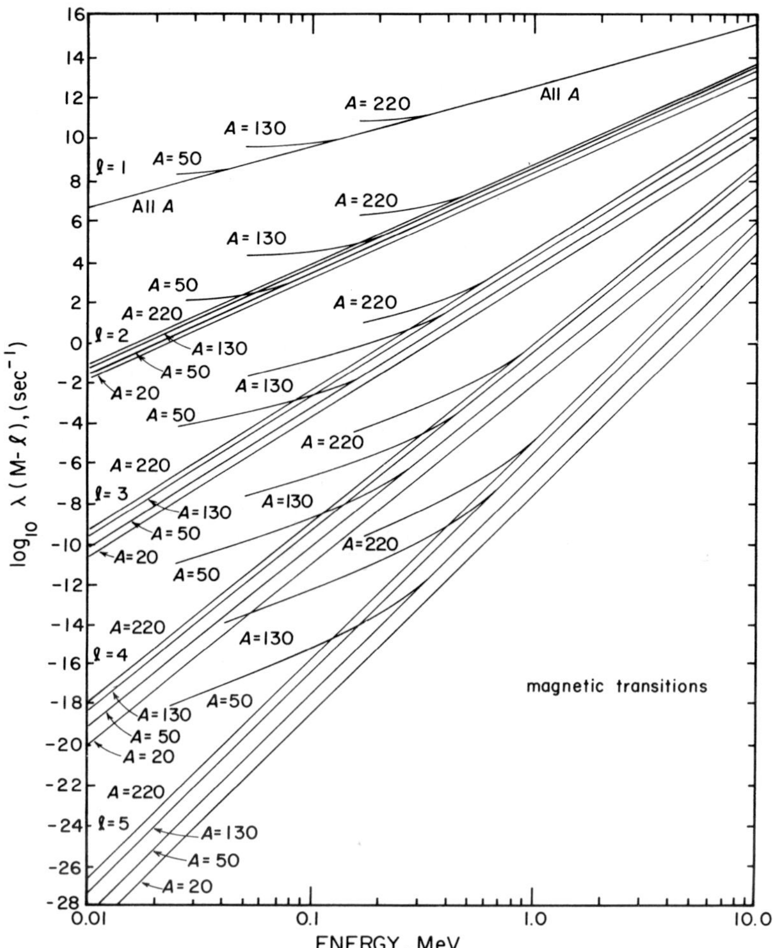

Fig. II-15. The theoretical single-particle transition probabilities for *magnetic* multipole radiation vs. photon energy. The results are shown for a range of mass numbers A for each value $l = 1 \rightarrow 5$. The lower branch is for γ-ray emission only, and the upper branch includes internal conversion. (From R.W. Hayward, in E.U. Condon and H. Odishaw (eds.), *Handbook of Physics*. Copyright 1967, McGraw-Hill Book Company. Used by permission.)

example, could be produced by (p, p′γ) reactions. We show in Figure II-16 the known experimental level schemes of ^{12}C and the population of the various levels by β^- and β^+ decay process, taken from Lederer *et al.* (1968). Consider the first two excited states of ^{12}C: Since the second excited and ground states are both $J = 0$, the $E0$ transition is strictly forbidden (although decay by two photon emission is possible). The second excited state in this case decays by α emission with a measured lifetime of 5×10^{-17}s as determined from experiment. The first excited state, however, is 2^+ so $2 \leqslant l \leqslant 2$ and $E2$ and $M2$ radiation is possible, but there is no parity change so the observed radiation must be $E2$. From the Weisskopf rates (Figure II-14) the lifetime of the level would be expected to be $\sim 10^{-12}$s for a 4.4 MeV γ-ray, and the experimental value is 4×10^{-14}s.

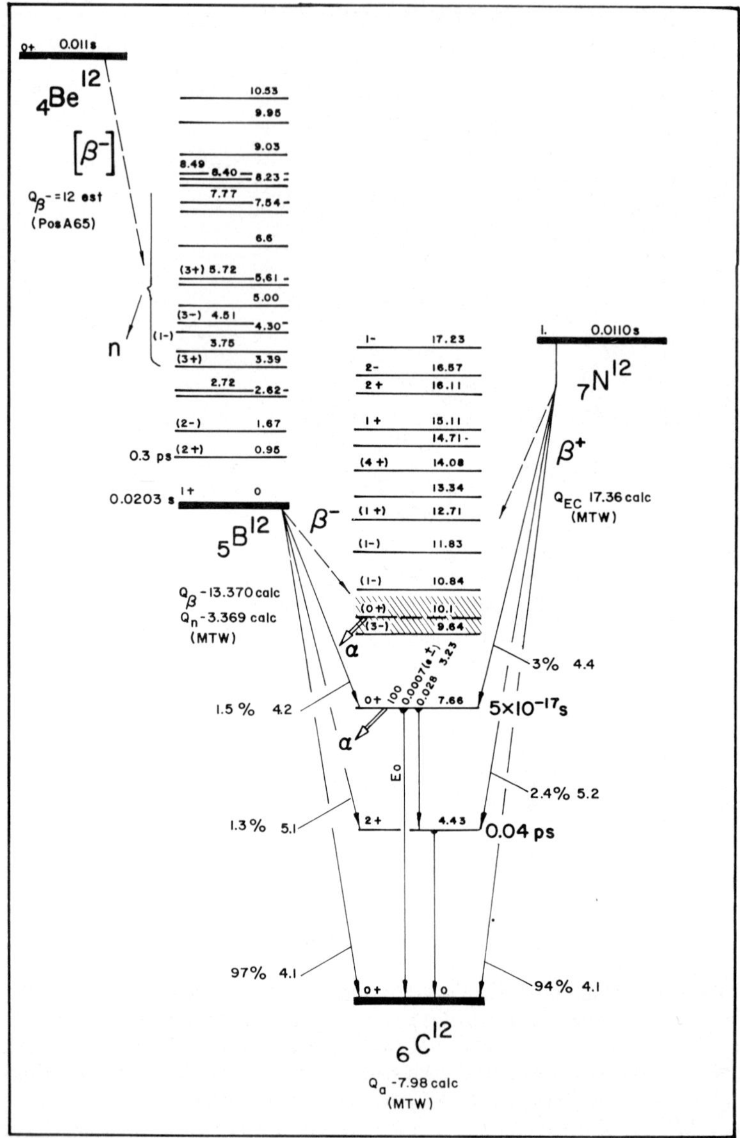

Fig. II-16. The known energy level scheme for ^{12}C is shown with experimentally determined spins and parities. The population of some of the ^{12}C states by β^- and β^+ decay of ground state ^{12}B and ^{12}N is shown. (From C. M. Lederer *et al.*, *Table of Isotopes*, John Wiley & Sons, New York, 1968, by permission of the U.S. Energy Research and Development Administration.)

This line is a strong emission from ^{12}C, which can be excited in many ways.

Now consider the levels of ^{16}O shown in Figure II-17, where the ground and first excited states are both determined to be 0^+, so a single γ-ray transition of energy 6.052 MeV is strictly forbidden. This level of ^{16}O, however, decays by two photon emission or

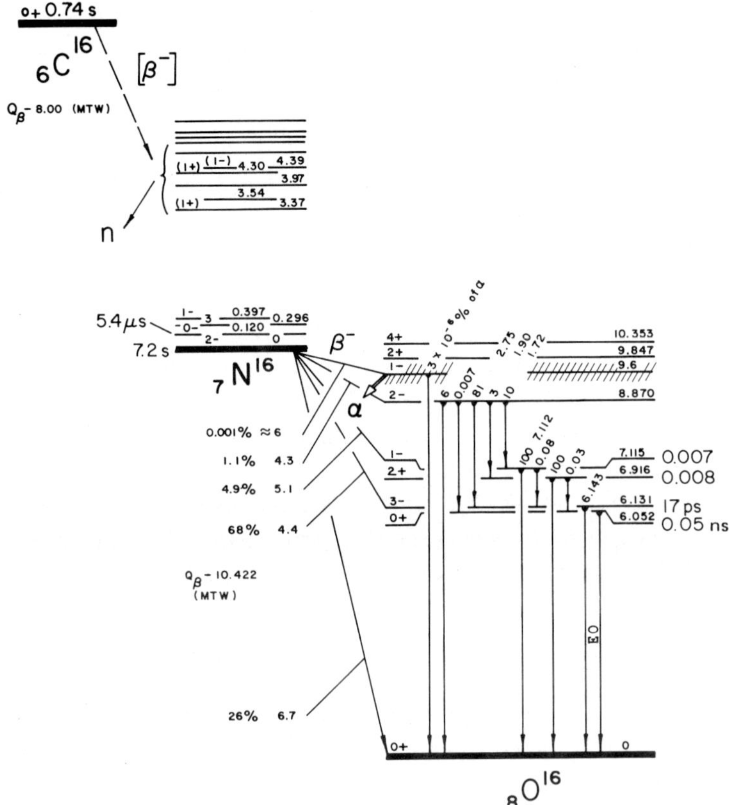

Fig. II-17. The known energy level scheme for ^{16}O is shown with experimentally determined spins and parities. The population of some of the ^{16}O states by β^- decay of ground state ^{16}N is shown. (From C. M. Lederer *et al., Table of Isotopes*, John Wiley & Sons, New York, 1968; by permission of the U.S. Energy Research and Development Administration.)

by a photon and a conversion electron (not of unique energy) with a lifetime of 5×10^{-11} s. Now the level just above the first excited state is 3^-, so we expect $3 \leqslant l \leqslant 3$ or $E3$ and $M3$ radiation, but since there is a parity change, the radiation must be $E3$ (octopole), and the measured lifetime of the state is $\sim 1.7 \times 10^{-11}$ s. A γ-ray from this level of ^{16}O is also a strong feature seen in experimental situations wherever the 3^- state is populated.

As stated in Equation (II.46) above, the intensity of the γ-rays from excited nuclei depends also on the instantaneous number of nuclei which are in a particular excited state. In the examples shown for ^{12}C and ^{16}O, the various excited levels were populated by β^- and β^+ decay of a parent nuclide. In this case, then, the instantaneous population is determined by the instantaneous number of parent nuclei and the radioactive half-life of the parent. There are, of course, several other ways that excited states can be produced, such as (p, p′γ) or (n, n′γ). Spallation reactions from, for example, protons on a target of higher atomic number may also produce the excited nuclide of interest. The γ-ray intensity expected then depends on $A_{i \to f}$ and on the total probability of exciting the level in question by all the different processes according to:

$$N_i = \sum_{j=1}^{N} N_{ij}$$

where j refers to the process which excites the atom to state i.

a. *Nuclear Excitation by Charged Particles:* (p, p$'\gamma$), (α, $\alpha'\gamma$), (p, γ), (α, γ), *Spallation, and Fission*

Direct production of γ-rays can occur in charged particle interactions with a nuclide $_Z N^A$ according to, for example, (p, p$'\gamma$), (p, γ), (α, γ), or similar reactions with heavier incident charged particles. Another process of γ-ray line production by charged particles occurs by Coulomb excitation in which the incident particle does not penetrate the Coulomb barrier as is required in the specific nuclear reactions mentioned here. The cross section for this process which excites surface oscillations of the nucleus (e.g. rotational levels) is typically small compared to (p, p$'\gamma$) reactions at comparable bombarding energies ~ 10 MeV (cf. Alburger, 1966 and also Marmier and Sheldon, 1970).

Consider a general nuclear reaction of the form

$$a + A \to b + B \qquad \text{or } A(a, b)B \tag{II.49}$$

where, in general, lower case letters represent light particles and capitals represent heavy particles. Conservation of energy for the reaction is designated by $E_a + E_A = E_b + E_B$ where E represents the total energy of the reacting particles (rest and kinetic). If the recoiling nucleus represented by B is left in an excited state E_B^* then $E_B^2 = P_B^2 c^2 + (M_B c^2 + E_B^*)^2$ where P_B is its momentum and M_B is its rest mass.

Such reactions are usually classified according to the Q_0 values where

$$Q_0 = (M_a + M_A - M_b - M_B)c^2. \tag{II.50}$$

If $Q_0 < E_B^*$ (endoergic) then energy is required in the center of mass system for the reaction to proceed, and if $Q_0 > E_B^*$ (exoergic) then no energy need be added for the reaction to proceed. Since any energy required must come from the initial kinetic energy of a and A, conservation of energy gives:

$$Q_0 + \epsilon_a + \epsilon_A = \epsilon_b + \epsilon_B + E_B^* \tag{II.51}$$

and

$$E_{\text{threshold}} = (E_B^* - Q_0)\left(\frac{M_b + M_B}{M_b + M_B - M_a}\right) \tag{II.52}$$

for the non-relativistic case (< 100 MeV) and $E_B^* \ll M_B c^2$, where ϵ refers to kinetic energies and in Equation (II.51) $\epsilon_A = 0$. Addition of the conservation of momentum relation provides all the information necessary to describe all the reaction parameters. A convenient summary giving the kinematical relations for the energies and emission angles of the secondary particles may be found in several places, for example: Marion *et al.* (1959), Marmier and Sheldon (1969, p. 601), Von Buttlar (1968, p. 51), and Evans (1955, p. 408).

(p, p$'\gamma$) *and* (α, $\alpha'\gamma$). In the case of inelastic scattering to an excited state [for example for (p, p$'\gamma$) or (α, $\alpha'\gamma$) reactions], $a \equiv b$ and $A \equiv B$, so $Q_0 = 0$. Therefore, in the laboratory system the threshold for the reaction is

$$(\epsilon_a)_{\text{threshold}} = \left(1 + \frac{M_a}{M_A}\right) E_B^* . \tag{II.53}$$

If the heavy particle is the projectile and the light particle is at rest then a and A should be exchanged; however, the threshold energy expressed in energy per nucleon of the projectile is the same for either case.

Figure II-18 shows the energy dependence of the (p, p$'\gamma$) and (α, $\alpha'\gamma$) cross sections for ^{12}C and ^{16}O as given by Ramaty and Lingenfelter (1975). The abscissa is given in units of MeV per nucleon so these cross sections may be used either for the case when the proton or α-particle is the projectile incident on a given target nucleus leaving the latter in a given excited state, or for the inverse case when the heavy nucleus is the

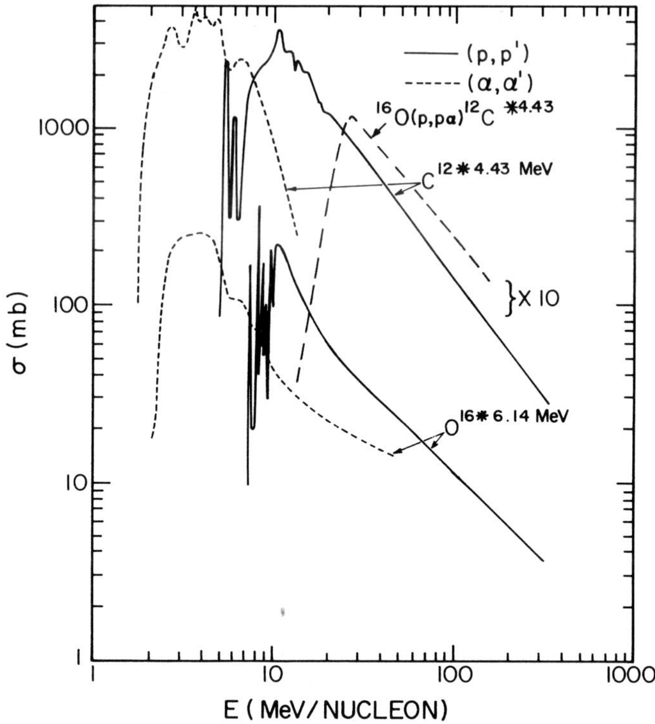

Fig. II-18. The energy dependence of the (p, p$'\gamma$) and (α, $\alpha'\gamma$) reaction cross sections on ^{12}C and ^{16}O plotted vs. the incident kinetic energy per nucleon. (From R. Ramaty and R.E. Lingenfelter: 1975, in S.R. Kane (ed.), *Solar Gamma-, X-, and EUV Radiation,* IAU COSPAR Symposium No. 68, D. Reidel Publishing Company. Used by permission.)

projectile and leaves in an excited state after scattering from a proton or α-particle at rest. The kinematics are different for these two cases and the most important effect for γ-ray line astronomy is to seriously broaden the line from its natural width in the case the heavy nucleus is the projectile. Even when the proton is the projectile, the excited nucleus receives enough kinetic energy to produce a slightly broadened spectral line. This point is discussed further in Sections II-2.5.1 and III-3.2.4.

(p, γ) *and* (α, γ). These processes are known as radiative proton or alpha capture reactions and, of course, can occur for any nuclide. They are similar to the slow neutron capture (n, γ) reactions to be discussed below, in which it is convenient to think of the compound state, $(p + {}_ZN^A) = {}_{Z+1}N^{A+1}$, as being excited with an amount of energy equal to the binding energy of the last proton in the nucleus ${}_{Z+1}N^{A+1}$ in the case for which the proton has zero kinetic energy. The nucleus, thus excited, can decay by γ-ray emission to all allowed lower states giving a rich spectrum of γ-ray lines. This type of reaction could proceed in principle with a charged particle of zero kinetic energy except for the Coulomb barrier; so the cross section at low energies behaves as $\sigma \propto 1/v_p \exp(-G_p)$ which rises monotonically from zero kinetic energy until the barrier transparency factor $\exp(-G_p)$ is sufficiently increased to give a significant cross section. For a high Coulomb barrier $G_p \propto 2\pi Z_A e^2/\hbar v_p$ where v_p is the velocity of the incident proton (see Evans, 1955, p. 874).

A more general discussion of (p, γ) and other reactions is given by Alburger (1966) and elucidates the significance of resonances, and we will follow that discussion. In Figure II-19 is shown an energy level diagram which illustrates the typical case when a competing reaction, such as (p, α), also can occur. On the left is shown the target nucleus

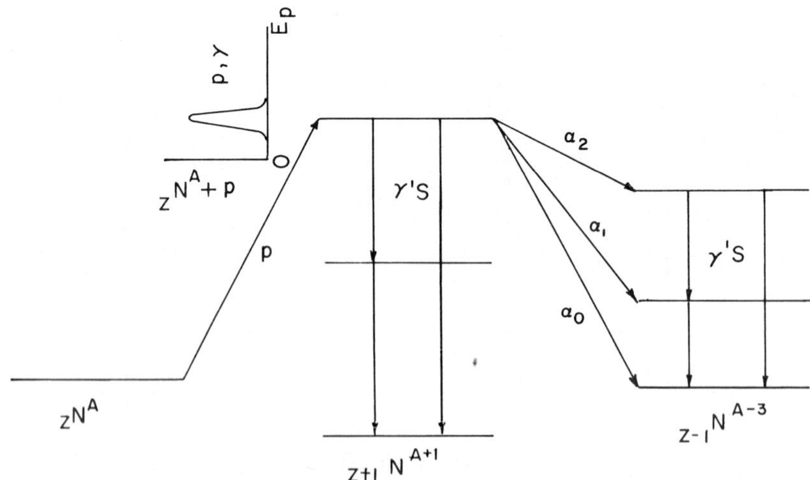

Fig. II-19. Energy level diagram illustrating (p, γ) and (p, α) reactions. (From D. Alburger: 1966, in K. Siegbahn (ed.), *Alpha-, Beta-, and Gamma-Ray Spectroscopy, Vol. 1.* Used by permission North-Holland Publishing Company.

$_ZN^A$ which absorbs a proton. The reaction can only occur if the proton kinetic energy E_p is such that the total excitation energy of the compound state is close to one of the (virtual) levels of the nucleus $_{Z+1}N^{A+1}$. When these levels are widely spaced and one is dealing with a thin target situation in a laboratory accelerator experiment and if the proton energy is steady, then a resonant capture can occur as the proton energy is varied. The resonance is identified by observing the yield of γ's and α's. This is shown at the inset in the Figure II-19. The resonance indicates the presence of a virtual level in the compound nucleus. Accelerators with well controlled energy (such as Van de Graaffs) have traced out the shape and widths of the resonances which can be related to the mean life of the virtual state by the uncertainty principle according to τ (sec) $= 6.6 \times 10^{-16}/\Delta E$ (eV). Since the state can decay by several paths, each denoted by i, the total width of the level is $\Delta E = \sum_{i=1}^{N} \Delta E_i$, where N is the total number of decay paths. Since we are interested in the gamma ray width, ΔE_γ, the lifetime of the level for emitting a particular gamma ray line will be larger than the total lifetime.

The Q of the reaction is, in the usual manner, for (p, γ) reactions

$$Q = {}_ZM^A + {}_1M^1 - {}_{Z+1}M^{A+1}$$

where $_1M^1 = m_p$, the mass of the proton, and the total energy added to the nucleus $_ZM^A$ at a resonance is

$$E_T = 7.5845\,\text{MeV} + (1 - m_p/{}_{Z+1}M^{A+1})E_p. \qquad \text{(II.54)}$$

The first term is the mass defect of the proton and the last term is the kinetic energy of the proton at resonance in the center of mass sytem. As indicated in Figure II-19, the γ-ray energies are determined by the allowed transitions in the product nucleus. A convenient tabulation by Butler (1959), gives the resonance energies for (p, γ) reactions on several elements as well as the γ-ray energies and the cross sections at resonance.

Some recent measurements of several (p, γ) reaction rates have been published by Roughton *et al.* (1974). This work gives the yield for (p, γ) reactions on ^{12}C, ^{29}Si, ^{46}Ti, ^{47}Ti and ^{56}Fe for proton kinetic energies up to ~4 MeV. Typically the reaction yields are $\sim 10^{-6}$ reactions per proton, which is small compared to typical (p, p$'\gamma$) reaction yields which are $\sim 10^{-2}$.

We have so far described some simpler cases of γ-ray line production by charged particles. In the practical case, spallation reactions or the fission process can leave residual nuclei in excited states which give γ-rays directly, and any radioactive by-products which decay by particle emission ($\beta^+, \beta^-, p, \alpha, \ldots$) can also populate levels in the daughter nuclei. Very little information is available, however, on the spallation cross sections leading to excited states of the product nuclei. We discuss briefly in Section VI-6.1.2 the calculations of spallation cross sections which are available. Concerning the γ-rays emitted in association with fission, usually about 5 MeV is emitted in the form of γ-rays per fission event, but the associated γ-ray spectrum is very complex, falling monotonically from keV energies to >10 MeV. Recent experiments show structure which possibly results from production of fission fragments in high spin states. A fairly recent review of this subject has been given by Johansson and Kleinheinz (1966).

We have discussed above several independent mechanisms by which charged particles can produce γ-rays. In practice, it is extremely difficult to calculate the γ-ray yield from a given target even at a single incident energy. Some experimental data are available from work at Oak Ridge, however, inspired by shielding studies for the Apollo program. Zobel *et al.* (1968) have reported γ-ray cross sections from the bombardment of various light and intermediate weight elements Be, C, O, Al, Co, and Bi by 16 to 160 MeV protons and 59 MeV α-particles. The target thicknesses were such that about 15 to 20% of the incident beam energy was removed by ionization loss so the targets were more thin than thick. Most of the yield measurements were made with the γ-ray spectrometer at 135° to the incident proton beam. In general, the angular distribution was isotropic to ~50% except for 14.5 MeV protons on C. Because of the importance of these results for γ-ray line astronomy, we reproduce in Table II-6 the resulting cross sections for producing different γ-ray lines for each target and each proton energy or α-particle energy as shown in the first two columns. In columns 3 and 4 are shown the possible reaction and transition energy in the product nucleus giving the experimentally observed γ-ray line in column 5. Finally, the production cross section for producing the particular line is shown in column 6, so no correction is needed for the transition probability. In general,

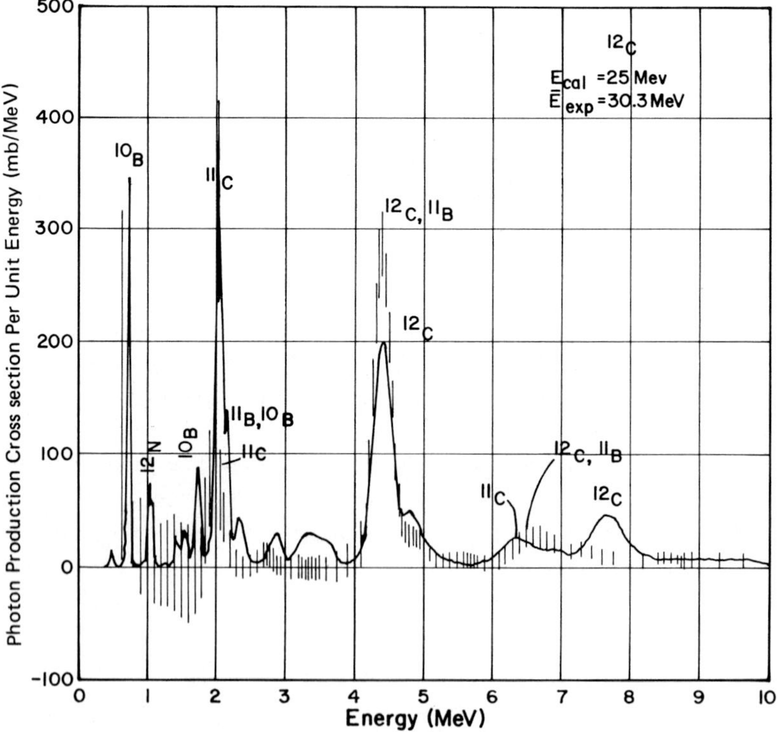

Fig. II-20. The measured γ-ray spectrum obtained when a thick C target is bombarded with 30.3 MeV protons is shown by the 1σ lines delineating the spectrum. The solid curve is an attempt at a theoretical fit to the experimental data of Zobel *et al.* (1968, *Nucl. Sci. Eng.* **32**, 392). (From Y. Shima and R.G. Alsmiller: 1970, *Nucl. Sci. Eng.* **41**, 47. Used by permission American Nuclear Society.)

the total measured γ-ray production depends on proton energy in a way similar to the stopping power due to ionization. The approximate total γ-ray yield per proton (for $E_\gamma > 700\,\mathrm{keV}$) is typically $\sim 10^{-2}$ for $E_p \sim 30\,\mathrm{MeV} \rightarrow 50\,\mathrm{MeV}$.

The detailed shape of the γ-ray spectrum from these measurements is also of considerable interest and some examples are shown in Figures II-20 and II-21 from Shima and Alsmiller (1970). In Figure II-20, the ordinate gives the photon intensity expressed as a cross section per unit γ-ray energy interval (MeV) where the γ-ray energy is shown on the abscissa. This spectrum is for 30.3 MeV protons incident on C. The actual experimental cross sections are shown by $\pm 1\sigma$ vertical lines for each channel of the spectrometer, and the solid curves are an attempt at a calculational fit to the measured values by Shima and Alsmiller (1970) using an intranucleus cascade model (see Section VI-6.1.2). As can be seen, the fit to the details of the spectrum is not very good. In Figure II-21 is shown a similar result for 28.2 MeV protons on ^{16}O. The importance of this data for our work is illustrated in Figure II-21 where bombardment of ^{16}O yields a very strong ^{12}C deexcitation line as well as the 6.1 MeV ^{16}O line! This latter type of information is of great value in γ-ray line astronomy for both theoretical predictions and interpretation of experimental results, since it is the simplest way to construct a γ-ray spec-

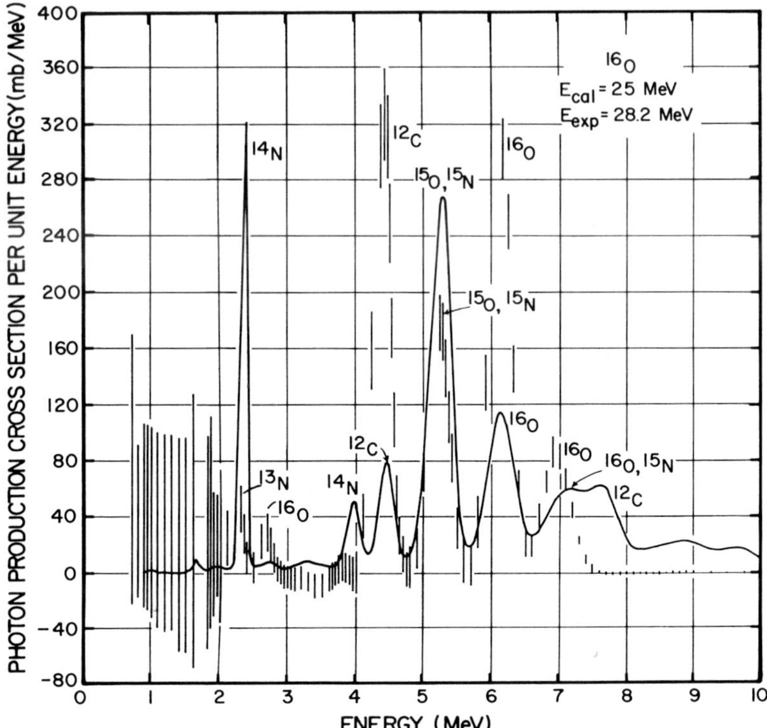

Fig. II-21. The measured γ-ray spectrum obtained when a thick O_2 target is bombarded with 28.2 MeV protons is shown by the 1σ lines delineating the spectrum. The solid curve is an attempt at a theoretical fit to the experimental data of Zobel *et al.* (1968, *Nucl. Sci. Eng.* **32**, 392). (From Y. Shima and R.G. Alsmiller: 1970, *Nucl. Sci. Eng.* **41**, 47. Used by permission American Nuclear Society.)

TABLE II-6

Cross sections and energies of γ-rays produced by protons and α-particles as measured at 135° from the incident beam axis (From W. Zobel *et al.*: 1968, *Nucl. Sci. and Eng.* **32,** 392. Used by permission American Nuclear Society.)

Target	$\bar{E}_p^a$ (MeV)	Possible Reaction	Transition Energy (keV)	$E_\gamma{}^b$ (keV)	$\sigma_{p,x\gamma}(135°)$ (mb/sr)
^7Li	30.8	$^7Li(p, pn)^6Li$	3560	3520 ± 83	0.63 ± 0.09
	30.8	Sum over all gamma-ray energies above 700 keV			2.1 ± 0.9
Be	[56.6]	$^9Be(\alpha, pn)^{11}B$	2900^c	2800 ± 60	0.8 ± 0.4
	[56.6]	$^9Be(\alpha, 2pn)^{10}Be$	3368	3300 ± 80	0.8 ± 0.3
	147	$^9Be(p, \alpha)^6Li$	3560	3575 ± 15	0.16 ± 0.7
	49.3			3475 ± 60	0.28 ± 0.11
	30.2			3450 ± 80	0.63 ± 0.11
	14.7			3475 ± 65	0.57 ± 0.14
	[56.6]	$^9Be(\alpha, n)^{12}C$	4430	4330 ± 60	1.9 ± 0.2
	147			6250 ± 35	0.037 ± 0.018
	49.3			6125 ± 135	--------
	30.2			6180 ± 85	0.40 ± 0.09
	14.7			6150 ± 90	0.14 ± 0.05
	[56.6]	$^9Be(\alpha, 2n)^{11}C$	6490	6570 ± 120	0.57 ± 0.11
^{11}B	50.6	$^{11}B(p,pn)^{10}B$	1033^d	1015 ± 50	3.1 ± 0.8
	31.0			1020 ± 50	3.2 ± 0.9
	50.6	$^{11}B(p, pn)^{10}B$	1433^d	1405 ± 65	1.4 ± 1.2
	31.0			1416 ± 65	1.7 ± 1.8
	50.6	$^{11}B(p,p')^{11}B$	2130	2100 ± 85	2.1 ± 0.8
	31.6	$^{11}B(p,pn)^{10}B$	2150	2205 ± 85	2.4 ± 0.5
	50.6	$^{11}B(p,p')^{11}B$	2900^c	2920 ± 50	1.0 ± 0.2
	31.0			2930 ± 80	1.4 ± 0.3
	50.6	$^{11}B(p, 2p)^{10}Be$	3368	3320 ± 55	1.0 ± 0.2
	31.0			3380 ± 100	1.2 ± 0.3
	50.6	$^{11}B(p,p')^{11}B$	4460	4390 ± 70	1.6 ± 0.2
	31.0			4400 ± 95	3.2 ± 0.3
	50.6	$^{11}B(p,p')^{11}B$	5030	4950 ± 80	0.4 ± 0.1
	31.0			4950 ± 115	0.7 ± 0.2
	50.6	$^{11}B(p,p')^{11}B$	6760	6720 ± 100	0.8 ± 0.1
	31.0			6640 ± 120	2.3 ± 0.2
	50.6	$^{11}B(p,p')^{11}B$	8570	8670 ± 140	0.3 ± 0.1
	31.0			8500 ± 170	0.5 ± 0.1
	50.6	Sum over above peaks			11.7 ± 1.7
	31.0				16.6 ± 2.1
	50.6	Sum over all gamma-ray energies above 700 keV			17 ± 3
	31.0				24 ± 3
C	145	$^{12}C(p, pn)^{11}C$	1990	1988 ± 22	0.50 ± 0.13
	50.3			2000 ± 40	2.3 ± 2.0
	30.3			1978 ± 30	1.8 ± 0.9
	14.6			2030 ± 35	< 1.0
	[56.3]	$^{12}C(\alpha, \alpha n)^{11}C$	2760^c	2800 ± 50	0.8 ± 0.4
	[56.3]	$^{12}C(\alpha, 2pn)^{13}C$	3850	3820 ± 70	2.3 ± 0.5
	145	$^{12}C(p,p')^{12}C$	4430	4474 ± 30	0.89 ± 0.23
	50.3	$^{13}C(p, 2p)^{11}B$	4460	4380 ± 70	3.5 ± 0.3

TABLE II-6 (continued)

	30.3			4375 ± 70	8.2 ± 0.5
	14.6			4340 ± 60	49 ± 4
	[56.3]	$^{12}C(\alpha,\alpha')^{12}C$	4430	4360 ± 70	12.7 ± 0.6
		$^{12}C(\alpha,\alpha p)^{11}B$	4460		
	[56.3]	$^{12}C(\alpha,n)^{15}O$	5240	5250 ± 70	1.9 ± 0.3
		$^{12}C(\alpha,p)^{15}N$	5270		
	145	$^{12}C(p,2p)^{11}B$	6760	~6750[f]	0.24 ± 0.09
	50.3	$^{12}C(p,pn)^{11}C$	6490	6600 ± 150[f]	0.9 ± 0.1
	30.3			6650 ± 150[f]	1.8 ± 0.3
	14.6			———————	< 0.1
	[56.3]	$^{12}C(\alpha,\alpha p)^{11}B$	6760	6520 ± 150	2.6 ± 0.2
		$^{12}C(\alpha,\alpha n)^{11}C$	6490		
	145				1.63 ± 0.28
	50.3				6.7 ± 2.0
	30.3	Sum over above peaks			11.8 ± 1.1
	14.6				49 ± 4
	[56.3]				20.3 ± 1.5
	145				3.3 ± 0.5
	50.3	Sum over all gamma-ray energies above 700 keV			11.1 ± 3
	30.3				15 ± 3
	14.5				52 ± 4
	[56.3]				37 ± 15
O[g]	145	$^{16}O(p,\alpha)^{13}N$	2365	2355 ± 25	0.26 ± 0.10
	48.3			2190 ± 25	< 1.0
	28.2			2230 ± 50	1.0 ± 0.4
	12.1				< 1.0
	145	$^{16}O(p,p')^{16}O$	2750[h]		———————
	48.3			2600 ± 80	0.9 ± 0.3
	28.2			2662 ± 60	0.6 ± 0.3
	12.1			2725 ± 60	2.0 ± 0.6
	[51.8]	$^{16}O(\alpha,\alpha')^{16}O$	2750[h]		< 0.7
	145	$^{16}O(p,2pn)^{13}C$	3850	3720 ± 30	0.22 ± 0.10
	48.3				0.6 ± 0.3
	28.2			———————	< 0.25
	12.1			3625 ± 115	0.8 ± 0.5
	[51.8]	$^{16}O(\alpha,2pn)^{13}C$	3850	3780 ± 100	1.7 ± 0.5
	145	$^{16}O(p,p\alpha)^{12}C$	4433	4430 ± 30	1.26 ± 0.46
	48.3			4270 ± 70	4.0 ± 0.4
	28.2			4335 ± 70	8.7 ± 0.7
	12.1			———————	< 0.2
	[51.8]	$^{16}O(\alpha,2\alpha)^{12}C$	4433	4370 ± 70	9.0 ± 0.7
	145	$^{16}O(p,pn)^{15}O$	5240	5260 ± 25	0.96 ± 0.39
	48.3	$^{16}O(p,2p)^{15}N$	5276	5090 ± 100	4.1 ± 0.5
	28.2			5188 ± 75	5.8 ± 0.5
	12.1			———————	< 0.2
	[51.8]	$^{16}O(\alpha,\alpha n)^{15}O$	5240	5250 ± 70	6.9 ± 0.6
		$^{16}O(\alpha,\alpha p)^{15}N$	5276		
	145	$^{16}O(p,2p)^{15}N$	6328	6290 ± 35	4.4 ± 1.6
	48.3	$^{16}O(p,p')^{16}O$	6140	6010 ± 100	6.8 ± 0.5
	28.2			6090 ± 80	9.9 ± 0.8
	12.1			6220 ± 75	10.5 ± 1.2
	[51.8]	$^{16}O(\alpha,\alpha p)^{15}N$	6328	6210 ± 80	4.7 ± 0.5
		$^{16}O(\alpha,\alpha')^{16}O$	6140		

TABLE II-6 (continued)

Target	$\bar{E}_p$[a] (MeV)	Possible Reaction	Transition Energy (keV)	E_γ[b] (keV)	$\sigma_{p,x\gamma}(135°)$ (mb/sr)
	145	$^{16}O(p,p')^{16}O$	7120	7100 ± 50	0.98 ± 0.35
	48.3			6800 ± 150	2.2 ± 0.2
	28.2			6950 ± 90	3.7 ± 0.3
	12.1			7080 ± 80	4.6 ± 0.6
	[51.8]	$^{16}O(\alpha,\alpha')^{16}O$	7120	6980 ± 100	2.5 ± 0.3
	145				8.1 ± 1.9
	48.3				19.7 ± 1.5
	28.2	Sum over above peaks			29.8 ± 2.3
	12.1				18.6 ± 1.6
	[51.8]				25.1 ± 2.0
	145				8.9 ± 1.9
	48.3				37 ± 1.5
	28.2	Sum over all gamma-ray energies above 700 keV			31 ± 5
	12.1				24 ± 5
	[51.8]				59 ± 30
Mg	30.3	$^{24}Mg(p,p')^{24}Mg$	1368	1393 ± 40	18 ± 9
	30.3	$^{24}Mg(p,2p)^{23}Na$	1650[i]	1675 ± 45	16 ± 9
	30.3	$^{24}Mg(p,2p)^{23}Na$	2640	2630 ± 105	4.7 ± 2.6
	30.3	$^{24}Mg(p,2p)^{23}Na$	3240[i]	3320 ± 125	< 0.1
	30.3	$^{24}Mg(p,2\alpha p)^{16}O$	6140	6250 ± 150	1.9 ± 1.2
	30.3	$^{24}Mg(p,2\alpha p)^{16}O$	7120	7070 ± 170	1.0 ± 1.1
	30.3	Sum over above peaks			42 ± 13
	30.3	Sum over all gamma-ray energies above 700 keV			98 ± 10
Fe	155[j]	$^{59}Co(p,n)$	870	857 ± 15	12.2 ± 4.8
	31.4	$^{56}Fe(p,p')^{56}Fe$	845	875 ± 35	30 ± 15
	15.7[k]			870 ± 20	44 ± 14
	[57.2]	$^{56}Fe(\alpha,\alpha')^{56}Fe$	845	830 ± 45	63 ± 34
	155[j]	$^{57}Co(p,p')^{59}Co$	1289	1264 ± 15	13.5 ± 5.2
	31.4	$^{56}Fe(p,p')^{56}Fe$	1240[l]	1340 ± 55	61 ± 18
	15.7[k]			1280 ± 30	25 ± 14
	[57.2]	$^{56}Fe(\alpha,\alpha')^{56}Fe$	1240[l]	1240 ± 40	131 ± 53
	155[j]				26 ± 7
	31.4[k]				91 ± 23
	15.7	Sum over above peaks			69 ± 20
	[57.2]				194 ± 61
	155[j]				84 ± 18
	31.4[k]				240 ± 20
	15.7	Sum over all gamma-ray energies above 700 keV			214 ± 17
	[57.2]				650 ± 52
Al	146	$^{27}Al(p,p)^{27}Al$	1015	1026 ± 10	1.1 ± 0.8
	50.6			950 ± 30	3 ± 7
	30.0			989 ± 25	6.5 ± 5.5
	14.4			1000 ± 25	9.8 ± 5.9
	[56.2]	$^{27}Al(\alpha,\alpha')^{27}Al$	1015	975 ± 30	7 ± 13
	146	$^{27}Al(p,pn)^{26}Al$	1340[m]	1392 ± 15	2.4 ± 0.9
	50.6	$^{27}Al(p,\alpha)^{24}Mg$	1369	1290 ± 30	6.8 ± 6
	30.0			-------	1.5 ± 6
	14.4			1375 ± 30	10.6 ± 9

TABLE II-6 (continued)

[56.2]	$^{27}Al(\alpha, \alpha n)^{26}Al$ 1340[m] / $^{27}Al(\alpha, \alpha p2n)^{24}Mg$ 1369		1290 ± 50	20 ± 20
146	$^{27}Al(p,p')^{27}Al$	1720[m]	1677 ± 15	1.1 ± 0.5
50.6	$^{27}Al(p,pn)^{26}Al$	1850 or	1700 ± 40	9.3 ± 7
30.0	$^{27}Al(p, 2p)^{26}Mg$	1830	1770 ± 25	20 ± 7
14.4			1825 ± 35	10 ± 12
[56.2]	$^{27}Al(\alpha,\alpha')^{27}Al$	1720	1740 ± 30	68 ± 21
146	$^{27}Al(p,p')^{27}Al$	2219	2250 ± 25	0.5 ± 0.7
50.6			2080 ± 50	5.8 ± 4
30.0			2185 ± 25	7.6 ± 3.4
14.4			2205 ± 30	21 ± 4
[56.2]	$^{27}Al(\alpha,\alpha')^{27}Al$	2219	~ 2200	11 ± 6
[56.2]	$^{27}Al(\alpha,\alpha p2n)^{24}Mg$	2753[m]	2770 ± 80	23.3 ± 3.4
146			———————	———————
50.6			———————	< 1.0
30.0			2925 ± 50	2.7 ± 2.0
14.4			2960 ± 50	12.6 ± 2.7
146	$^{27}Al(p, 2pn)^{25}Mg$	3410	3400 ± 20	0.2 ± 0.4
50.6			3320 ± 60	< 1.0
30.0			3380 ± 55	1.0 ± 1.0
14.4			———————	———————
146	$^{27}Al(p, 2pn)^{25}Mg$	3920[m]	3975 ± 25	0.5 ± 0.5
50.6			3810 ± 70	< 1.0
30.0			3850 ± 70	0.8 ± 1.0
14.4			———————	———————
[56.2]	$^{27}Al(\alpha, \alpha pn)^{25}Mg$		3930 ± 70	9.1 ± 1.8
146	Sum over above peaks			5.8 ± 1.6
50.6	Sum over above peaks			26 ± 14
30.0	Sum over above peaks			40 ± 12
14.4	Sum over above peaks			64 ± 17
[56.2]	Sum over above peaks			129 ± 33
146	Sum over all gamma-ray energies above 700 keV			35 ± 8
50.6	Sum over all gamma-ray energies above 700 keV			88 ± 7
30.0	Sum over all gamma-ray energies above 700 keV			107 ± 9
14.4	Sum over all gamma-ray energies above 700 keV			140 ± 11
[56.2]	Sum over all gamma-ray energies above 700 keV			263 ± 22

[a] $\bar{E}_p$ = average proton energy in target, in MeV. Values in brackets refer to incident α-particles.
[b] E_γ = measured gamma-ray energy, in keV.
[c] 2900 = 5030 – 2130.
[d] 1033 = 1740 – 717; 1433 = 2150 – 717.
[e] 2760 = 4750 – 1990.
[f] Multiple peaks are observed at about this energy.
[g] An H_2O target was used at 160-, 56-, and 33-MeV incident proton energies; a BeO target at other energies.
[h] 2750 = 8800 – 6130.
[i] 1650 = 2090 – 440; 3240 = 3680 – 440.
[j] A Co target was used at 155 MeV.
[k] Measurement made at 90° rather than 135°.
[l] 1240 = 2085 – 845.
[m] 1340 = 1760 – 418; 1720 = 2730 – 1013; 2753 = 4122 – 1369; 3920 = 7331 – 3410.

trum for a given situation. The γ-ray measurements cited above have been made with NaI (Tl) spectrometers. An example of recent high resolution γ-ray yield measurements has been made by Chang *et al.* (1974) using a Ge(Li) spectrometer in studying 100 MeV proton interactions in thin targets of ^{56}Fe and ^{58}Ni. This result shows an extremely rich spectrum of γ-ray lines. Detailed measurements of this type for several bombarding energies for different projectiles and targets will be necessary in interpreting any observed cosmic γ-ray line spectra.

Production of γ-ray lines by strongly interacting cosmic ray secondaries, such as π mesons, is also possible; however, this is a second order process. One interesting example of this process is the capture of slow π^- mesons in a dense medium according to the reaction p(π^-, γ) n giving a γ-ray line at $\sim$129 MeV. This reaction also competes with p(π^-, π^0)n; however, the π^- meson slowing-down time in the medium must be $\leqslant 10^{-8}$ s, the half life of the π^- mesons. This requirement is met in many high energy accelerator experiments. Ullrich *et al.* (1974) have described measurements of the γ-ray spectra resulting from the capture of π^- mesons (initial energy 220 to 380 MeV) in several light elements such as O, F, Mg, P, Ca, etc. Again, a very rich γ-ray line spectrum is observed. This reference also refers to much earlier work in the field.

b. *Nuclear Excitation by Neutrons:* (n, n′γ), (n, γ), *and Activation*

Neutrons are secondary products of the interaction of energetic particles ($\gtrsim$ 10 MeV) with matter; however, the half-life of a free neutron is only 11 min. Neutrons that are produced in astrophysical situations by a steady flux of charged particles can be an important source of γ-ray lines from several reactions. Also, with neutron production in transient events such as solar flares, γ-ray producing reactions will give the most intense flux within about 2 mean lives of the free neutron after production in a burst, that is, $\sim$2000s (see Sections III-3.1 and V-5.1).

A convenient summary of the processes by which neutrons can produce γ-ray lines has been given recently by Gorenstein and Gursky (1970) in connection with γ-ray production in planetary atmospheres and planetary surfaces. There are essentially three processes of interest:

(a) Inelastic scattering of fast neutrons denoted by (n, n′γ) or (n, n′xγ) reactions, where in the latter case γ-rays can result from a *residual* excited nucleus.

(b) Capture γ-rays, denoted by (n, γ).

(c) Activation γ-rays, which result if the nuclide produced following neutron capture is radioactive or if fast neutron reactions, such as (n, p), (n, α), (n, αn), etc., produce the activation.

The γ-rays from each of these processes occur on a differing time scale following the production of a neutron. The first process is essentially prompt and would occur on a time scale after production, determined by the transit time of the neutron from its production region to a dense medium ($\sim 10^{-1}$ s for 10 MeV neutrons produced at 500 km above the solar photosphere); but once there, the reaction time would characteristically be dependent on the specific process. The second process occurs after the characteristic thermalization time in a dense medium (in H the capture time $\tau_c \sim 10^{19}/n_H$ s

TABLE II-7

Principal γ-rays from the interaction of 14 MeV neutrons. The differential cross section given pertains to γ-radiation emitted perpendicular to the neutron beam. (From P. Gorenstein and H. Gursky: 1970, *Space Sci. Rev.* **10,** 770, D. Reidel Publishing Company. Used by permission.)

Element	E_γ (MeV)	Probable reaction	$(d\sigma/d\Omega)$ 90° (mb sr^{-1})
C	4.43	$^{12}C(n,n')^{12}C^*$	13.1 ± 1.3
N	7.03	$^{14}N(n,n')^{14}N^*$	4.8 ± 1.6
	5.1	$^{14}N(n,n')^{14}N^*$	4.3 ± 1.4
	4.46	$^{14}N(n,\alpha')^{11}B^*$	5.0 ± 1.0
O	7.12	$^{16}O(n,n')^{16}O^*$	5.0 ± 1.0
	6.93	$^{16}O(n,n')^{16}O^*$	3.8 ± 0.9
	6.13	$^{16}O(n,n')^{16}O^*$ $^{16}O(n,p)^{16}N \xrightarrow{\beta^-} {}^{16}O^*$	22.6 ± 1.7
	3.68	$^{16}O(n,\alpha)^{13}O^*$	5.5 ± 1.1
	2.75	$^{16}O(n,n')^{16}O^*$	3.8 ± 0.4
Na	2.24	$^{23}Na(n,n')^{23}Na$	3.2 ± 1.1
	1.63	$^{23}Na(n,n')^{23}Na$ $^{23}Na(n,\alpha)^{20}F \xrightarrow{\beta^-} {}^{20}Ne^*$	18.8 ± 1.9
	1.27	$^{23}Na(n,d)^{22}Ne^*$	14.6 ± 2.5
	0.64	$^{23}Na(n,\alpha)^{20}F^*$ $^{23}Na(n,n')^{23}Na^*$	6.4 ± 1.3
	0.44	$^{23}Na(n,n')^{23}Na^*$ $^{23}Na(n,p)^{23}Ne \xrightarrow{\beta^-} {}^{23}Na^*$	78.5 ± 5.7
Mg	2.75	$^{24}Mg(n,n')^{24}Mg^*$	4.1 ± 0.8
	1.83	$^{26}Mg(n,n')^{26}Mg^*$ $^{26}Mg(n,p)^{26}Na \xrightarrow{\beta^-} {}^{26}Mg^*$	11.1 ± 0.8
	1.37	$^{24}Mg(n,n')^{24}Mg^*$	30.9 ± 3.1
	1.14	$^{26}Mg(n,n')^{26}Mg^*$ $^{26}Mg(n,p)^{26}Na \xrightarrow{\beta^-} {}^{24}Mg^*$	5.9 ± 1.4
	0.47	$^{24}Mg(n,p)^{24}Na^*$	10.4 ± 2.1
	0.35	$^{24}Mg(n,\alpha)^{21}Ne^*$	11.6 ± 3.8
Al	3.0	$^{27}Al(n,n')^{27}Al^*$	7.9 ± 1.6
	2.21	$^{27}Al(n,n')^{27}Al^*$	10.8 ± 1.1
	1.81	$^{27}Al(n,d)^{26}Mg^*$	13.7 ± 2.7
	1.72	$^{27}Al(n,n')^{27}Al^*$	4.3 ± 1.4
	1.01	$^{27}Al(n,n')^{27}Al^*$ $^{27}Al(n,p)^{27}Mg \xrightarrow{\beta^-} {}^{27}Al^*$	20.5 ± 1.5
	0.84	$^{27}Al(n,n')^{27}Al^*$ $^{27}Al(n,p)^{27}Mg \xrightarrow{\beta^-} {}^{27}Al^*$	12.0 ± 2.0
Si	7.4	$^{28}Si(n,n')^{28}Si^*$	2.7 ± 1.0
	6.8	$^{28}Si(n,n')^{28}Si^*$	3.7 ± 1.3
	5.10	$^{28}Si(n,n')^{28}Si^*$	3.9 ± 0.8
	2.84	$^{28}Si(n,n')^{28}Si^*$	5.3 ± 0.6
	1.78	$^{28}Si(n,n')^{28}Si^*$ $^{28}Si(n,p)^{28}Al \xrightarrow{\beta^-} {}^{28}Si$	75.9 ± 8.5
	0.96	$^{28}Si(n,\alpha)^{25}Mg^*$	4.4 ± 1.5
	0.60	$^{28}Si(n,\alpha)^{25}Mg^*$	4.7 ± 1.5

TABLE II-7 (continued)

Element	E_γ (MeV)	Probable reaction	$(d\sigma/d\Omega)$ 90° (mb sr^{-1})
K	2.83	$^{39}K(n,n')^{39}K^*$	8.1 ± 1.6
	2.16	$^{39}K(n,d)^{38}Ar^*$	19.3 ± 2.0
	1.66	$^{41}K(n,n')^{41}K^*$	8.9 ± 0.9
	0.78	{ $^{39}K(n,\alpha)^{36}Cl^*$; $^{41}K(n,2n)^{40}K^*$ }	6.3 ± 1.3
Ca	3.73	$^{40}Ca(n,n')^{40}Ca$	9.0 ± 1.8
	1.61	{ $^{40}Ca(n,p)^{40}K^*$; $^{40}Ca(n,\alpha)^{37}Ar^*$ }	5.4 ± 0.6
	0.89	$^{40}Ca(n,p)^{40}K$	4.8 ± 1.0
	0.77	$^{40}Ca(n,p)^{40}K$	5.6 ± 1.2
Fe	1.82	$^{56}Fe(n,n')^{56}Fe^*$	5.4 ± 0.6
	1.33	{ $^{56}Fe(n,n')^{56}Fe^*$; $^{56}Fe(n,2n)^{55}Fe^*$ }	7.4 ± 2.5
	1.24	$^{56}Fe(n,n')^{56}Fe^*$	23.0 ± 2.3
	1.02	$^{56}Fe(n,n')^{56}Fe^*$	6.3 ± 1.3
	0.92	$^{56}Fe(n,2n)^{55}Fe^*$	11.7 ± 3.9
	0.85	{ $^{56}Fe(n,n')^{56}Fe^*$; $^{56}Fe(n,p)^{56}Mn \xrightarrow{\beta^-} {}^{56}Fe^*$ }	114.5 ± 8.4

where n_H is the proton number density). The third process gives γ-ray lines on a time scale (following activation) that is determined by the half-life of the activated nuclide.

In Table II-7 are shown the principal prompt γ-rays resulting from the interaction of fast neutrons with the most abundant naturally occurring elements. The cross sections given in column 4 are from experimental results with 14MeV neutrons. These reactions are energy dependent (similar to (p, p$'\gamma$) except for the Coulomb barrier); however, a large body of data is available from the 14MeV neutrons resulting from D-T reactions using Van de Graaff accelerators. Marmier and Sheldon (1969) discuss the general energy dependence of fast neutron reactions. This table is provided to show representative fast neutron cross sections, and the same qualitative behavior is expected for neutrons with energies considerably below or above 14MeV. In general, the predominate (n, n$'\gamma$) reactions are those where the γ-ray results from decay of the lowest lying energy levels of the target nucleus. Reactions such as (n, n$'$pγ) and (n, dγ) are also important in some cases. The γ-ray lines resulting from these reactions may also in general be expected to be Doppler broadened (see Section II-2.5.1). For example, a nucleus of $A = 20$ recoiling from a collision with a 14MeV neutron will have a maximum velocity of $\sim 0.016c$ and the γ-ray line will have a typical half width of 1.6%. This becomes important if measurements are made with high resolution γ-ray spectrometers such as Ge(Li) which typically have resolutions of $\sim$ 1% or less depending on the γ-ray energy.

Production of γ-rays from neutron capture, especially in H, is of great importance in astrophysical situations because of the predominant abundance of this element. The capture cross section of neutrons generally follows a behavior $\sigma(n, \gamma) \propto 1/v_n$, except when capture at resonance occurs. For example, in H, $\sigma_H(n, \gamma) = 7.26 \times 10^4/v_n$ (barns), where v_n is in cm s^{-1}. Observation of the 2.23 MeV capture γ-ray in H is discussed in Sections III-3.1 and V-5.1. Table II-8 shows the most prominent capture γ-rays for the

TABLE II-8

Capture γ-rays from the abundant elements†. (From P. Gorenstein and H. Gursky: 1970, *Space Sci. Rev.* **10,** 770, D. Reidel Publishing Company. Used by permission.)

Element	A	Fractional natural isotopic abundance	$\sigma(n, \gamma)^a$	E_γ (MeV)	$\eta(E_\gamma)^b$
H	1	~ 1	.33	2.23	1
C	12	.99	.0034	4.95	.75
				3.68	.25
				1.26	.25
N	14	~ 1	.075	2.52	.06
				1.88	.21
				1.68	.12
				3.53	.09
				3.68	$<.23$
				4.51	.15
				5.30	.21
				5.27	.32
				5.53	.21
				5.56	.11
				6.32	.18
				10.82	.14
O	16	~ 1	$<.0002$		
Na	23	1	.534	6.40	.26
				3.98	.22
				3.59	.17
				2.52	.17
				2.03	.20
				2.75	1.0 (^{24}Mg)
				1.38	(^{24}Mg)
				.87	.25
				.47	.69
Mg	(Natural)		.063	3.92	.47
				2.28	.24
Al	27	~ 1	.235	7.72	.16
				1.78	.88 (^{28}Si)
Si	28	.92	.16	4.94	.61
				3.54	.64
Ca	40	.97	.44	6.42	.42
				4.42	.16
				3.61	.07
				2.00	.29
				1.94	.99
Fe	56	.92	2.55	7.64	.44
				7.28	.05
				6.02	.085
				5.92	.08

† Bartholomew *et al.* (1967)

[a] Thermal capture cross section in barns

[b] Fractional yield of gamma of energy E_γ

TABLE II-9

Long-lived γ-ray emitters produced by cosmogenic activation.[a] (From P. Gorenstein and H. Gursky: 1970, *Space Sci. Rev.* **10,** 770, D. Reidel Publishing Company. Used by permission.

Isotope	$T_{1/2}$	E_γ (MeV)
Be^7	53 d	0.48
Na^{22}	2.6 yr	1.28
Al^{26}	7.4×10^5 yr	1.83
A^{42}	33 yr	1.52
Sc^{46}	84 d	1.12, 0.89
Mn^{54}	291 d	0.84

[a] These include beta emitters whose half-lives exceed 10 d.

most abundant naturally-occurring elements through Fe. The cross section given in column 4 is for a neutron of thermal velocity at 300 K (2200 m s^{-1}). It should be noted that generally the capture γ-spectra in elements above H are quite complex. A detailed discussion may be found in Motz and Backström (1966), and a fairly recent tabulation of capture γ-ray energies and cross sections is given by Bartholomew *et al.* (1967).

The last process (activation) is of most importance when the active nuclide is a β emitter. For activation of naturally-occurring nuclides by cosmic ray bombardment, this process is known as 'cosmogenic activation', and in Table II-9 some activation nuclides with half-lives longer than ten days are listed. The radioactive nuclides listed here may be produced in excited states by cosmic ray spallation and, in some cases, by fast neutron activation. Several fast neutron reactions giving radioactive species which β^- decay to excited states of their daughters are given in Table II-7.

c. *Thermonuclear (Exoergic) Reactions*

In general, we should consider as potential γ-ray sources the nuclear reactions possible under conditions of high temperatures which can produce either neutrons, positrons, or γ-rays directly. Such reactions normally have $Q > 0$ and are strongly temperature dependent. The brief discussion we give here will follow that given by Clayton (1968) which is appropriate for thermonuclear reactions. Complete treatments are also given by Fowler *et al.* (1967) and Chiu (1968). Consider a system of particles a and X of concentration N_a (cm^{-3}) and N_X (cm^{-3}) interacting according to the reaction

$$a + X \rightarrow Y + b.$$

In general the reaction rate may be expressed as

$$r_{aX} = N_a N_X \int_0^\infty v\sigma(v)\, \phi(v) dv \tag{II.55}$$

where v is the relative velocity between the reacting pair, $\sigma(v)$ is the cross section for the reaction, and $\phi(v)dv$ is the probability that the relative velocity of the pair of particles is of a magnitude v to $v + dv$. The reaction rate may also be expressed as

$$r_{aX} = (1 + \delta_{aX})^{-1}\, N_a N_X \langle\sigma v\rangle \tag{II.56}$$

where $\langle\sigma v\rangle$ is the average value given by Equation (II.58) and δ is the Kronecker delta introduced to account for the number of unique pairs (a, X) in case a = X. In general, $\langle\sigma v\rangle$ may be the average over a Maxwellian velocity distribution giving $\langle\sigma v\rangle_{th}$, and if averaged over a nonthermal velocity distribution, this becomes $\langle\sigma v\rangle_{NT}$. In Equation (II.55), for a monoenergetic group of ions of velocity v_0

$$\phi(v)\mathrm{d}v = \delta(v - v_0)\mathrm{d}v, \qquad \text{so} \quad \langle\sigma v\rangle_{NT} = v_0\,\sigma(v_0).$$

It is useful to consider the lifetime $\tau_a(X)$ of particle (a) against reactions with species X, since this gives a qualitative measure of the importance of the reaction for a given situation. Following Clayton (1968, p. 293),

$$\tau_a(X) = [\langle\sigma v\rangle N_a]^{-1} \ (\mathrm{s}). \tag{II.57}$$

The case where $\phi(v)\,\mathrm{d}v$ is based on a Maxwellian gas composed of species a and X has, of course, in the theory of stellar interiors received considerable attention, and we will summarize briefly the important results. The important formula is (Clayton, 1968, p. 296).

$$\langle\sigma v\rangle_{th} = 4\pi\left(\frac{\mu}{2\pi kT}\right)^{3/2}\int_0^\infty v^3\,\sigma(v)\exp\left(\frac{-\mu v^2}{2kT}\right)\mathrm{d}v \tag{II.58}$$

where μ is the reduced mass of the reacting pair and all other terms have been previously defined. In order to complete the reaction rate calculation, $\sigma(v)$ must be known and the possibility of resonant states in the compound system (a + X) must also be considered. To avoid complexity only nonresonant reactions will be described here. Since the Coulomb barrier limits the reaction rate, the cross section is usually expressed as

$$\sigma(E) \equiv \frac{S(E)}{\mathrm{E}}\exp\left[\frac{-2\pi Z_1 Z_2 e^2}{\hbar v}\right]. \tag{II.59}$$

In the case of low energies (< 100 keV), $S(E)$ is often found to be a slowly varying function of energy so, in general,

$$S(E) = S(E_0) + \left(\frac{\mathrm{d}S}{\mathrm{d}E}\right)_{E_0}(E - E_0) \tag{II.60}$$

where E_0 is the energy of the 'Gamow peak.' The 'Gamow peak' is a concept appropriate to a thermal distribution of particle energies and a cross section controlled by the Coulomb barrier and is illustrated in Figure II-22. Here the folding of the exponential term in the velocity distribution (Equation (II.58)) and the Coulomb barrier factor in Equation (II.59) gives the sharp peak at energy E_0 where generally $E_0 \gg kT$. Thus in a purely thermonuclear setting, the small number of particles in the tail of the Maxwellian distribution is primarily responsible for the reaction rate. $S(E)$ in general depends on purely nuclear parameters which can be either calculated or experimentally obtained.

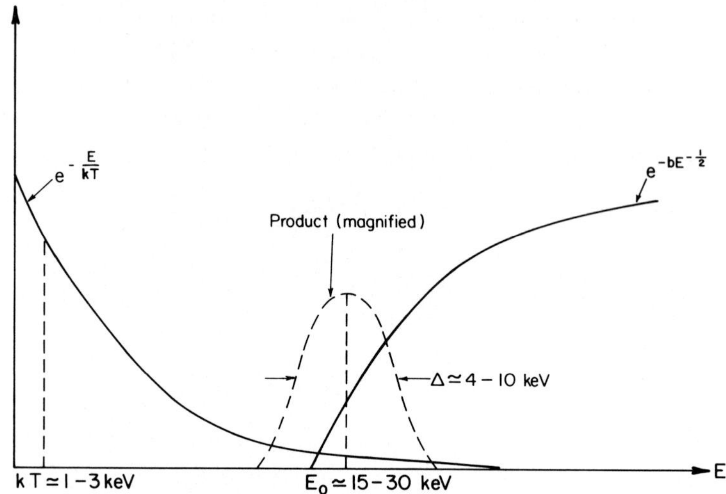

Fig. II-22. The concept of the "Gamow Peak" is illustrated. The dominant energy dependent factors in thermonuclear reactions are shown. (From D.D. Clayton, *Principles of Stellar Evolution and Nucleosynthesis.* Copyright 1968, McGraw-Hill Book Company. Used by permission.)

The above brief discussion is sufficient to understand the reactions most commonly discussed for thermal settings. The quantity $\langle\sigma v\rangle$ is conveniently expressed for approximate numerical work (for particles 1 and 2) as

$$\langle\sigma v\rangle_{12} = [7.2 \times 10^{-19} S(E_0)]\,(AZ_1Z_2)^{-1}\,\tau^2 e^{-\tau}\ (\mathrm{cm^3\,s^{-1}}) \tag{II.61}$$

where $A = (A_1A_2)/(A_1 + A_2) = \mu/M_u$ (M_u is the mass of 1 amu, so A is the reduced weight, expressed in mass units). Also

$$\tau = 42.48\left(\frac{Z_1^2Z_2^2A}{T_6}\right)^{1/3}$$

where T_6 is the temperature in units of 10^6 K. The latter is also written as $\tau = BT_6^{-1/3}$. The units of $S(E_0)$ in Equation (II.61) is (keV barns).

Table II-10 is a list of the parameters for several exoergic nuclear reactions important in the PP chains and CNO cycle for stellar energy generation. All of the quantities have been defined in the discussion above and they are sufficient to estimate the thermonuclear reaction rates for any temperature using Equations (II.61) and (II.56). In the upper part of the table, the reaction time scale (Equation (II.57)) is given for the assumed conditions at the center of the Sun shown in the central part of the table.

It should be noted that the weakest nuclear reaction, ^{1}H (p, $\beta^+\nu$) ^{2}D, cannot be expected to be of any importance in producing the positrons needed for an annihilation line at 0.51 MeV, since even for the conditions at the center of the Sun, the reaction time scale is $\sim 10^{10}$ yr as shown in column 7 of Table II-10. Extremely bizarre cirumstances would be necessary to make this reaction important, since even for 1 MeV protons the cross section is $\sim 10^{-47}$ cm^2 and cannot be measured in the laboratory! As can be seen

TABLE II-10

Reaction rate parameters for thermonuclear reactions in the PP chains and CNO cycle. (From D. D. Clayton, *Principles of Stellar Evolution and Nucleosynthesis,* pp. 380 and 392. Copyright 1968, McGraw-Hill Book Company. Used by permission.)

Reactions of the PP chains

Reaction	Q value MeV	Average ν loss MeV	$S(E=0)$, keV barns	dS/dE barns	B	τ_{12} yr[a]
$H^1(p,\beta^+\nu)D^2$	1.442	0.263	3.7×10^{-22}	4.2×10^{-24}	33.81	7.9×10^9
$D^2(p,\gamma)He^3$	5.493		2.5×10^{-4}	7.9×10^{-6}	37.21	4.4×10^{-8}
$He^3(He^3, 2p)He^4$	12.859		5.0×10^3		122.77	2.4×10^5
$He^3(\alpha,\gamma)Be^7$	1.586		4.7×10^{-1}	2.8×10^{-4}	122.28	9.7×10^5
$Be^7(e^-,\nu)Li^7$	0.861	0.80				3.9×10^{-1}
$Li^7(p,\alpha)He^4$	17.347		1.2×10^2		84.73	1.8×10^{-5}
$Be^7(p,\gamma)B^8$	0.135		4.0×10^{-2}		102.65	6.6×10^1
$B^8(\beta^+\nu)Be^{8*}(\alpha)He^4$	18.074	7.2				3×10^{-8}

[a] Computed for $X = Y = 0.5$, $\rho = 100$, $T_6 = 15$ (Sun)

Reactions in the CNO cycle

Reaction	Q value MeV	Average ν loss MeV	$S(E=0)$, keV barns	dS/dE barns	B	τ_{12} yr
$C^{12}(p,\gamma)N^{13}$	1.944		1.40	4.26×10^{-3}	136.93	
$N^{13}(\beta^+\nu)C^{13}$	2.221	0.710				
$C^{13}(p,\gamma)N^{15}$	7.550		5.50	1.34×10^{-2}	137.20	
$N^{14}(p,\gamma)O^{15}$	7.293		2.75		152.31	
$O^{15}(\beta^+\nu)N^{15}$	2.761	1.00				
$N^{15}(p,\alpha)C^{12}$	4.965		5.34×10^4	8.22×10^2	152.54	
$N^{15}(p,\gamma)O^{16}$	12.126		2.74×10^1	1.86×10^{-1}	152.54	
$O^{16}(p,\gamma)F^{17}$	0.601		1.03×10^1	2.81×10^{-2}	166.96	
$F^{17}(\beta^+\nu)O^{17}$	2.762	0.94				
$O^{17}(p,\alpha)N^{14}$	1.193		Resonant reaction		167.15	

from inspection of the parameter values in the table, the strongest reaction rate is expected from $^2D(p, \gamma)\,^3He$ which gives a capture γ-ray at ~5.5 MeV.

Several other exoergic reactions involving light isotopes of hydrogen and helium are of interest in the production of γ-rays, since neutrons are produced which can give γ-ray lines by the several processes discussed in Section II-2.4.3.b. The two reactions of interest are $^2D(d, n)\,^3He$ and $^2D(t, n)\,^4He$ which are exoergic with Q values 3.25 and 17.6 MeV, respectively. The reaction $^2D(d, p)\,^3H$ competes with the first reaction but the branching ratio is ~1/2. In Figure II-23 is shown the quantity $\langle\sigma v\rangle_{th}$ plotted versus the deuteron kinetic temperature for these two neutron-producing reactions when the reacting particles are in thermal equilibrium. The kinetic temperature is $T_k = kT$, so at 1 keV, $T = 1.16 \times 10^7$K.

2.4.4. POSITRON-ELECTRON ANNIHILATION

The earliest astrophysical considerations on the γ-ray spectra produced by positron-electron annihilation had tacitly assumed that a simple two photon spectrum results,

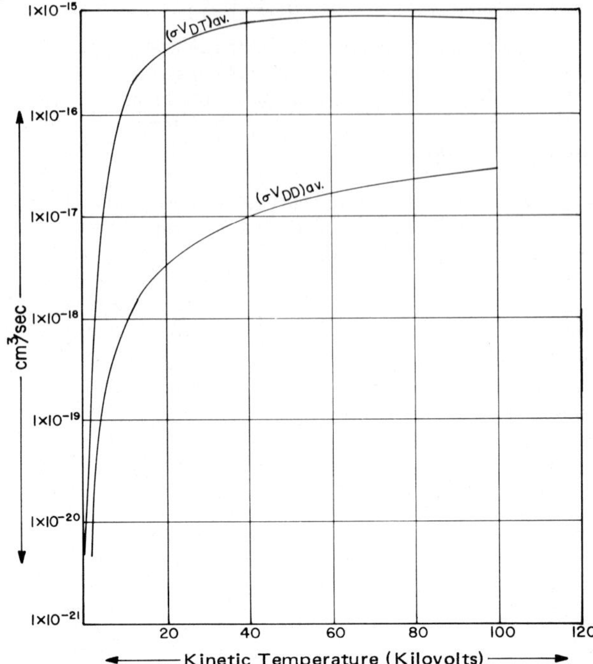

Fig. II-23. The average value $\langle \sigma v_{th} \rangle$ is shown for D-D and D-T reactions. (From R.F. Post: 1956, *Rev. Mod. Phys.* **28**, 338. Used by permission.)

promptly giving a distinct line at 0.511 MeV when the center of mass of the annihilating pair is at rest relative to the observer. This is the basic spectrum observed in unsophisticated laboratory experiments where one is usually dealing with positron annihilation in solids, and the most elementary arguments suffice to explain the situation. In the astrophysical case, a variety of physical situations may be present and various possible fates of a positron must be carefully considered. A vast amount of literature is available on this problem, and we will summarize several aspects of this literature. The following factors must then be considered:

(a) The initial energy of the positron.

(b) The slowing down process in media such as a plasma, a neutral gas, a dense gas, or a solid.

(c) The annihilation process itself through either free annihilation, positronium formation, or bound state annihilation.

Depending on the particular situation, the annihilation process, which is a positron's ultimate fate, can lead to either single or multiple (2 or more) photon emission.

As will become clear, either 2 or 3 photon emissions are the most predominant emission modes (cf. Heitler, 1954).

a. *Single Photon Annihilation*

If the annihilation takes place with an electron bound in an atom, then the momentum

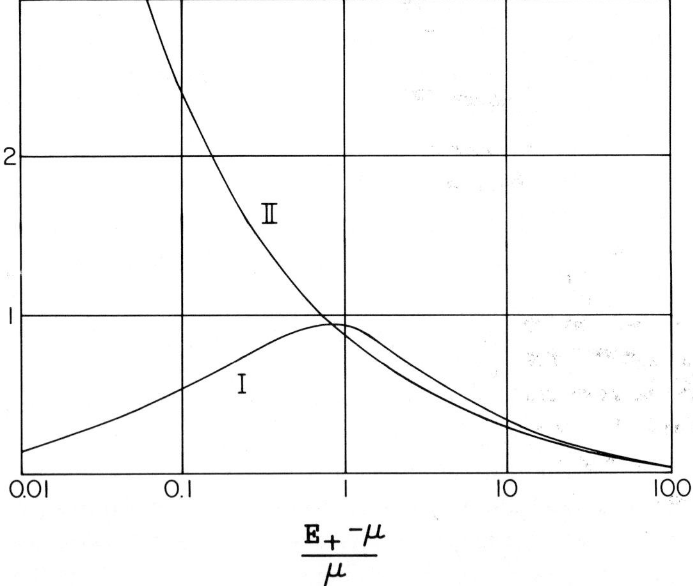

Fig. II-24. The cross sections for positron annihilation giving (I) single photon emission or (II) two photon emission vs. positron kinetic energy in units of electron rest mass. Note the different units for I and II described in the text. (From W. Heitler, *The Quantum Theory of Radiation,* Oxford University Press, London, 1954. Used by permission.)

of the single photon can be balanced by the nucleus or lattice in which it resides. As a concrete example Heitler (1954) treats the case of annihilation with a K-electron. The single quantum emitted has an energy E given by

$$E_1 = E_+ + \mu - \alpha^2 \mu/2 \tag{II.62}$$

where E_+ is the total energy of the positron, μ is the electron rest energy, and $\alpha = 1/137$. The quantity $\alpha^2 \mu/2$ is the binding energy of the K-electron. The cross section for this process is strongly dependent on the atomic number and is given approximately by:

$$\sigma_K(1\gamma) = \frac{4\pi r_0^2}{3} Z^5 \alpha^4 P_+/\mu \tag{II.63}$$

$$\sigma_K(1\gamma) = 4\pi r_0^2 Z^5 \alpha^4 \frac{\mu}{E_+} . \tag{II.64}$$

Equation (II.63) is for non-relativistic positrons of momentum P_+ and Equation (II.64) is for relativistic positrons. Here r_0 is the classical radius of the electron.

Figure II-24 shows the cross section for the single photon emission (I) and the cross section for the two photon emission (II) (see below) versus the positron kinetic energy in electron rest mass units $(E_+ - \mu)/\mu$. Note that the units for the two cross sections are different. In the one-quantum case, the ordinate should be multiplied by $\pi r_0^2 Z^5 \alpha^4$ and in the 2-quanta case, $\pi Z r_0^2$. The positron energy for which the single photon to two photon yield is the largest is at a kinetic energy of $(10\mu - \mu) \simeq 5$ MeV. For $Z = 82$

(Pb), the ratio is 20% and is probably an upper limit (Heitler, 1954); the single photon energy would also be about 5 MeV. Since the two photon annihilation cross section increases rapidly as the positron energy falls, there is probably no astrophysical situation where the single spectrum would have practical significance.

b. *Free Two and Three Photon Annihilation*

The more general case usually discussed is that of a positron with total energy $E_+ = k_+ + \mu$ annihilating with a free negatron. In the case of an annihilating collision in an S-state with electron and positron spins opposite (a singlet collision), two photons are emitted, and when the spins are parallel (a triplet collision), three photons are emitted (see Yang, 1950). The relative probability of three photon annihilation to two photon annihilation in the unbound case has been shown to be $\alpha/\pi \simeq 1/372$ in Heitler (1954, p. 272), when the collisions are averaged over electron spins. It is therefore sufficient to consider only the case of the two photon emission if no bound state is formed. In the laboratory case where positrons are slowed down rapidly enough (i.e., in a solid), the annihilation takes place essentially at rest, in which case the two annihilation photons must have equal and opposite moments. Under these conditions a sharp γ-ray line at 0.511 MeV is observed.

The more general situation appropriate to astrophysics, where free annihilation may take place in flight, has been given by Stecker (1971) in detail. We will summarize the important results of this treatment. First, the differential cross section for γ-ray production in the center of mass system of the colliding electron pair is expressed as

$$d\sigma = \frac{\sigma_0}{2\gamma_c^2\beta_c}\phi(\chi\,|\,\gamma)d\chi = \frac{\sigma_0}{2\gamma_c^2\beta_c}\left[\frac{1+\beta_c^2(2-\chi^2)}{(1-\beta_c^2\chi^2)} - \frac{2\beta_c^4(1-\chi)^2}{(1-\beta_c^2\chi^2)}\right]d\chi \quad \text{(II.65)}$$

where $\sigma_0 = \pi r_0^2 = 2.5 \times 10^{-25}$ cm^2, and

$$\beta_c = \left(\frac{\gamma-1}{\gamma+1}\right)^{1/2}, \qquad \gamma_c = \left(\frac{\gamma+1}{2}\right)$$

are the center of mass velocity of the positron and its Lorentz factor, respectively, and the laboratory Lorentz factor is

$$\gamma = \frac{E_+}{\mu} = \frac{k_+ + \mu}{\mu} = (1-\beta^2)^{-1/2}.$$

χ is the cosine of the angle between the incoming positron and the outgoing γ-ray in the center of mass system (see Figure II-25).

The energy of a two photon annihilation γ-ray in the laboratory system depends on the angle of emission of the photons in the center of mass system. This is given as

$$E_\gamma = \mu\gamma_c^2(1+\beta_c\chi) \quad \text{(II.66)}$$

for

$$1 \leqslant \gamma \to \infty \;; \quad -1 \leqslant \chi \leqslant 1.$$

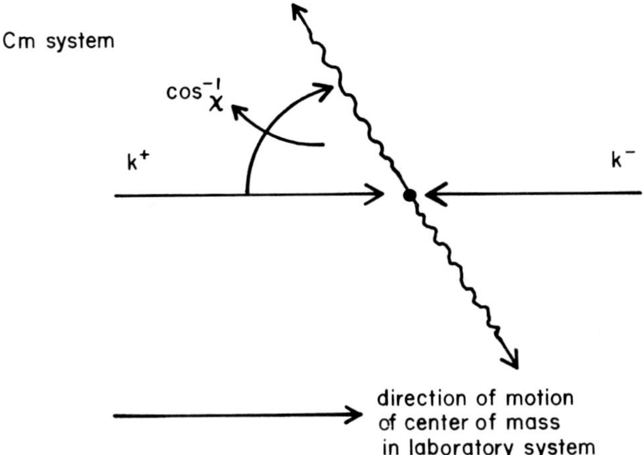

Fig. II-25. The annihilation of an electron-positron pair in the center of mass system leading to two photon emission.

Stecker (1971) derives the production spectrum of γ-rays for a positron density spectrum $n(\gamma)$ cm^{-3} interacting with an electron density n_{e^-}. The result is

$$Q(\eta) = 4H_+(1/2)n_{e^-}\,\sigma_0 c \int_{G(\eta)}^{\infty} \frac{d\gamma\; n(\gamma)}{\gamma(\gamma^2-1)^{1/2}}\,\phi[\chi_0(\eta,\gamma),\gamma]\;(\mathrm{s}^{-1}) \qquad \text{(II.67)}$$

where $\eta(\gamma) = E_\gamma/\mu$, H_+ is the Heaviside step function (see Stecker, 1971, Equation 2-18), and the lower limit in the integral is

$$G(\eta) = [\eta^2 + (\eta-1)^2]/(2\eta-1).$$

$\phi[\chi_0(\eta,\gamma),\gamma]$ is found from the angular distribution function part of the differential cross section given in Equation (II.65) by inserting $\chi_0(\eta,\gamma) = ((2\eta-1)-\gamma)/(\gamma^2-1)^{1/2}$ instead of χ and noting that the limits on the center of mass angles give limits on $\chi_0 = \pm 1$ which in turn give limiting values for η such that

$$\eta_\pm = \gamma_c^2(1 \pm \beta_c).$$

Stecker (1971) has shown that η_+ and η_- are roots of the equation

$$\eta^2 - 2\gamma_c^2\eta + \gamma_c^2 = 0.$$

Using the relation between γ_c and γ then gives

$$\gamma = \frac{\eta^2 + (\eta-1)^2}{2\eta-1}\,. \qquad \text{(II.68)}$$

Figure II-26 from Stecker (1971) shows a plot of this equation versus $\eta(\gamma) = E_\gamma/\mu$ which gives the annihilation γ-ray energy.

The physical meaning of this figure is that the allowable values for the laboratory energy of an annihilation photon for a given positron energy are contained in the double

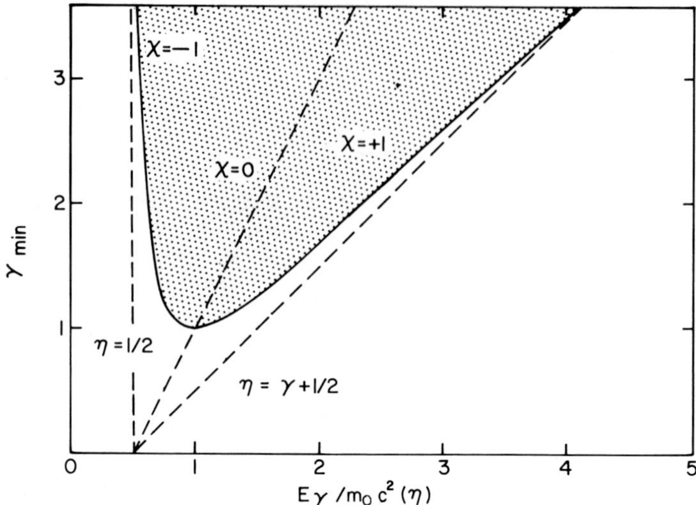

Fig. II-26. The parameters are shown from which the energy of each annihilation photon can be found. (From F.W. Stecker: 1971, *NASA SP-249* and Monobook Company, Baltimore, Maryland, 1971.)

shaded region above the γ_{min} curve given by Equation (II.68). Thus, if the two annihilation photons are emitted at a center of mass angle of 90° ($\chi = 0$), then the photons have equal energy given by the value of η where a horizontal line at γ intersects the line $\chi = 0$. If the two photons are emitted along the direction of motion of the center of the mass, then $\chi = +1$ for one photon and $\chi = -1$ for the other.

The photon energies are then found from the $\chi = -1$ and $\chi = +1$ branches of the curve of Equation (II.68). For intermediate values of χ, the photon energies lie in the shaded region mentioned.

For the trival case where the annihilation takes place at rest, $\gamma = 1$, and therefore the line intersects the curve, Figure II-26, at the minimum where $\eta = 1$, so both photons have an energy of $\mu = m_0c^2 = 0.511$ MeV.

For the general case of annihilation in flight, it is readily seen from Figure II-26 that there is a lower limit to the energy of an annihilation photon since the asymptotic lower bound level is $E_{\gamma 1} = \eta_{-}\mu \to \mu/2$ for $\gamma \to \infty$ or $\gamma_c \to \infty$, as read from the $\chi = -1$ branch in Figure II-26.

The other photon has in this case an energy, read off the $\chi = +1$ branch, of $E_{\gamma 2} = \eta_{+}\mu = \gamma_c^2(1 + \beta_c)\mu = \mu/(1 - \beta_c)$ which $\to \infty$ as $\gamma, \gamma_c \to \infty$.

Of direct experimental interest is the differential annihilation γ-ray spectrum for annihilation in flight. Stecker (1971) points out that in the ultrarelativistic limit ($\gamma \gg 1$), the angular distribution of the emitted photons, given by $\phi(\chi | \gamma)$ in Equation (II.65), is strongly peaked at $\chi = \pm 1$, so the principal emission of gamma rays in the laboratory lies along the velocity vector the center of mass, and the two γ-rays have nearly the asymptotic values $E_{\gamma 1} = \mu/2$ and $E_{\gamma 2} = \mu(\gamma + 1/2)$. The forward moving photon ($\chi = +1$) carries away most of the energy available and the backward moving photon ($\chi = -1$) is shifted down in energy, but no lower than 0.25 MeV.

In this case, the total cross section for a high energy positron of energy $E_+ = \gamma\mu$ to produce a high energy photon of energy $E_\gamma \gg \mu$ is approximately

$$\sigma_A(E_+|E_\gamma) \simeq \sigma_A(E_\gamma)\,\delta(E_\gamma - E_+) \tag{II.69}$$

where

$$\sigma_A = \frac{\sigma_0}{\gamma+1}\left[\frac{\gamma^2+4\gamma+1}{(\gamma^2-1)^{1/2}}\ln\left[\gamma+(\gamma^2-1)^{1/2}\right] - \frac{\gamma+3}{(\gamma^2-1)^{1/2}}\right] \tag{II.70}$$

from Dirac (1930).

When $\gamma \gg 1$,

$$\sigma_A \simeq \sigma_0\left[\frac{\ln(2\gamma)-1}{\gamma}\right]. \tag{II.71}$$

For the nonrelativistic case ($\gamma \sim 1$), the approximate form of the annihilation cross section is

$$\sigma_A \simeq \frac{\sigma_0}{\beta}. \tag{II.72}$$

As a final remark, the free annihilation process may also produce three or more γ-rays, but the relativistic cross section for the process is usually down by a factor $\alpha^{n-2} = (1/137)^{n-2}$ from free two photon annihilation, where n is the number of photons emitted. Specifically, for the case of annihilation of nonrelativistic positrons, the ratio of the three photon annihilation cross section to the two photon annihilation cross section is $\sigma_{A,3\gamma}/\sigma_{A,2\gamma} \sim 1/372$ according to Ore and Powell (1949).

Thus, the positron spectrum in the slowing down medium itself determines the annihilation inflight spectrum with a low energy cutoff at 0.25 MeV and an upper limit determined by E_+.

In general, the inflight production rate of γ-rays, $Q(\eta)$, must be compared with the slowing down rate of the positrons to see where inflight production may be important astrophysically. The former is inferred from Ginzburg and Syrovatskii (1964b, p. 381) as

$$Q(\eta) \sim n_e \sigma_0 c n_+(\eta)\;[\ln(\eta) - 1]/\eta \tag{II.73}$$

where all quantities have been defined previously.

c. *Bound State (Positronium) Annihilation*

The discussion above has indicated that the free annihilation of positrons will lead primarily to two photon emissions and, if the annihilation occurs near rest, then two γ-rays of equal energy (0.511 MeV) are emitted. This is the situation in laboratory experiments where the positrons are emitted in a solid; however, formation of a bound state of the electron and positron, known as positronium and denoted by Ps, is a very likely occurrence in a gas. The discovery of positronium by Deutsch (1951) was recently reviewed (Maglich, 1974), and a thorough discussion of laboratory work may be found in Green and Lee (1964). Since the bound system is very similar to the H atom (Rydberg energy = 6.8 eV), and both particles are fermions, two bound states are most likely: (a) para-positronium or singlet Ps ($^1S_0$) where the electrons have anti-parallel spins, or (b) ortho-positronium or triplet Ps ($^3S_0$) where the electrons have parallel spins. Formation of Ps in the spherically symmetric, $l = 0$, ground state is the most probable mode of formation. However, even in the bound state the lifetime against annihilation is very short ($\leqslant 10^{-7}$ s).

Stecker (1971) has reviewed the consequences of annihilation through either the singlet or triplet state. Basically the result is that the number of photons emitted depends on the angular momentum of the original state, so an *even* number of photons (2) is emitted for the (even) singlet annihilation and an *odd* number of photons (3) is emitted for (odd) triplet annihilation. This is due to the fact that each photon carries off only one unit of angular momentum. Thus, since the multiplicity of the triplet state is three (with three magnetic substates $m = 0, \pm 1$), in 3/4 of the total annihilations 3 photons are emitted, giving a continuum; and in only 1/4 of the annihilations is there produced a γ-ray line at 0.511 MeV. The triplet annihilation γ-ray spectrum has been calculated by Ore and Powell (1949) and is expressed as a normalized differential spectrum $P_T(E_\gamma)$ which is found to be a monotonically rising spectrum for $0 < E_\gamma < 0.511$ MeV as shown in Figure II-27. The positronium annihilation γ-ray spectrum may be constructed from this according to

$$P(E_\gamma) = 0.75 \times 3 \times P_T(E_\gamma) + 0.25 \times 2 \times P_s(E_\gamma) \tag{II.74}$$

where P_T and P_s are triplet and singlet γ-ray distributions normalized to unity. $P_s(E_\gamma)$ is a delta function, $\delta(E_\gamma - 0.511\,\text{MeV})$. The lifetimes of the two P_s states are $\tau(^1S_0) = 1.25 \times 10^{-10}$ s and $\tau(^3S_1) = 1.4 \times 10^{-7}$ s (Heitler, 1954, p. 275).

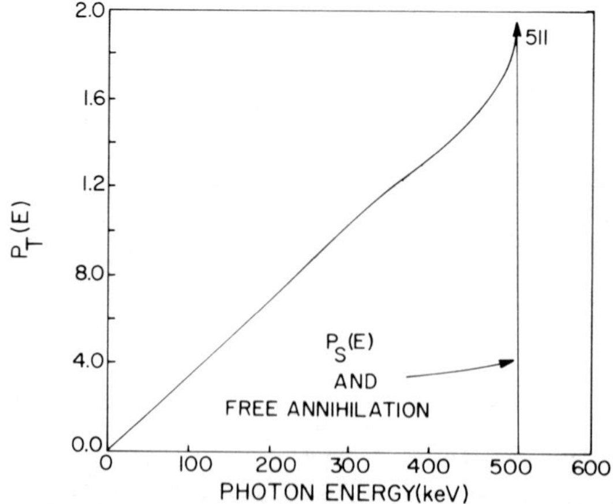

Fig. II-27. The normalized positronium γ-ray spectrum for triplet and singlet annihilation. (From M. Leventhal, *Astrophys. J. (Letters)* **183**, L147. Copyright 1973, The American Astronomical Society. Used by permission of the University of Chicago Press.)

Leventhal (1973a) has pointed out that if the density of the gaseous medium is sufficiently low, then positronium, once formed, can survive breakup by ionizing collisions where the critical density of the neutral gas is given by $n_c \leqslant 1/T_c \sigma_I v$. Here, T_c is the ionizing collision time, σ_I is the P_s ionization cross section, and v is the positron velocity. If $T_c > \tau(^3S_1)$ then the bound state annihilation can proceed giving the spectrum above. Using values for atomic H from Massey and Mohr (1954) of $\sigma_I = 7 \times 10^{-17}\,\text{cm}^2$ at $v = 2 \times 10^8\,\text{cm s}^{-1}$, Leventhal (1973a) finds $n_c \lesssim 10^{15}\,\text{cm}^{-3}$. Since the matter density is this high only when near condensed objects, it is expected that the predominant annihilation spectrum from positrons near rest will be a continuous spectrum. This conclusion was reached by Stecker (1969) for the interstellar medium. He calculated the ratio of positronium formation to free annihilation for atomic H and found that most of the positrons annihilating near rest do so through Ps formation at an average energy of ~ 35eV. Besides collisional breakup of positronium, a strong magnetic field ($\sim$5000G) will cause almost all the $^3S_1$ (m = 0) orthopositronium to mix with the singlet state and annihilate quickly into 2 photons (Green and Lee, 1964). The more general astrophysical problem of positronium formation in a plasma has received little treatment. Ramaty and Lingenfelter (1975) however, have pointed out that if the temperature is sufficiently high ($\sim 10^6$ K), then the rate of free annihilation just equals the rate of positronium formation.

2.5. Special Effects

Several effects can influence γ-rays in subtle ways before they reach the observer. We will discuss here the effect of Doppler shifts on the line widths and energies of γ-ray lines and, briefly, the attentuation of the flux of the γ-rays over large cosmic distances.

2.5.1. DOPPLER SHIFTS

In many γ-ray producing reactions such as inelastic scattering of protons on heavier targets or spallation reactions where the products are left in excited states, the γ-ray source may be moving with velocities between 10^8 and $10^{10}\,\text{cm s}^{-1}$ while emitting γ-rays. The gamma ray energy emitted is given by the usual Doppler formula, $E_\gamma = E_\gamma^0(1 + (v/c)\cos\theta)$, where θ is the angle between the direction of the γ-ray and the velocity vector of the nucleus giving a γ-ray of energy E_γ^0. (The shift of the γ-ray energy from E_γ^0 in the nuclide's rest frame due to the nuclear recoil on emission is discussed in Section II-2.4.3.) In a case where the nuclear velocity vectors are distributed at random, the observed γ-rays have energies over the range $E_\gamma^0(1 - (v/c)) < E_\gamma < E_\gamma^0(1 + (v/c))$, thus the shift in the energy of the photon is $\Delta E_\gamma = E_\gamma^0(v/c)\cos\theta$, and so the average width (FWHM) of the Doppler shifted line is $\Delta E_{\pm} \cong (4v/\pi c)\, E_\gamma^0$. Note that this is not the correct line width since, in general, v will depend on the angle of emission of the excited nucleus, and the line width must be found by also taking into account the angular distribution of the scattering cross section. This has been done by Meneguzzi and Reeves (1973), who have estimated the magnitude of this effect. As an example, consider the reaction $^{16}O(p, p'\gamma)^{16}O$ for exciting the 6.13 MeV level. The maximum velocity attainable for ^{16}O when $E_p = 10$ MeV is $v(^{16}O) \cong 2.8 \times 10^8$ cm s^{-1}, in which case the Doppler width of the line is $\Delta E_\gamma \sim 70$ keV.

This width is small compared with the resolution of a NaI(Tl) spectrometer which is typically 130 keV at 6 MeV; however, with a Ge(Li) detector the best resolution attainable ranges from 1 to 5 keV so the advantage of such an instrument would be lost. There is a compensating effect, however, since this Doppler shift can be attenuated in the same way as was used in the method of measuring nuclear lifetimes (cf. Bell, 1966). If the excited recoil nuclei are moving in a medium with a stopping time T, for the particle type and energy involved, and τ is the lifetime of the excited level, then for $\tau \gg T$, there is no Doppler shift since the excited nucleus has come to rest. If $\tau \ll T$, the full Doppler shift occurs.

In solids, the stopping time for ions (e.g., C, O) with characteristic nuclear recoil energies is $\lesssim 10^{-12}$ s, so usually the first condition results; however, in the astrophysical case the latter situation can be expected except near condensed objects. As an example, consider the 6.13 MeV level of ^{16}O for which $\tau = 1.7 \times 10^{-11}$ s. We can estimate the stopping time of ^{16}O in a typically light element gaseous medium from the fact that the stopping time for 8.8 MeV α-particles ($v_\alpha \sim 2 \times 10^8$ cm s^{-1}) in air at sea level is $\sim 8.5 \times 10^{-9}$ s. The stopping time in H would be half this since the energy loss in light elements is $\propto Z/A$. The rate of stopping of the ^{16}O should be ($Z^2 = 64$) times faster than for the α-particles of the same velocity, so the stopping time at a density of $\sim 3 \times 10^{19}$ cm^{-3} would be $\sim 1.3 \times 10^{-10}$ s, and should be inversely proportional to the density. Even at sea level then, $\tau(^{16}O) \ll T$. Therefore, the atmospheric ^{16}O line predicted by Peterson *et al.* (1973b) from (n, n$'\gamma$) at 6.13 MeV should appear much broader than the instrument width if measured with a high resolution detector such as Ge(Li), but this Doppler broadening would be barely detectable with a NaI(Tl) spectrometer. Such a measurement should prove whether a line at 6.13 MeV is actually an atmospheric emission or a local background effect.

These considerations should be taken into account when evaluating the predictions of gamma ray line fluxes from different sources in Chapter III and the design of experiments in Chapter VI. Since this important effect has only been discussed in the literature for cosmic γ-ray lines by Meneguzzi and Reeves (1973, 1975), we refer the reader to the table of nuclear lifetimes given by Lindskog *et al.* (1966).

The other information required is the collision time T of fast ions and atoms in astrophysical plasmas, but, unfortunately, this is not readily available in the literature. Rough estimates can be made using the elementary theory of slowing down of fast heavy particles in various materials (see, e.g., Evans, 1955).

2.5.2. GRAVITATIONAL AND COSMOLOGICAL RED SHIFTS

Gamma ray lines produced near strong gravitational fields or at earlier epochs of the Universe are expected to have their energies shifted downward. In the former case, the gravitational red shift ΔE of a γ-ray emitted from a source at rest is given by

$$z_g = \Delta E/E_\gamma^0 = GM/c^2 R \tag{II.75}$$

where M and R are the mass and radius, respectively, of a condensed object (see McVittie, 1965). For example, a neutron star of one solar mass and $R \simeq 10^6$ cm will produce a gravitational red shift of ~0.15 for an emission line produced at the surface of the star (see Section III-3.2.2 for further discussion). The cosmological red shift is defined as

$$z = \Delta\lambda/\lambda_0 = \Delta E/E_0 \tag{II.76}$$

and is expressed as

$$(1 + z) = \frac{R(t_0)}{R(t)} \tag{II.77}$$

where $R(t_0)$ is the scale of the Universe at t_0, the present epoch, and $R(t)$ is the scale of the universe at the time, t, when the photon was emitted. Thus, $t_0 - t$ is the travel time of the radiation. In general $R(t)$ is undetermined unless the general relativistic model of the Universe is known. A discussion of the cosmological red shift may also be found in McVittie (1965) and in Sciama (1971). In Section III-3.3., we discuss examples corresponding to $z \simeq 70$ for π_0 γ-rays and $z \simeq 2$ for the n-p capture line. The velocity, v, of the source when the photon was emitted can be found from the relativistic formula

$$1 + z = \sqrt{\frac{1 + v/c}{1 - v/c}}\,. \tag{II.78}$$

2.5.3. ABSORPTION OF GAMMA RAYS IN GALACTIC AND METAGALACTIC SPACE

Normally it is assumed that absorption of γ-rays of MeV energies is not a serious problem (see Chapter IV and Appendix A for a discussion of absorption coefficients for γ-rays). For example, the $1/e$ attenuation lengths (optical depth = 1) of γ-ray photons in neutral H at 100 keV, 1 MeV, and 100 MeV are 3.3, 10, and 100 g cm^{-2}, respectively. If the matter in the Universe had the current particle density value of ~10^{-5} cm^{-3} (~1.6×10^{-29} g cm^{-3}) for all time, then the above mean free paths would correspond to distances ~10^{30} cm for rectilinear travel of photons. Of course, near condensed objects, these absorption lengths are accumulated in very short distances. For example, 10g cm^{-2} is the integrated mass cm^{-2} down to a solar photospheric depth of ~1000 km. However according to our present evolutionary cosmological picture (cf. Sciama, 1971) the existence of the Hubble law and the isotropic microwave background radiation requires that the density of matter at an earlier epoch be $n(z) = n_0(1 + z)^3$. In addition, a photon received now at energy $E_\gamma(t_0)$ was emitted with an energy $E(t) = zE_\gamma(t_0)$, and the temperature of the Universe varies as $T = T_0(1 + z)$.

In the cosmic medium, the presence of low energy photons such as starlight, cosmic black body radiation, etc., also gives the possibility of absorption by the reaction

$$\gamma + (h\nu) \rightarrow e^+ + e^- .$$

These cosmological effects have now received a fairly thorough theoretical treatment and, in general, the problem is quite complex. Arons (1971a, b) has solved the general γ-ray propagation problem by solving a cosmological photon transport (CPT) equation in order to arrive at a γ-ray spectrum from a cosmological distribution of sources. Stecker (1975) has also recently reviewed the work on the absorption of γ-rays as a function of red shift z. The important result for our discussion is shown in Figure II-28, which plots the critical red shift as a function of the γ-ray photon energy received. This figure is based upon work by both Arons and McCray (1969) and Fazio and Stecker (1970).

The graph should be interpreted as follows. First, the critical red shift is determined from the condition that the optical depth $\tau = 1$ where, in general,

$$\tau(E_\gamma, z) = \int_0^{z_{max}} \kappa(E_\gamma, z) \left(\frac{dl}{dz}\right) dz .$$

Here, $\kappa(E_\gamma, z)$ is the absorption coefficient (cm^{-1}) for a particular process, and dl/dz is

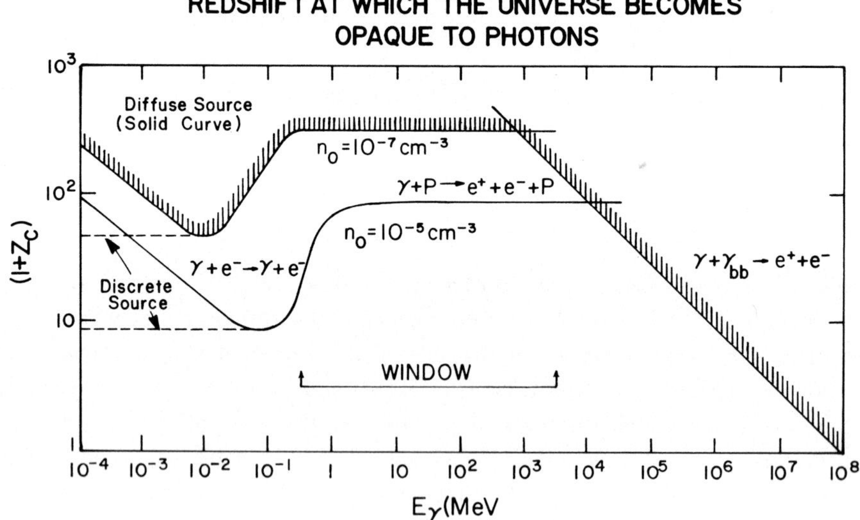

Fig. II-28. The redshift at which the Universe becomes opaque to photons given as a function of *observed* γ-ray energy. γ-rays originating at all redshifts below the curve can reach us unattenuated with the energy indicated. The two curves on the left side of the figure are for attenuation by Compton scattering with intergalactic electrons having the densities indicated and for pair production and are based on the calculations of Arons and McCray (1969). The right hand curve results from attenuation of γ-rays by interactions with the microwave black body radiation and is based on the discussion of Fazio and Stecker (1970). (From F.W. Stecker: 1975, in J.L. Osborne and A.W. Wolfendale (eds.), *Origin of Cosmic Rays.* Used by permission D. Reidel Publishing Company, Dordrecht, Holland.)

found from general relativity (see Stecker, 1975). The γ-rays originating from all red shifts below a given curve reach us unattenuated with the energy indicated. The density values $n_0 = 10^{-5}$ cm^{-3} and $n_0 = 10^{-7}$ cm^{-3} correspond to the range of values for a closed and open Universe according to the evolutionary picture.

The two curves on the left are based on the work of Arons and McCray (1969), who have considered only scattering to lower energies by the Compton process and absorption by pair production. The intergalactic medium is assumed to be fully ionized so no photoelectric absorption is possible. The significance of the low energy portion of the curves (<100keV) is that the steeply rising part of the curve gives the red shift necessary to produce a serious distortion in a cosmological diffuse radiation field due to Compton scattering. The dotted extension below 100 keV, for both density values, is based on using the Thomson cross section in the equation for the optical depth $\tau = 1$. This gives the criterion for a significant obscuration of a low energy discrete source. In the energy region above 1.02 MeV, the pair production process completely absorbs photons; the curve here refers to both distortion of a diffuse background and obscuration of a discrete source. It is of interest to consider the implications of these curves for γ-ray astronomy. As Arons and McCray (1969) point out, "the energy range $E_{\gamma 0} > 1$ MeV offers the best chance of observing discrete sources at very large red shifts where at our present state of understanding, anything is likely to happen!" Also, a cutoff of discrete source counts in the 100 keV to 1 MeV region may provide a measure of the density of the intergalactic medium, but of course we need to find any extragalactic discrete sources in this energy range first. With the assumption that the Universe is fully ionized, the windows out to large red shifts for a diffuse background fall in the energy range $E_\gamma < 1$ keV and $E_\gamma >$ 100keV. From Figure II-28 it is seen that a high density Universe becomes transparent above $E_\gamma \sim 500$keV.

The last process to consider is the absorption of γ-ray photons by interaction with ambient photons to produce electron pairs as mentioned above. The threshold for pair production by this process is $E_\gamma = (m_0 c^2)^2/\epsilon$ where ϵ is the energy of the ambient photon (Fazio, 1967; Stecker, 1974). The cross section for this is a maximum ($\sim 10^{-25}$ cm^2) near the threshold energy. However, since $\epsilon \sim 10^{-6}$ MeV for starlight and even lower for the 2.7K radiation, this process is reserved for absorption of γ-rays with energies $E_\gamma \lesssim (0.5)^2/10^{-6} \sim 10^{11}$ eV.

The right hand curve in Figure II-28 gives the critical red shift for this process when the ambient photons are microwave background at $T = 2.7$ K. A more complete discussion is given by Fazio (1967) and Stecker (1971, 1974). Jelley (1966) has also suggested that this absorption process is significant for γ-rays above 10^{11} eV in QSO's and above 10^8 eV (100 MeV) in point X-ray sources. In order that the process be important for MeV γ-rays, the target photons must have energies of $\sim 1/4$ MeV corresponding to a temperature of $\sim 3 \times 10^9$ K. The absorption of γ-rays by this process in photon fields around pulsars and QSO's has also been studied recently by Pollack *et al.* (1971). This work shows that photons emitted from the surface of a neutron star, such as in the Crab Nebula, will be strongly attenuated if the energy is $\gtrsim 1$ MeV.

TABLE II-11

Summary of differential photon spectral shapes from various mechanisms (see the text for the equation references).
For Electrons $k_c E^{-\alpha}$; $\alpha > 1$. For Protons $k_p E^{-m}$; $m > 1$

Mechanisms	Spectral shapes		References	Eqn. no.
Nonthermal electron bremsstrahlung	$I_b(E_\gamma) \propto E_\gamma^{-\alpha}$	$v_e \sim c$	Fazio (1967); Stecker (1974)	(II.32)
Nonthermal proton bremsstrahlung	$I_b(E_\gamma) \propto E_\gamma^{-(m+1)}$	$v_p \sim c$	Jones (1971)	
Nonthermal electron bremsstrahlung (nonrelativistic electron < 200 keV)	$I(E_\gamma) \propto E_\gamma^{-(\alpha+1/2)}$	for Thin Target	Brown (1971)	
	$I(E_\gamma) \propto E_\gamma^{-(\alpha-1)}$	for Thick Target	Brown (1971)	
Inverse compton (electrons)	$I_c(E_\gamma) \propto E_\gamma^{-(\alpha+1)/2}$		Ginzburg and Syrovatskii (1964); Felten and Morrison (1966)	(II.12)
Synchrotron radiation (electrons) (Magnetobremsstrahlung)	$I_s(E_\gamma) \propto B^{(\alpha+1)/2} E_\gamma^{-(\alpha+1)/2}$		Fazio (1967); Ginzburg and Syrovatskii (1964); Felten and Morrison (1966)	(II.20)
Thermal bremsstrahlung				
Optically thin[a]	$P_{ff}(E_\gamma) \propto 1/E_\gamma \exp(-E_\gamma/kT)$		Hayakawa (1969)	(II.33)
Optically thick	$I(E_\gamma) \propto E_\gamma^2 [\exp(E_\gamma/kT) - 1]^{-1}$		Hayakawa (1969)	(II.3)

[a] Note: Edges and lines from free-bound and bound-bound transitions.

2.6. Summary of γ-Ray Spectra

For convenience, a summary of formulae for the various possible differential γ-ray spectral shapes is given in Table II-11. In general, only the dependence of the spectra on the photon energy is indicated, and other physical parameters are suppressed in the relations. The complete form of a given equation may be found by the equation number shown at the right in the table.

CHAPTER III

THEORETICAL ESTIMATES OF γ-RAY EMISSION

In Table II-2 we have listed several astrophysical sites where γ-ray line production might be expected. In the case of the galactic disk, core, and the Sun, the calculation of the expected γ-ray fluxes from the different production mechanisms is on a reasonably firm basis since experimental results are available (cf. Sections V-5.1 and V-5.2). In addition, the evidence for a γ-ray line feature near 500 keV from the direction of the Galactic Center (Section V-5.2.3) has given rise to a flurry of theoretical predictions as to the origin of this feature. The same is true for the γ-ray bursts discovered on the Vela satellites (Section V-5.4). We will review here the predicted fluxes from several sources listed in Table II-2. Since the case of γ-ray production in the solar atmosphere has received considerable treatment, we will review these calculations in more detail than for other sources, since the results are generally applicable to any astrophysical source where a solar abundance is a reasonable approximation. Only those theoretical predictions for other sources will be discussed here that seem most pertinent to γ-ray line astronomy. The experimental evidence for γ-ray fluxes in the nuclear transition region is described in Chapter V, and reference to the discussion there should be made when considering such material in this chapter.

3.1. Solar γ-Rays

The interest in the emission of γ-rays and neutrons from the Sun has been closely coupled almost from the time it was realized that the Sun could produce particles with cosmic ray energies. This latter evidence was obtained in 1942 independently by Ehmert (1948) and Lange and Forbush (1942). This evidence appeared in the form of an intensity increase as registered by a coincidence detector and a shielded ion chamber, respectively, detecting charged cosmic rays in the atmosphere at sea level. In 1949 Adams and Braddick (1951) detected, in association with a solar flare, an anomalously large increase in intensity in a neutron monitor, while the corresponding charged particle monitor showed a much smaller relative increase in intensity. A possible explanation of the large neutron intensity increase was advanced in 1951 by Biermann *et al.* (1951), who pointed out that protons accelerated at the Sun could interact with the solar atmosphere and produce nuclear reactions with the emission of neutrons. It was soon realized, however, that charged cosmic rays produced these solar cosmic ray events.

It was several years later that Severnyi (1957), in the study of the shape of the Balmer emission line in large flares, suggested evidence for a Dα (deuterium) line. He was thus led to hypothesize a flare model in which the production of neutrons in D-D thermonuclear reactions occurred behind shock fronts converging on one another at the site of a flare. Severnyi also hypothesized at this time that these thermonuclear reactions would lead to the near simulataneous production of X-rays, γ-rays, and the charged solar cosmic rays. These ideas were discussed in several papers (Severnyi, 1958, 1964; Severnyi and Shabanskii, 1961). Goldberg *et al.* (1958) have studied the profile of Hα immediately after the large east limb flare of 1956, February 10 (the large cosmic ray burst was associated with the February 23 flare). Since this line, from a loop prominence, was observed to be asymmetric toward the violet, Goldberg *et al.* (1958) suggested that the D might have been produced by nuclear reactions during the flare.

Direct evidence for the occurrence of nuclear reactions at the Sun in association with solar flares was obtained during November 1960 when ^{3}H and ^{3}He were found by radiochemical means in recovered casings from a Discoverer satellite by Fireman *et al.* (1961) and Schaeffer and Zähringer (1962), respectively. Flamm *et al.* (1962) made estimates of the solar production of ^{3}H and ^{3}He from solar proton spallation of ^{4}He in the solar atmosphere. These reactions also give neutrons as a by-product, and Fireman (1963) and Chupp (1963) made estimates of the fluxes of neutrons and γ-rays expected at the Earth following large flares. They pointed out that detection of either of these components would provide confirmation for the ^{3}H and ^{3}He observations at the Earth and also enhance our understanding of solar flares. Elliot (1964) proposed that the flare process itself might be a result of the slow ($\sim$1 day) acceleration of ambient chromospheric and coronal protons with the flash phase of a flare being a direct result of the release of high energy protons ($>$10 MeV) into the dense solar atmosphere. As a test of this idea, Elliot (1964) proposed a search for the emission of high energy γ-rays (π^0) and neutrons at low levels before the flare. These ideas have been further developed (Elliot, 1969, 1973).

Dolan and Fazio (1965) made the first extensive study of the γ-ray spectrum expected during solar flares. They concluded that in the energy region 10 keV to 1 MeV the spectrum would be predominantly continuous due to solar flare electron bremsstrahlung. Flare protons could also produce copious γ-ray lines from ^{12}C(p, p$'\gamma$), ^{14}N(p, p$'\gamma$), and ^{16}O(p, p$'\gamma$); and neutron capture, via the reaction ^{1}H(n, γ)D, would give a strong 2.22 MeV γ-ray line. The required neutron source was provided by spallation reactions due to energetic proton ($\lesssim$ 30 MeV) reactions on He, C, N, and O. It was also concluded that the positrons (β^+) emitted by radioactive spallation products of CNO and π^+ meson decay would give the 0.51 MeV annihilation line. In addition, higher energy γ-rays ($>$10 MeV) from π^0 meson decay were also expected. Dolan and Fazio (1965) had also investigated the neutron production rate from the thermonuclear reaction ^{2}H(d, n) ^{3}He and concluded that any contribution from this source would be negligible compared to neutron production by spallation reactions which was estimated to be $\sim 10^{-1}$ neutrons cm^{-3} s^{-1}. This conclusion was based on the assumption that the D density $n_d \sim 10^{-5}\, n_p$, $n_p \sim 3 \times 10^{13}$ cm^{-3}, and that the flare temperature is $\sim 3 \times 10^5$ K. Severnyi and Shabanskii (1961) on the other hand have predicted that the above reaction could produce as many as 1 neutron cm^{-3} s^{-1}, which would exceed the contributions from spallation reactions given above. The question of thermonuclear reactions in flares will be mentioned later in connection with solar flare γ-ray observations (Section V-5.1); however, it has been tacitly assumed that only spallation reactions by energetic solar particles are effective in producing neutrons, although special circumstances may also allow a contribution from thermonuclear reactions.

These early estimates were comprehensively extended by Lingenfelter and Ramaty (1967), who treated the complete problem (except for thermonuclear reactions) of the production of all charged and neutral nuclear reaction secondaries assuming the differential solar cosmic ray spectrum is an exponential rigidity spectrum as seen at the Earth. This work gave generalized fluxes for all the above γ-ray lines, and when applied specifically to the large 1960, November 12 flare implied extremely large γ-ray and neutron fluxes ($\gg 1\ \mathrm{cm}^{-2}\ \mathrm{s}^{-1}$). This work greatly stimulated experimental activity to measure solar γ-ray and neutron fluxes. More recently, Kuzhevskii (1969) gave estimates of γ-ray line fluxes from several $(p, p'\gamma)$ reactions. Cheng (1972), using the Lingenfelter and Ramaty (1967) calculations, evaluated the expected γ-ray spectrum for two different approximate flare models. Lingenfelter (1969) assumed that the solar flare optical emission was a result of the ionization loss of protons in the chromosphere and obtained a ratio between the expected γ-ray line yield and the ionization loss rate. Chupp (1971) reviewed the evidence for the acceleration of charged particles in the solar atmosphere with specific reference to the production of γ-rays and neutrons, and the status of the experimental efforts to detect these radiations. Reference to other theoretical work on solar neutron and γ-ray fluxes may be found in a recent review by Ramaty *et al.* (1975).

Even though some evidence had been given for solar neutron fluxes and γ-ray line fluxes (Daniel *et al.*, 1967; Hirasima *et al.*, 1969), the observation of distinct γ-ray lines at 0.51 MeV and 2.2 MeV by the OSO-7 γ-ray detector (Chupp *et al.*, 1973a, b) in close time coincidence with the large solar flares on 1972, August 4 and 7 provided a stimulus for more detailed calculations of the production of neutral secondaries in association with solar flares. This need was reinforced by the observation in space of charged secondaries ^{3}He and ^{3}H, following several flares before the August 1972 events. This work was reported by Anglin *et al.* (1973) at the same time as the first report of the γ-ray lines (see Ramaty and Stone, 1973). We will briefly review here the most relevant results of these new calculations.

Ramaty and Lingenfelter (1973) have extensively revised their earlier calculations of 1967 on the production of solar flare nuclear reaction secondaries. It should be noted, however, that there is still no complete geometrical flare model available at this time which would allow more refined calculations, so one should keep this in mind when interpreting any results such as the OSO-7 results discussed in Section V-5.1.

3.1.1. CALCULATION OF γ-RAY YIELD

The detailed calculations of the solar flare yield of the different γ-ray lines made originally by Lingenfelter and Ramaty (1967) assumed the accelerated solar cosmic rays had an exponential rigidity spectra of the form $\exp(-P/P_0)$, where P_0 is the characteristic rigidity parameter which determines the hardness of the spectrum. These calculations have now been extended to include power law spectra by Ramaty and Lingenfelter (1975); a recent review of the theoretical basis for these calculations has been given by Ramaty *et al.* (1975). This work also gives up-to-date results based on the latest cross section data. We will review here only the basis of the new γ-ray yield calculations. They considered two limiting cases:

(a) A *thin target* model in which the spectrum of accelerated particles is *not* modified during the time in which nuclear reactions occur. This implies that either the total path length traversed by the particles at the Sun is small in comparison with their nuclear interaction length, or that the particle energy loss from ionization and nuclear interactions is just balanced by energy gain by a continuously operating acceleration mechanism.

(b) A *thick target* model in which the particles undergo nuclear reactions as they slow down and stop.

The composition of the ambient solar atmosphere used in the recent Ramaty and Lingenfelter (1975) calculations is that given by Cameron (1973). The power law and exponential rigidity differential spectra assumed for the *thin target* case are

$$N_i(E) = k_i E^{-s} \qquad \text{(particles (MeV per nucleon)}^{-1}\text{)} \tag{III.1}$$

and

$$N_i(P) = k_i' \exp(-P/P_0) \qquad \text{(particles MV}^{-1}\text{)} \tag{III.2}$$

where in both cases the particle spectra give the *instantaneous* number of particles of species i in the interaction region. The k_i and k_i' are found by the normalization $\int_{30}^{\infty} N_i(E)\,dE = \alpha_i$ and $\int_{P_{30}}^{\infty} N_i(P)\,dP = \alpha_i$, where α_i is the solar abundance relative to H. The spectral index, s, and the characteristic rigidity, P_0, are assumed to be the same for all species. In the *thick target* case, the instantaneous number is replaced by the total numbers, $\bar{N}_i(E)$ or $\bar{N}_i(P)$ particles that enter the thick target per unit energy or rigidity interval, respectively. The normalization is as before. The secondary production for the *thin target* case may be expressed as

$$q_{ij} = n_j \int_0^{\infty} N_i(E)\, c\beta\, \sigma_{ij}(E)\,dE \quad (\text{s}^{-1}) \tag{III.3}$$

where the subscripts refer to accelerated species (i) incident on target species (j) with a number density n_j, and $c\beta$ is the velocity of the projectile i. $\sigma_{ij}(E)$ is the cross section as a function of energy per nucleon for a particular reaction.

In the thick target case the *total* number of secondaries, Q_{ij}, for a particular reaction is calculated from

$$Q_{ij} = \eta_j \int_0^{\infty} \bar{N}_i(E') \left[\int_0^{E} \left(\frac{dE}{dx}\right)_i^{-1} \sigma_{ij}(E)\,dE \right] dE' \tag{III.4}$$

where $(dE/dx)_i$ (g cm^{-2}) is the total stopping power of the projectile in solar material and η_j is the number of target nuclei g^{-1} of solar material. Equation (III.4) is rewritten by Ramaty and Lingenfelter (1975) to give

$$Q_{ij} = \eta_j \int_0^{\infty} \bar{N}_i(>E) \left(\frac{dE}{dx}\right)_i^{-1} \sigma_{ij}(E)\,dE \tag{III.5}$$

where $\bar{N}_i(>E)$ is the normalized integral spectrum or the number of particles, *i*, with energy greater than *E*. Similar formulae are used for exponential rigidity spectra. The basic results of these calculations give the secondary yields of positrons, neutrons, and excited nuclei.

3.1.2. Positron and Neutron Production

First, let us briefly review possible sources of the positrons and neutrons that are required to give rise to the observed line features at 0.5 and 2.2 MeV. In the first case, we consider the positron emitters produced as a result of the interactions of accelerated protons, α-particles, and other solar constituents interacting in the solar atmosphere. The reactions which can produce positrons are either exoergic ($Q > 0$) or endoergic ($Q < 0$). Table III-1 shows several positron emitters from exoergic and endoergic reactions having thresholds of a few MeV to 41 MeV for the light isotope reactions, to several hundred MeV for π^+ production. The fifth column gives the threshold in the laboratory system in MeV per nucleon for cases where $Q < 0$. Note that the thresholds are the same whether the light particle or the heavy particle is the projectile when the projectile energy is expressed in MeV per nucleon. The basic reference to the cross sections used for the β^+ calculations is Lingenfelter and Ramaty (1967) with some recent revisions (see Ramaty *et al.*, 1975).

TABLE III-1

Principal positron emitters from energetic particle reactions. (From R. Ramaty, *et al.*: 1975, 'Solar Gamma Rays', *Space Sci. Rev.* **18**, 341, D. Reidel Publishing Company, by permission.)

Positron emitter	Half-life	Maximum positron energy (MeV)	Production mode	Threshold (MeV)
μ^+	1.5×16^{-6} s	53	$p + {}^1H \rightarrow \pi^+ \ldots$	292.3
			$p + {}^4He \rightarrow \pi^+ \ldots$	185
			$\pi^+ \rightarrow \mu^+ + \nu$	–
^{10}C	19 s	1.9	$p + {}^{16}O \rightarrow {}^{10}C + \ldots$	41.4
			$p + {}^{14}N \rightarrow {}^{10}C + \ldots$	17.1
			$p + {}^{12}C \rightarrow {}^{10}C + \ldots$	34.4
^{11}C	20.5 min	0.92	$p + {}^{16}O \rightarrow {}^{11}C + \ldots$	27.5
			$p + {}^{14}N \rightarrow {}^{11}C + \ldots$	3.1
			$p + {}^{12}C \rightarrow {}^{11}C + \ldots$	17.9
^{12}N	0.011 s	16.4	$p + {}^{12}C \rightarrow {}^{12}N + n$	19.6
^{13}N	10 min	1.19	$p + {}^{16}O \rightarrow {}^{13}N + \ldots$	5.5
			$p + {}^{14}N \rightarrow {}^{13}N + \ldots$	9.0
^{14}O	71 s	1.8(99.4%)	$p + {}^{16}O \rightarrow {}^{14}O + \ldots$	30.7
		4.1(0.6%)	$p + {}^{14}N \rightarrow {}^{14}O + n$	6.3
^{15}O	2.06 min	1.74	$p + {}^{16}O \rightarrow {}^{15}O + \ldots$	14.3
			$\alpha + {}^{12}C \rightarrow {}^{15}O + n$	2.8
^{19}Ne	17.4 s	2.2	$\alpha + {}^{16}O \rightarrow {}^{19}Ne + n$	3.75

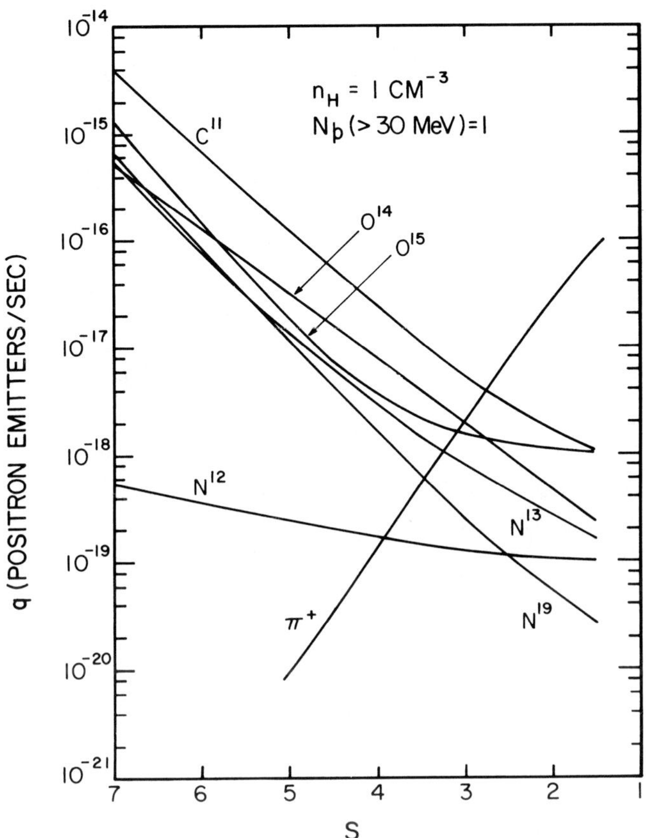

Fig. III-1. The normalized production rate of positron emitters for a *power law* (E^{-s}) differential spectrum of solar cosmic rays under *thin target* conditions. (From R. Ramaty *et al.*: 1975, 'Solar Gamma Rays', *Space Sci. Rev.* **18**, 341, D. Reidel Publishing Company, by permission.)

Figure III-1 shows the normalized production rate of several secondary positron emitters from Equation (III.1) for a thin target and a power law incident spectrum. In order to find the absolute production rate the ordinate must be multiplied by $n_{\mathrm{H}} \times N_{\mathrm{p}}$ (>30 MeV), the product of the H density and the integral number of protons.

Figure III-1 shows that the yield of positrons through the $\pi^+ \rightarrow \mu^+$ decay chain is predominant over positron emitters (see Table III-1) for flat spectra (s-small) and is negligible compared to the latter for very steep spectra. The corresponding result for a thin target using an exponential rigidity spectrum is shown in Figure III-2. In this case the absolute yield is found by multiplying by $n_{\mathrm{H}} \times N_{\mathrm{p}}$ $(P > 0)$ where now N_{p} $(P > 0)$ is the number of protons with a rigidity greater than zero. Similar curves have been deduced for the thick target case. An example is shown in Figure III-3 for an exponential rigidity spectrum where the normalization is for 1 proton with rigidity greater than $P = 0$. One general conclusion from these calculations is that the efficiency of the production of π^+ mesons

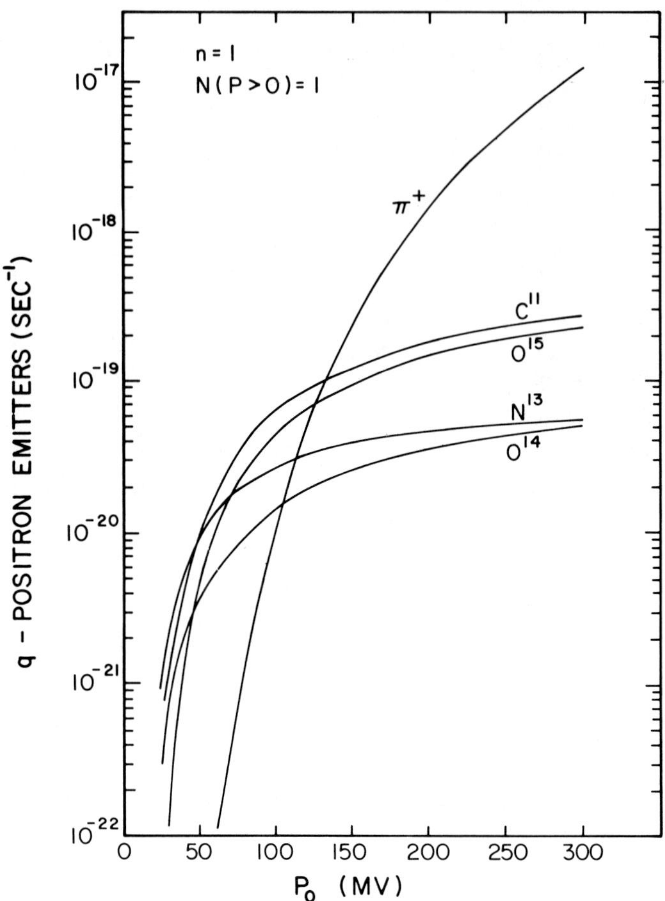

Fig. III-2. The normalized production rate of positron emitters for an exponential rigidity spectrum ($\propto e^{-P/P_0}$) of solar cosmic rays under *thin target* conditions. (From R. Ramaty and R.E. Lingenfelter: 1973, *NASA SP–342.*)

compared to total positron emitters is always greater in the thick target case than for the thin target case for any spectral shape (Ramaty and Lingenfelter, 1975).

The predominant sources of neutrons from exoergic and endoergic reactions are shown in Table III-2, here with laboratory reaction thresholds in MeV per nucleon for the case $Q < 0$. For neutron production Ramaty and Lingenfelter (1975) have revised the neutron production cross sections originally used by Lingenfelter and Ramaty (1967). These are shown in Figure III-4 for several neutron production processes such as protons on H, He, and CNO of solar abundance and α-particles on He or CNO. Since the cross sections are plotted vs. energy per nucleon, they are also appropriate for the case of the heavy particle as the projectile; however, the kinematics for the reaction products are different. In the case of (pCNO) and (αCNO) processes, the cross sections shown give the total neutron production per nucleus ($A \geqslant 12$) where the normalization is to one such nucleus based on the abundances of Cameron (1973). It is of particular significance to note the very low

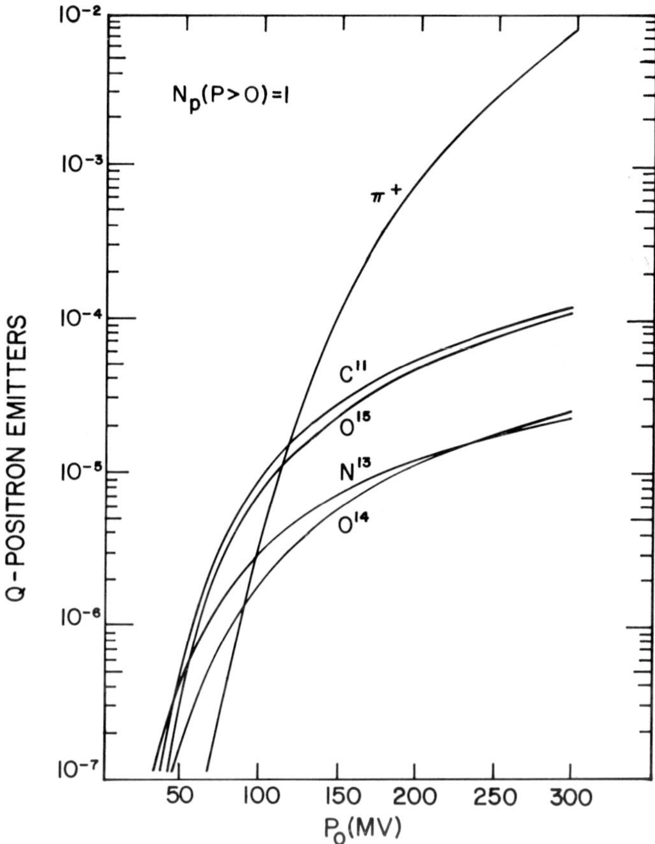

Fig. III-3. The normalized production rate of positron emitters for an exponential rigidity spectrum ($\propto e^{-P/P_0}$) of solar cosmic rays under *thick target* conditions. (From R. Ramaty and R.E. Lingenfelter: 1973, *NASA SP-342.*)

threshold of the αCNO processes. In Figures III-5 and III-6 are shown the normalized thin target neutron production rates for both power law and exponential rigidity spectra. The absolute neutron production rates are found from these curves by multiplying by $n_H N_p$ (>30 MeV) as described above for β^+ emitter production. Several important observations should be noted about these results. For very flat primary spectra (small s or large P_0), the neutrons are predominantly from the breakup of He by protons; i.e. pα and αp processes. In the case of a very steep *power law spectrum* the predominant neutron production is by α-particles interacting with heavy nuclei ($A \geqslant 12$), (αCNO) and (CNOα) processes. This is a direct result of the rapidly increasing number of low energy nucleons and lower thresholds (<10 MeV per nucleon) for these reactions whereas the threshold for neutron production via pα and αp reactions is ~30 MeV per nucleon. Also from Figure III-5 it can be seen that α-α reactions make a predominant contribution only in a very limited range of spectral shapes ($s \sim 4$). In the case of an exponential rigidity spectrum, the relative number of accelerated particles levels off at unity as $P \to 0$ and does not

TABLE III-2

Neutron production from energetic particle reactions. (From R. Ramaty *et al.*: 1975, 'Solar Gamma Rays', *Space Sci. Rev.* 18, 341, D. Reidel Publishing Company, by permission.)

	Reaction		Threshold (MeV/nucleon)
1.	$p + {}^{1}H$	$\rightarrow n + p + \pi^{+}$	292.3
2.	$p + {}^{4}He$	$\rightarrow {}^{3}He + p + n + (\pi)$	25.7
		$\rightarrow {}^{2}H + 2p + n + (\pi)$	32.6
		$\rightarrow 3p + 2n + (\pi)$	35.4
3.	$p + {}^{12}C$	$\rightarrow n + \ldots$	19.6
	$p + {}^{13}C$	$\rightarrow n + \ldots$	3.2
4.	$p + {}^{14}N$	$\rightarrow n + \ldots$	6.3
5.	$p + {}^{16}O$	$\rightarrow n + \ldots$	16.6
	$p + {}^{18}O$	$\rightarrow n + \ldots$	2.5
6.	$p + {}^{20}Ne$	$\rightarrow n + \ldots$	15.9
7.	$p + {}^{56}Fe$	$\rightarrow n + \ldots$	5.5
8.	$\alpha + {}^{4}He$	$\rightarrow {}^{7}Be + n$	9.5
9.	$\alpha + {}^{12}C$	$\rightarrow n + \ldots$	2.8
	$\alpha + {}^{13}C$	$\rightarrow n + \ldots$	$Q = +$ 2.2
10.	$\alpha + {}^{14}N$	$\rightarrow n + \ldots$	1.5
11.	$\alpha + {}^{16}O$	$\rightarrow n + \ldots$	3.8
	$\alpha + {}^{18}O$	$\rightarrow n + \ldots$	0.21
12.	$\alpha + {}^{20}Ne$	$\rightarrow n + \ldots$	2.16
	$\alpha + {}^{22}Ne$	$\rightarrow n + \ldots$	0.15
13.	$\alpha + {}^{56}Fe$	$\rightarrow n + \ldots$	1.37
14.	$\alpha + {}^{25}Mg$	$\rightarrow n + \ldots$	$Q = +$ 2.6
15.	$\alpha + {}^{26}Mg$	$\rightarrow n + \ldots$	$Q = +$ 0.04
16.	$\alpha + {}^{29}Si$	$\rightarrow n + \ldots$	0.43
17.	$d + {}^{2}D$	$\rightarrow {}^{3}He + n$	$Q = +$ 3.25
18.	$d + {}^{3}H$	$\rightarrow {}^{4}He + n$	$Q = +$ 17.6

approach ∞ as in the case of a power law spectrum. Therefore, for essentially all values of P_0 the pα neutron production predominates. Also, as pointed out by Ramaty and Lingenfelter (1975), this is a result of the fact that a high energy particle of $Z \geqslant 2$ has a *greater* rigidity than a proton of the *same energy per nucleon*. Thus, in an exponential rigidity spectrum the flux of particles ($Z \geqslant 2$) relative to protons is lower at the same energy/nucleon than for the power law spectrum, while the cross sections are the same (see Figure III-4).

It is of interest to compare the total neutron production rates for all the possibilities. These results have been calculated by Ramaty and Lingenfelter (1975) and are shown in Figure III-7. Thin target yields are shown on the left ordinate and thick target yields on the right ordinate. It is interesting to note that for *flat spectra* in both thick and thin target cases the neutron production rates are about the same regardless of the type of spectra. This is a consequence of the fact that in flat spectra most of the neutrons are produced in pα reactions, which have a threshold at ~30 MeV per nucleon, and that all spectra are normalized to 1 proton >30 MeV in these recent calculations. Finally, in a

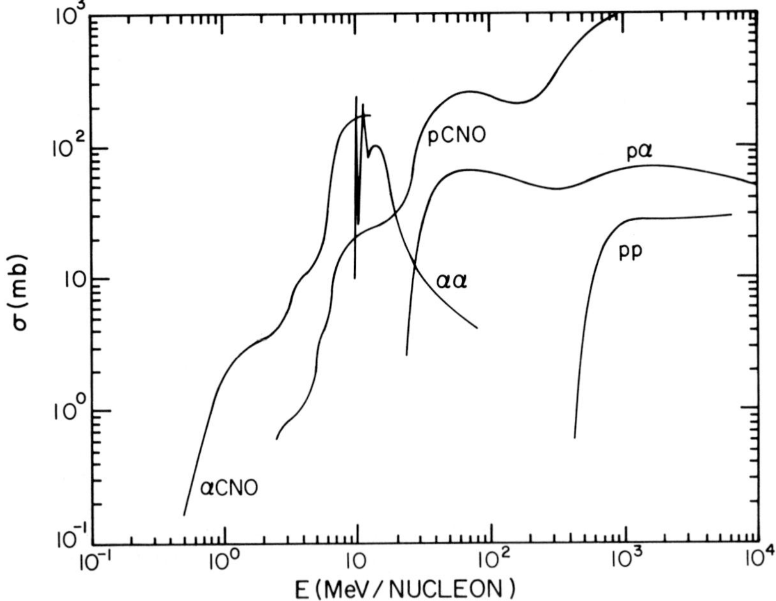

Fig. III-4. The neutron production cross section for several reactions. (From R. Ramaty and R.E. Lingenfelter: 1975, in S.R. Kane (ed.), *Solar Gamma-, X-, and EUV Radiation*, IAU Symposium No. 68, D. Reidel Publishing Company, by permission.)

power law spectrum with this same normalization, there are many more low energy particles, so αCNO and CNOα processes provide orders of magnitude more neutrons as was shown in Figure III-5.

The total yield of positron emitters compared to the total neutron yield has also been computed by Ramaty and Lingenfelter (1975) and the result is shown in Figure III-8. When used with Figure III-7, these two curves give the *total* yields of neutrons and positron emitters for the two extreme target conditions and the two spectral shapes. This basic information thus allows one to estimate the expected 0.51 MeV and 2.22 MeV γ-ray fluxes. It should be realized, however, that these calculations have not taken into account the geometry of the particle motions due to the specific accelerating mechanism and the solar magnetic field configuration. Neutron production will be highly anisotropic, especially for hard spectra, and local magnetic fields may strongly control the densities of the target medium and the trapping of positron emitters and the positrons.

a. *γ-Rays from Positrons and Neutrons*

The production of γ-ray lines from positrons and neutrons of course requires that these fast particles slow down and either annihilate or be captured, so there is a time delay between production and capture. This is very important in the case of solar flares, where the production appears to be in a rarified gas during a burst lasting ~1000 s. The positron capture time is $t_a \simeq 1.6 \times 10^{14} n_e^{-1}$ (s), and for neutrons $t_c \sim 1.4 \times 10^{19} n_p^{-1}$ (s), where in the first case n_e is the electron density (cm^{-3}), and in the second, n_p is the proton

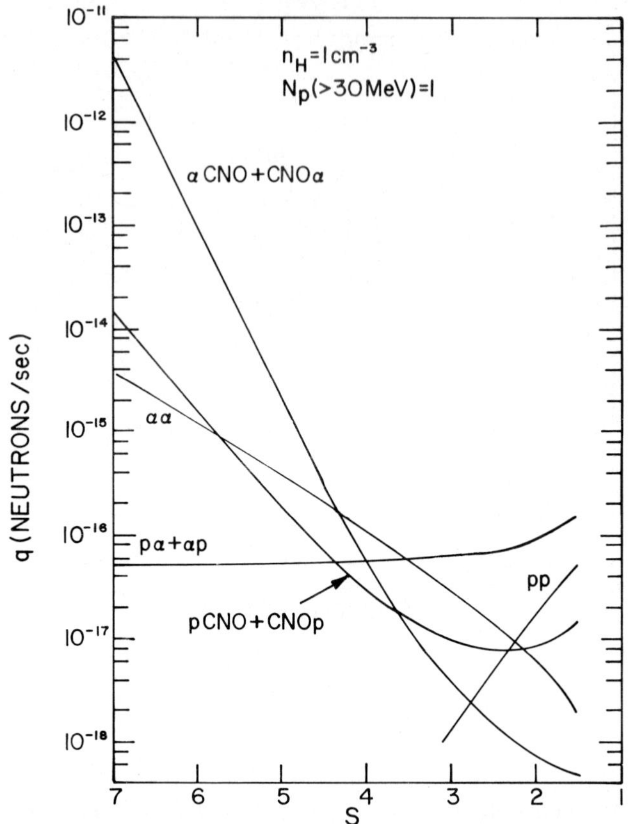

Fig. III.5. The normalized neutron production rate for a power law spectrum of solar cosmic rays under *thin target* conditions. (From R. Ramaty and R.E. Lingenfelter: 1975, in S.R. Kane (ed.), *Solar Gamma-, X-, and EUV Radiation*, IAU Symposium No. 68, D. Reidel Publishing Company, by permission.)

density (cm^{-3})(cf. Lingenfelter, 1969). Typical densities may be $10^{11} < n_e < 10^{15}$ cm^{-3} and $10^{16} < n_p < 10^{19}$ cm^{-3}, so it is clear that the time history of the 0.51 MeV and 2.22 MeV lines may not correspond closely to the impulsive reaction time scale in solar flares. For this reason it is of great importance to consider nuclear reactions that give γ-rays promptly such as direct nuclear excitation as discussed in Section II-2.4.3 (see also Section III-3.1.3).

The problem of determining the γ-ray spectrum resulting from the annihilation of positrons has not received a very thorough treatment, as yet, for the solar flare case. The reason for this is that the time history of the annihilation spectrum will depend on several factors which include: 1) the relative yields of the positron emitters (Table III-1) and their half-lives, 2) the slowing down time of the positrons, and 3) the relative yield of positrons from the π^+ meson channel. In addition the amount of positronium formation must be known since this determines the three photon continuum contribution from this source. *Ps* formation depends on the density and temperature of the medium. Because of

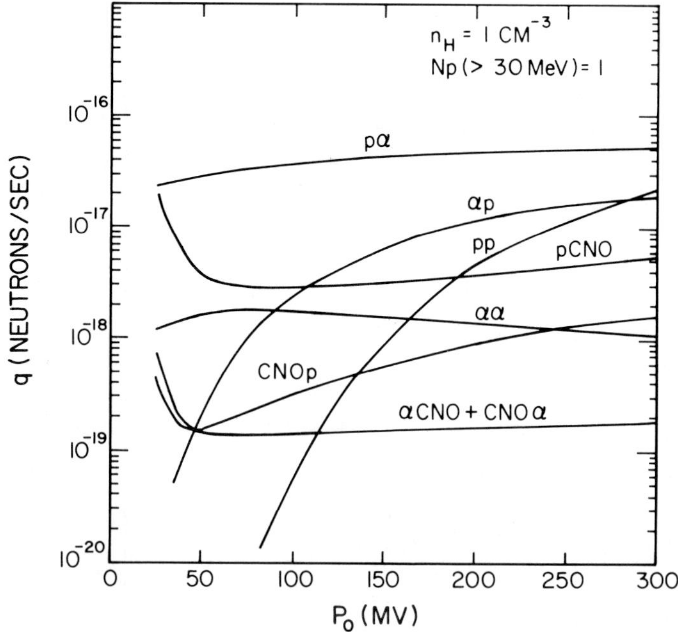

Fig. III-6. The normalized neutron production rate for an exponential rigidity spectrum of solar cosmic rays under thin target conditions. (From R. Ramaty and R.E. Lingenfelter: 1975, in S.R. Kane (ed.), *Solar Gamma-, X-, and EUV Radiation*, IAU Symposium No. 68, D. Reidel Publishing Company, by permission.)

the incomplete nature of the results, we will not describe them here other than to quote the main conclusions made by Ramaty and Lingenfelter (1975) in reference to interpreting the 1972, August 4 OSO-7 observations (see Section V-5.1). According to these observations the rise time of the 0.51 MeV line is approximately 100 s. This implies that the electron density of the medium where annihilation takes place is $\geqslant 10^{12}$ cm^{-3}. If the positrons are of nuclear origin (and not from pair-production of hard electron bremsstrahlung), then the positrons result mainly from π^+ mesons and short-lived radioactive nuclei such as ^{15}O, ^{14}O, and ^{12}N (see Table III-1).

Considerably more thought has been given to the fate of the neutrons following their production and their ultimate capture in H to give the 2.22 MeV γ-ray. Wang and Ramaty (1974) have recently carried out a detailed study on the time history of neutrons produced in nuclear reactions above the solar photosphere. Using Monte Carlo calculations, they explore the fate of monenergetic groups of neutrons produced isotropically, half of which go into the photosphere. This work takes into account several factors including an ambient ^{3}He/^{1}H ratio in the solar photosphere, the radioactive decay of neutrons, the escape of neutrons directly to the Earth, the escape of neutrons which scatter in the photosphere and then leave the Sun, and finally the capture of neutrons on protons which yields 2.2 MeV γ-rays. Attenuation and Compton scattering of the photons as they leave the region of capture is also considered. It is important to note that loss of neutrons in the nonradiative capture reaction ^{3}He(n, p) ^{3}H can reduce significantly the predicted

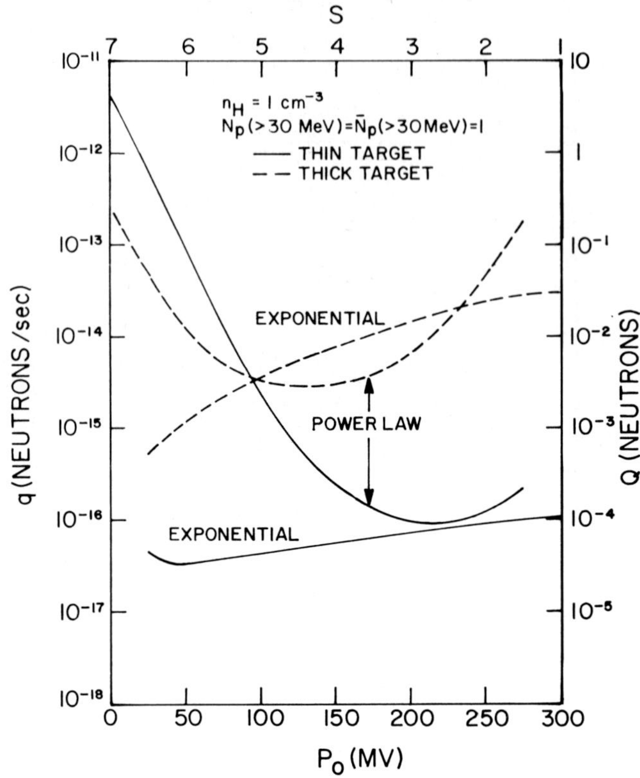

Fig. III-7. The normalized thin target and thick target neutron yields for power law (upper abscissa) and exponential rigidity spectra (lower abscissa). (From R. Ramaty and R.E. Lingenfelter: 1975, in S. Kane (ed.), *Solar Gamma-, X-, and EUV Radiation*, IAU Symposium No. 68, D. Reidel Publishing Company, by permission.)

2.2 MeV line intensity. The resulting Monte Carlo probabilities for the neutrons and the relative photon yields for various initial neutron energies are shown in Figure III-9, which shows the case when the $^3He/^1H$ abundance is 5×10^{-5} which is comparable to that in the solar wind. The ordinate on the left gives the probability that a neutron of energy E_n will suffer either of the four fates as shown by the solid curves; that is, spontaneous decay $n \to p + e^- + \bar{\nu} + 0.772$ MeV, nonradiative capture on ^3He, escape from the Sun, and radiative capture on protons. It is interesting to note the high probability of escape; however, any neutrons escaping with energies below 2 keV will be gravitationally trapped at the Sun and have another chance for capture by ^1H or ^3He. This refinement is not considered in these calculations. The sum of the probabilities equals one for each E_n.

The dashed curves (right ordinates) give the number of 2.2 MeV photons, $f(\theta, E_n)$, for each incident neutron of the indicated energy, which escape from the photosphere along the Earth-Sun line, which is at an angle θ to the vertical to the solar surface. Thus, if the flare is at central meridan then $\theta = 0$, and the 2.2 MeV photons encounter the smallest photospheric path length in escaping from the capture depth; whereas for a flare

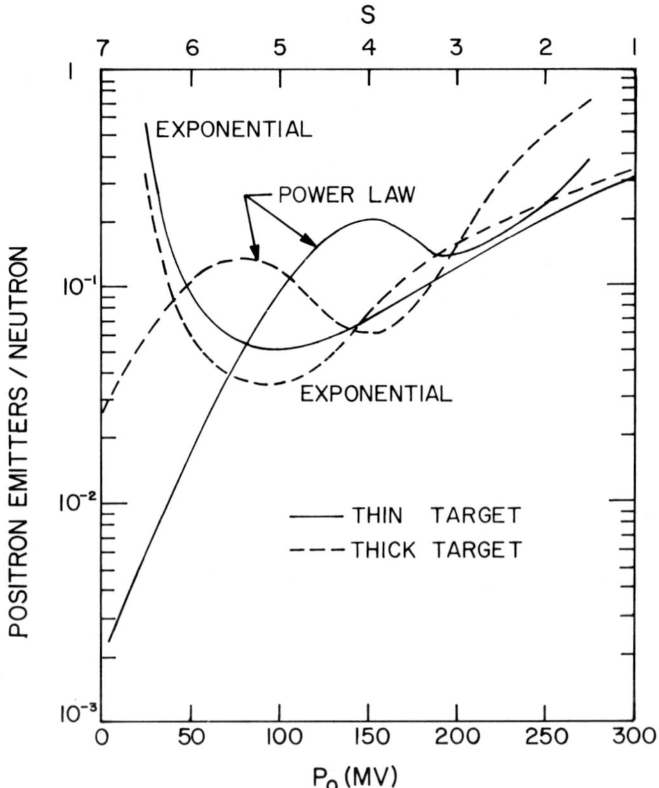

Fig. III-8. The ratios of the total positron yield to the total neutron yield for the thin and thick target models and for power law (upper abscissa) and exponential spectra (lower abscissa). (From R. Ramaty and R.E. Lingenfelter: 1975, in S.R. Kane (ed.), *Solar Gamma-, X-, and EUV Raidation*, IAU Symposium No. 68, D. Reidel Publishing Company, by permission.)

near either limb of the Sun, the photospheric depth along the Earth-Sun line may severely attenuate the photon flux. In these calculations it was assumed that an isotropic distribution of monenergetic neutrons of energy E_n was released over the photosphere, so the total number of neutrons can be found from the neutron yield calculations discussed previously to predict the 2.2 MeV flux. The assumption of isotropic production of neutrons may not be valid for all primary particle spectra, however. In Figure III-10, the resulting calculations are shown for a $^3He/^1H$ ratio of zero, in which case the relative γ yields rise.

In the previous Lingenfelter and Ramaty (1967) calculations, it was assumed that all downward moving neutrons are captured and all upward moving photons escape from the Sun. In this case the relative γ flux per neutron should be 1/2. However, from Figures III-9 and III-10 we see that, depending on the energy of the neutrons and the location of the flare on the Sun, one can overestimate the γ yield by at least a factor of 2.5. Furthermore, if the flare occurs close to the limb of the Sun, the 2.2 MeV line could become essentially unobservable. The time history of the 2.2 MeV line is also determined in these

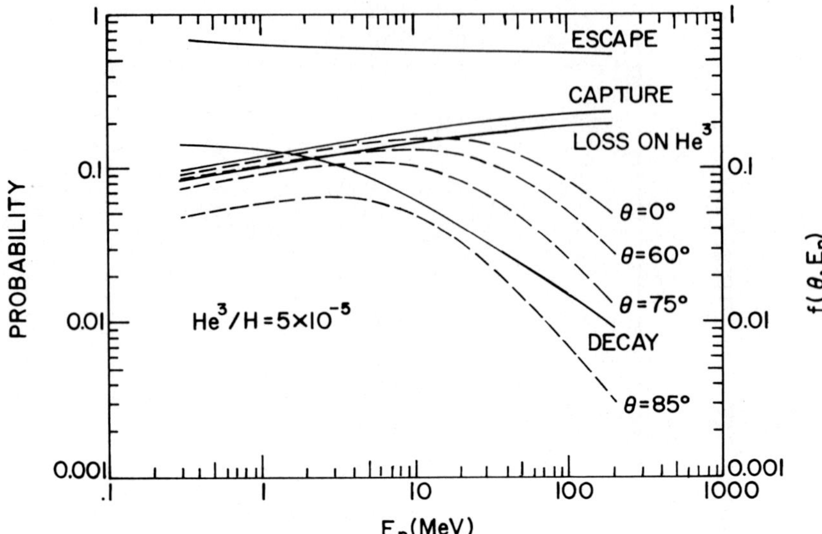

Fig. III-9. Monte Carlo probabilities for the fate of neutrons in the photosphere for ^{3}He/H = 5 × 10^{-5} are shown on the left ordinate versus the neutron energy (solid lines). The fraction, f, of the neutrons which give a 2.23 MeV capture γ-ray leaving the photosphere surface in the direction of the Earth and making an angle θ with the normal to the photosphere is on the right ordinate (dashed lines). $\theta = 0$ corresponds to the case of a flare at central meridian. (From W.T. Wang and R. Ramaty: 1974, *Solar Phys.* **36**, 129, D. Reidel Publishing Company, by permission.)

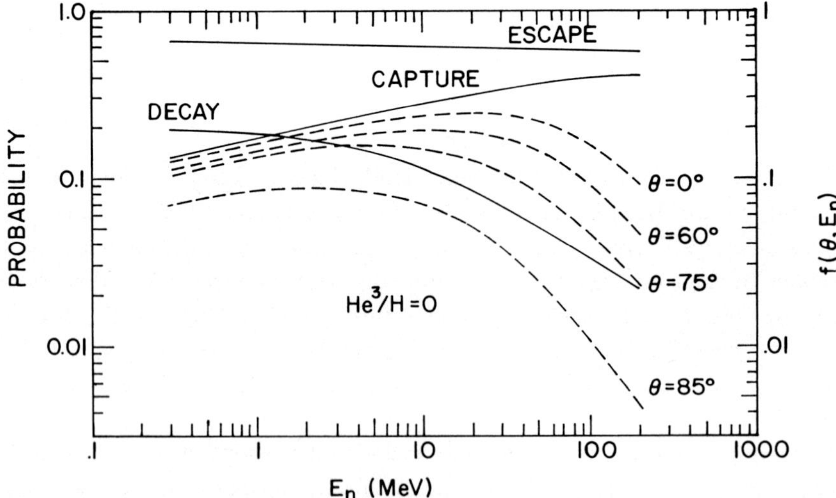

Fig. III-10. Monte Carlo probability for the fate of neutrons in the photosphere for ^{3}He/H = 0 is shown on the left ordinate versus the neutron energy (solid lines). The fraction, f, of the neutrons which give a 2.23 MeV capture γ-ray leaving the photosphere surface in the direction of the Earth and making an angle θ with the normal to the photosphere is on the right ordinate (dashed line). (From H.T. Wang and R. Ramaty: 1974, *Solar Phys.* **36**, 129, D. Reidel Publishing Company, by permission.)

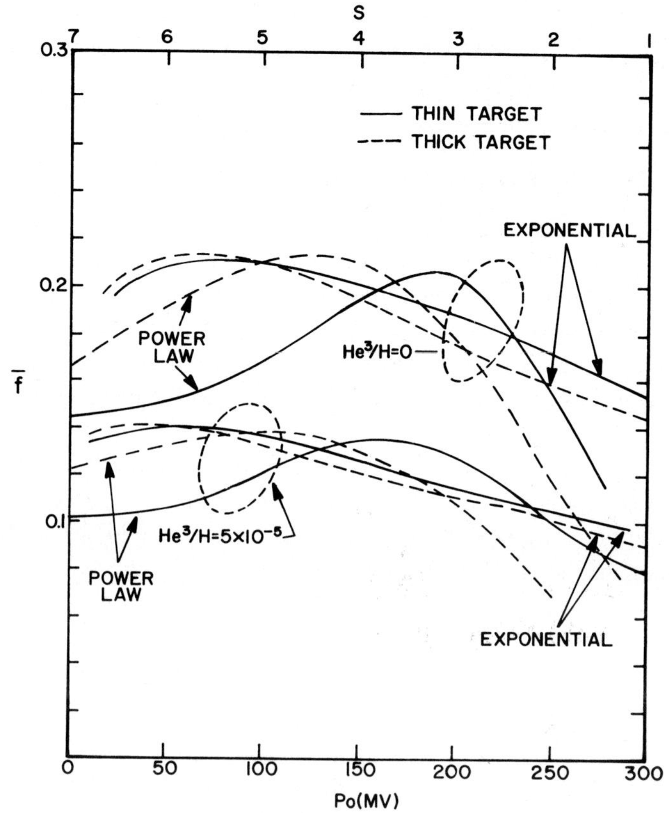

Fig. III-11. The average 2.2 MeV photon yield per neutron, $\bar{f}$, in the thin and thick target models, with power law and exponential spectra, and with $^3He/H = 5 \times 10^{-5}$ and $^3He/H = 0$ in the photosphere. (From R. Ramaty *et al.*: 1975, 'Solar Gamma Rays', *Space Sci. Rev.* 18, 341, D. Reidel Publishing Company, by permission.)

new calculations for each neutron energy. These Monte Carlo calculations will be valuable in the future when experiment and theory lead us to a more detailed acceleration model which gives the spectrum and time history of the solar cosmic rays at the Sun. The calculations also suggest that a limit may be placed on the $^3He/^1H$ abundance, and if a sufficient number of γ-ray events are recorded, the structure of the photosphere could be probed.

By using Figures III-9 and III-10, Wang (1975) has calculated the average 2.23 MeV photon yield per neutron, $\bar{f}$, by integrating $f(E_n)$ over the neutron production energy spectra for the different interaction models and energy spectra discussed above. The neutron energy spectra originally calculated by Lingenfelter and Ramaty (1967) were revised by Wang (1975) for calculating the quantity $\bar{f}$ shown in Figure III-11. It should be pointed out that this curve corresponds only to a flare at central meridian ($\theta = 0°$) and $\bar{f}$ must be reevaluated for $\theta \neq 0$ (see Figures III-9 and III-10).

In the case of a solar flare in which the neutrons are expected to be produced in an impulsive burst lasting ~100 s, it is interesting to consider the time history of a γ-ray burst. First the fraction $f(\theta, E_n)$ given in Figures III-9 and III-10, which is the fraction of captures per neutron, is used to find $\bar{f}$ which is related to the average 2.2 MeV flux at the Earth by

$$\begin{aligned} \phi(2.2\ \text{MeV}) &= \bar{q}\ \bar{f}/4\pi R^2 \qquad \text{thin target} \\ \bar{\phi}(2.2\ \text{MeV}) &= Q\ \bar{f}/4\pi R^2 \qquad \text{thick target} \end{aligned} \tag{III.6}$$

where $\bar{q}$ is the average neutron production rate (s^{-1}) and R is the Earth-Sun distance. Here it is assumed that the neutrons are produced impulsively for a time duration Δt, and are then captured at a later time. Therefore, $\bar{q}$ is a fictitious average production rate of neutrons obtained from experiment by $\bar{q} = (q\ \Delta t)/(T - t_0)$ where $T - t_0$ is the time duration of the observation of a 2.22 MeV γ-ray burst at the Earth and $(q\ \Delta t)$ is the total number of neutrons produced at the Sun. The capture of the neutrons actually lags their production.

If the neutron capture time T_c (s) $\cong 1.4 \times 10^{19}/n(\text{cm}^{-3})$ and T_d (s) is the mean life of the neutron against decay and He^3 loss is neglected, then the loss probability $\lambda(s^{-1}) = T_c^{-1} + T_d^{-1}$. Reppin *et al.* (1973) and Wang and Ramaty (1974) have used an approximate time history for the 2.2 MeV line given by

$$\phi(2.2\ \text{MeV}) \propto \exp[-\lambda(t - t_0)]\,. \tag{III.7}$$

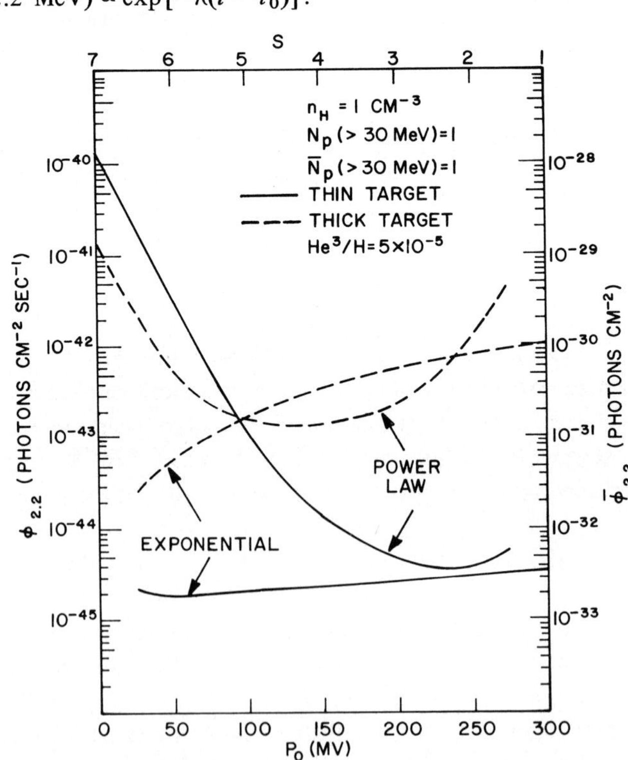

Fig. III-12. The 2.2 MeV line intensity at Earth in the thin and thick target models for power law and exponential spectra and $^3He/H = 5 \times 10^{-5}$. (From R. Ramaty *et al.*: 1975, 'Solar Gamma Rays', *Space Sci. Rev.* **18**, 341, D. Reidel Publishing Company, by permission.)

The γ-ray flux at the Earth can now be readily evaluated by using Equation (III-6). The basic secondary yields, $\bar{q}$ and Q, are found from Figure III-7 and $\bar{f}$ from Figure III-11. The results are given in Figure III-12 for all target and spectral conditions and for a photospheric abundance ratio ${}^{3}He/{}^{1}H = 5 \times 10^{-5}$. According to Ramaty *et al.* (1975) these results are not too strongly dependent on the ${}^{3}He$ abundance ratio. It should be emphasized that this figure applies only to the case of a central meridian flare. By use of this result, the actual γ-ray flux ($cm^{-2}\ s^{-1}$) at Earth (for a thin target production) is given by the product $[n_H \times N_p(> 30\ \text{MeV})]$ times the normalized ordinate value $\phi_{2.2}$ ($cm^{-2}\ s^{-1}$) on the left for the appropriate spectral shape. n_H is the actual H density in the production region for a steady state number of protons > 30 MeV ($N_p > 30$ MeV). For a thick target geometry, the time-integrated γ-ray flux at the Earth is given by the product $\bar{N}_p$ (> 30 MeV) times the normalized ordinate value on the right for the appropriate spectral shape.

3.1.3. Excited Nuclear States

Since excited nuclear levels have lifetimes $\ll 1$ s (except for isomeric states), the study of the intensity of the resulting γ-ray lines from the Sun during solar flares (or for that matter any cosmic source) will provide direct knowledge of the time history of the nuclear reactions producing the excited states. This time history therefore directly reflects the time dependent intensity of the energetic solar particles in their interaction with the solar atmosphere. Therefore it is expected that the problems of the solar flare particle acceleration phenomena can be greatly elucidated by study of prompt γ-ray lines. Ramaty *et al.* (1975) have also reviewed in detail the theoretical results for this problem. They have tabulated the prompt γ-ray lines in order of increasing energy as given in Table III-3. The γ-ray transition in a particular nuclide responsible for the line is shown in column 2 and the charged particle induced reaction leading to the excited nuclide is given in column 3. The appropriate relative intensities of the different γ-ray lines apply to the case of a typical solar flare and are based on the calculations discussed below. These result from two classes of interactions: 1) the case of accelerated protons and α-particles interacting with ambient solar material giving excited target nuclides leading to the γ-ray lines, and 2) the production by spallation reactions of excited fragments which are normally rare in the solar atmosphere. The first group of excited nuclei consists of isotopes of ${}^{12}C$, ${}^{14}N$, ${}^{16}O$, ${}^{20}Ne$, ${}^{24}Mg$, ${}^{28}Si$, ${}^{56}Fe$, and the second group consists of isotopes of ${}^{6}Li$, ${}^{7}Li$, ${}^{7}Be$, ${}^{10}B$, ${}^{11}B$, ${}^{11}C$, ${}^{13}C$, and ${}^{15}N$. The calculation of the relative intensities of the γ-ray lines given in the last column of Table III-3 is based on the secondary production rates given by Equations (III.3) and (III.4) for the energetic particle spectral shapes described above. In Section II-2.4.3 we have discussed some of the basic cross section data used in the calculations. Ramaty *et al.* (1975) have used this information, as well as several other cross section measurements, for the first group of interactions for incident protons and α-particles in the energy range $10\ \text{MeV} < E < 150\ \text{MeV}$ (see Ramaty *et al.*, 1975) for detailed references to these data). For the second group of reactions, the line intensities in Table III-3 at 0.431 MeV and 0.478 MeV from ${}^{7}Be$ and ${}^{7}Li$, respectively,

TABLE III-3

Prompt γ-ray lines. (From R. Ramaty *et al.*: 1975, 'Solar Gamma Rays', *Space Sci. Rev.* **18**, 341, D. Reidel Publishing Company, by permission.)

Photon energy (MeV)	Origin	Production modes	Approximate relative intensity
0.431	$^{7}Be^{*0.431} \to$ g.s.	$^{4}He(\alpha, n)^{7}Be^{*0.431}$	1
0.478	$^{7}Li^{*0.478} \to$ g.s.	$^{4}He(\alpha, p)^{7}Li^{*0.478}$	
		$^{4}He(\alpha, n)^{7}Be \xrightarrow{12\%} {}^{7}Li^{*0.478}$	1
0.72	$^{10}B^{*0.72} \to$ g.s.	$^{12}C(p, 2pn)^{10}B^{*0.72}$	
		$^{16}(O(p, -)^{10}B^{*0.72}$	0.1
0.845	$^{56}Fe^{*0.845} \to$ g.s.	$^{56}Fe(p, p')^{56}Fe^{*0.845}$	0.2
1.24	$^{56}Fe^{*2.08} \to {}^{56}Fe^{*0.845}$	$^{56}Fe(p, p')^{56}Fe^{*2.08}$	0.2
1.38	$^{24}Mg^{*1.37} \to$ g.s.	$^{24}Mg(p, p')^{24}Mg^{*1.37}$	0.1
1.63	$^{20}Ne^{*1.63} \to$ g.s.	$^{20}Ne(p, p')^{20}Ne^{*1.63}$	
	$^{14}N^{*3.94} \to N^{*2.31}$	$^{14}N(p, p')^{14}N^{*3.94}$	
		$^{16}O(p, 2pn)^{14}N^{*3.94}$	0.2
	$^{23}Na^{*2.07} \to {}^{23}Na^{*0.44}$	$^{24}Mg(p, 2p)^{23}Na^{*2.67}$	
1.78	$^{28}Si^{*1.78} \to$ g.s.	$^{28}Si(p, p')^{28}Si^{*1.78}$	0.2
1.99	$^{11}C^{*1.99} \to$ g.s.	$^{12}C(p, pn)^{11}C^{*1.99}$	0.07
2.31	$^{14}N^{*2.31} \to$ g.s.	$^{14}N(p, p')^{14}N^{*3.94} \to {}^{14}N^{*2.31}$	
		$^{14}N(p, n)^{14}O \to {}^{14}N^{*2.31}$	0.3
		$^{16}O(p, 2pn)^{14}N^{*2.31}$	
2.75	$^{16}O^{*8.88} \to {}^{16}O^{*6.14}$	$^{16}O(p, p')^{16}O^{*8.88}$	0.07
~3.62	$^{13}C^{*3.68} \to$ g.s.	$^{16}O(p, 3pn)^{13}C^{*3.68}$	0.07
	$^{6}Li^{*3.56} \to$ g.s.	$^{12}C(p, -)^{6}Li^{*3.56}$	
3.84	$^{13}C^{*3.84} \to$ g.s.	$^{16}O(p, 3pn)^{13}C^{*3.84}$	0.07
4.43	$^{12}C^{*4.43} \to$ g.s.	$^{12}C(p, p')^{12}C^{*4.43}$	
		$^{12}C(\alpha, \alpha')^{12}C^{*4.43}$	1
		$^{16}O(p, p\alpha)^{12}C^{*4.43}$	
~5.3	$^{15}O^{*5.26} \to$ g.s.	$^{16}O(p, pn)^{15}O^{*5.26}$	
	$^{15}N^{*5.28} \to$ g.s.	$^{16}O(p, 2p)^{15}N^{*5.28}$	0.3
	$^{15}N^{*5.31} \to$ g.s.	$^{16}O(p, 2p)^{15}N^{*5.31}$	
6.14	$^{16}O^{*6.14} \to$ g.s.	$^{16}O(p, p')^{16}O^{*6.14}$	0.5
		$^{16}O(\alpha, \alpha')^{16}O^{*6.14}$	
6.33	$^{15}N^{*6.33} \to$ g.s.	$^{16}O(p, 2p)^{15}N^{*6.33}$	0.5
~6.7	$^{11}B^{*6.76} \to$ g.s.	$^{12}C(p, 2p)^{11}B^{*6.76}$	
		$^{12}C(p, 2p)^{11}B^{*6.81}$	0.07
		$^{12}C(p, pn)^{11}C^{*6.50}$	
7.12	$^{16}O^{*7.12} \to$ g.s.	$^{16}O(p, p')^{16}O^{*7.12}$	0.2

have been inferred from theoretical calculations of the cross sections for the following three reactions: $^{4}He(\alpha, n)\ {}^{7}Be^{*}$, $^{4}He(\alpha, p)\ {}^{7}Li^{*}$, and $^{4}He(\alpha, n)\ {}^{7}Be$.

The first class of reactions mentioned above can also lead to spallation products which are left in excited states. An example of this, which is very important in solar flares, was discussed in Section II-2.4.3. There we pointed out that the results of Zobel *et al.* (1968) for protons on O give a much stronger ^{12}C line at 4.4 MeV than for a direct excitation ^{16}O line at 6.13 MeV (see Figure II-21). Thus, in the case of the solar atmosphere where O is nearly twice as abundant as C, the reaction $^{16}O(p, p\alpha)\ {}^{12}C^{*}$ makes a very important contribution to the line at 4.43 MeV (cf. Ramaty *et al.*, 1975).

One other important factor which has been taken into account in this work is the Doppler broadening of nuclear lines as we discussed in Section II-2.5.1. In the case of $^{12}C^*$ and $^{16}O^*$ the Doppler widths are smallest (< 100 keV) when energetic protons and alphas are incident on ambient C and O, whereas if fast C and O nuclei are excited in collisions with ambient H, then the emitted γ-rays are Doppler broadened severely. In fact, Ramaty *et al.* (1975) do not consider the γ-ray line intensities from this reverse process since it is not expected that these broad lines can be resolved experimentally from the background. We wish now to briefly describe the results of these calculations for general application. In Section V-5.1 the interpretation of solar flare γ-ray line observations will be discussed which makes use of these results.

The work on the prompt lines by Ramaty *et al.* (1975) has concentrated on the calculation of the production rates of ambient $^{12}C^*$ and $^{16}O^*$ giving lines at 4.43 MeV and 6.13 MeV, respectively. For flat power law spectra the excited states are produced mainly by proton reactions whereas for steep power law spectra the excitation by α-particles

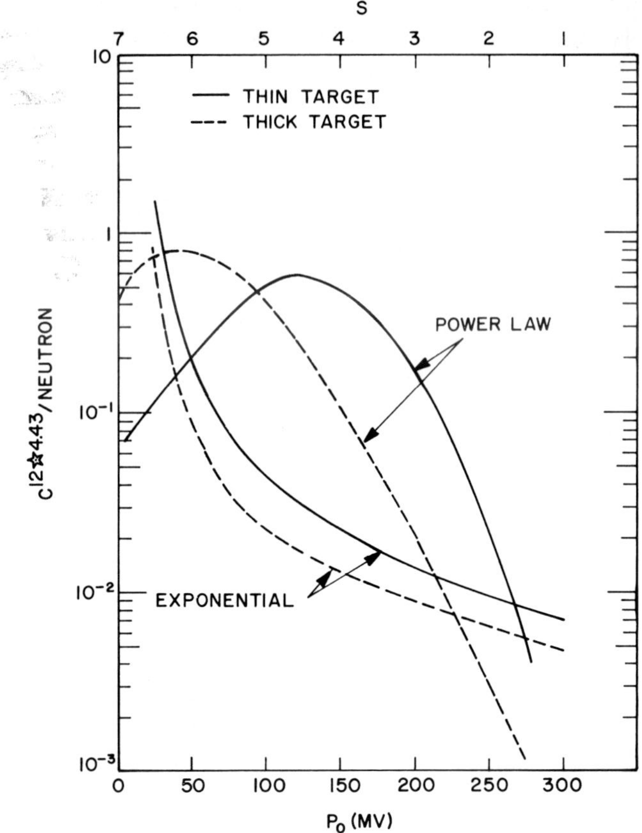

Fig. III-13. Ratios of the $^{12}C^{*4.43}$ yield to the total neutron yield for the thin and thick target models, and power law and exponential spectra. (From R. Ramaty *et al.*: 1975, 'Solar Gamma Rays', *Space Sci. Rev.* **18**, 341, D. Reidel Publishing Company, by permission.)

becomes important. On the other hand, for exponential spectra the proton induced reactions are the main source of excited states. Spallation production of $^{12}C^{*(4.43)}$ is also important as mentioned above, unless the spectrum is a steep power law for which direct excitation by inelastic scattering near threshold dominates.

Since there are many production modes for these excited states and there is a direct relation between the secondary production of neutrons and excited nuclear states, Ramaty *et al.* (1975) have determined the yield of the γ-rays per neutron produced. This ratio ($^{12}C^{*(4.43)}$ per neutron) is plotted in Figure III-13 for all target geometries and particle spectra considered. The absolute γ-ray line emission rate (s^{-1}), or total yield, at the Sun may be found by determining the number of neutrons produced from Figure III-7 for a thin or thick target geometry and a particular spectral shape.

It should be pointed out that the ratio given in Figure III-13 could be subject to change, if all modes of either neutron production or $^{12}C^{*}$ production are not included. The behaviour of Figure III-13 can be understood as follows: The ratio ($^{12}C^{*(4.43)}$ per neutron) decreases with P_0 for both targets since the neutron production cross section increases with energy and the excitation cross section is decreasing with energy. The same is true for power law spectra when $s < 4.5$ (thin target) or $s < 6$ (thick target).

When s is larger the ratio falls because neutrons are produced by αCNO and CNOα reactions which have thresholds below those for the $^{12}C^{*(4.43)}$ threshold. Similar detailed calculations could be carried out for all other prompt lines; however, Ramaty *et al.* (1975) have concentrated on the ^{12}C and ^{16}O lines at 4.43 MeV and 6.13 MeV, respectively, since there is evidence for these emissions in flares. From Table III-3 it is seen that the ^{16}O line is expected to be nearly half as intense as the ^{12}C line but intense lines are expected at 431 keV and 478 keV from $^{7}Be^{*}$ and $^{7}Li^{*}$, respectively. These arise from the following (α-α) reactions:

$$^{4}He\,(\alpha, n)\ ^{7}Be^{*}\ ; \quad 431\ keV$$

$$^{4}He(\alpha, n)\ ^{7}Be \xrightarrow[\epsilon]{12\%} {}^{7}Li^{*}\ ; \quad 478\ keV$$

$$^{4}He\,(\alpha, p)\ ^{7}Li^{*}\ ; \quad 478\ keV$$

The first two reactions have thresholds at 9.7 MeV per nucleon and 9.5 MeV per nucleon, respectively, and the last reaction has a threshold at 8.5 MeV per nucleon. Kozlovsky and Ramaty (1974) argue that the cross section for all these reactions is ~ 100 mb at 10 MeV per nucleon. For the first two reactions they have assumed that the production cross section of ^{7}Be in the ground state and first excited state is the same. Kozlovsky and Ramaty (1974) also point out that the Doppler width of these lines would be ~30 keV and may be difficult to observe as individual lines especially since there is a strong solar flare continuum radiation in this energy region (cf. Section V-5.1).

Table III-3 also lists the approximate relative intensities expected for lines at 0.72, 1.99, 3.62, 3.84, 5.3, 6.33 and 6.7 MeV which are produced by spallation products as indicated in the table. The intensities were scaled from experimental cross sections normalized to the spallation yield of $^{12}C^{*4.43}$ for $P_0 = 150$ MV. A similar procedure was used for finding the relative intensities of the other lines listed. It is of interest to consider the processes leading to the 2.31 MeV line listed in Table III-3, since it is close in energy to the neutron-proton capture line at 2.22 MeV. This line originates from the decay of the first excited state in ^{14}N, but at least three reactions (as shown) populate this level. Ramaty *et al.* (1975) have concluded that the main contribution to this line ($\sim$70%) comes through the channel $^{14}N(p, n)$ $^{14}O \xrightarrow[71\ \mathrm{sec}]{\beta^+} {}^{14}N^{*(2.31)}$. Since ^{14}O has a half-life of $\lesssim$ 1 min, the 2.31 MeV line is delayed with respect to the other prompt lines in Table III-3. It should be remembered that if a solar flare occurs far from the central solar meridian, then the intensity of the neutron capture line (2.22 MeV) could become unobservable (see Figure III-9) but the ^{14}N line could be observable.

a. *The Flux Ratio* $\phi(4.43)/\phi(2.22)$

In our discussion above concerning the intensity of the 2.2 MeV line expected at the Earth (see Figure III-12), Ramaty *et al.* (1975) give results for a flare at central meridian.

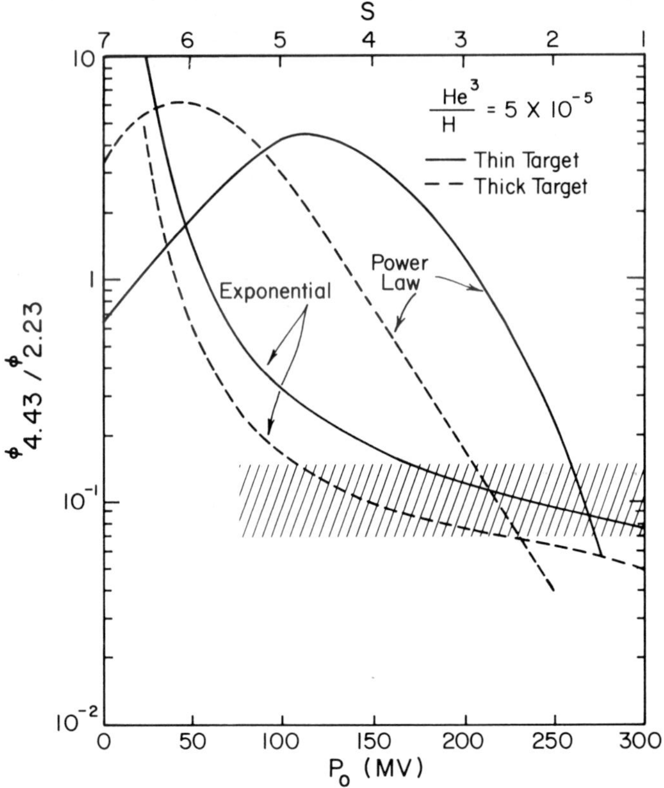

Fig. III-14. Ratios of the 4.4 MeV line intensity to the 2.2 MeV line intensity in the thin and thick target models, with power law and exponential spectra and with $^3He/H = 5 \times 10^{-5}$ in the photosphere. The shaded area is the data of Chupp *et al.* (1975). (From R. Ramaty *et al.*: 1975, 'Solar Gamma Rays', *Space Sci. Rev.* **18**, 341, D. Reidel Publishing Company, by permission.)

It should be remembered that the value of $\bar{f}$ (Figure III-11) required for this calculation depends on the longitude of the flare in the model used. It is useful, nevertheless, to review the results of Ramaty *et al.* (1975) for the central meridian case, obtained by comparing the expected ratio of the intensities of the 4.4 MeV and 2.2 MeV lines at the Earth since this can be easily compared to experiment (Section V-5.1) for the different models and spectral shapes assumed. This ratio is given by:

$$\frac{\phi_{4.43}}{\phi_{2.23}} = \left(\frac{^{12}\mathrm{C}^{*4.43}}{n}\right)(\bar{f})^{-1} \tag{III.8}$$

where $(^{12}\mathrm{C}^{*4.43}/n)$ is the yield ratio given in Figure III-13 and $\bar{f}$ is given in Figure III-11. It is understood, of course, that this ratio refers to the average fluxes since the 2.22 MeV line is delayed in its production relative to the 4.43 MeV line.

The result is shown in Figure III-14 for the case of a photospheric ^{3}He abundance given by $^3\mathrm{He/H} = 5 \times 10^{-5}$ only (cf. Ramaty *et al.* (1975) for a similar curve for $^3\mathrm{He/H} = 0$). The observed ratio was (0.11 ± 0.04) for the central meridian flare on 4 August, 1972 (Chupp *et al.*, 1975 and Section V-5.1), and is shown as the cross-hatched band in Figure III.14. The overlap between the experimental band and the four curves gives the spectral shape for the thin and thick target models. Therefore, for a power law spectrum for the accelerated particles, the spectral index must be near $s = 2$ if the target is thin and $s = 3$ if the target is thick. On the other hand, if an exponential rigidity spectral shape is appropriate, then Ramaty *et al.* (1975) have deduced that $P_0 = 165 \pm 55$ MV or $P_0 = 250 \pm 80$ MV for the thick and thin target cases, respectively.

The aim of our discussion here is to describe the γ-ray calculations and not discuss the solar flare problem; however, it is clear that further information would be needed to determine from γ-ray measurements which spectral shape or target geometry is appropriate. Also, Ramaty *et al.* (1975) have pointed out that these values depend significantly on the photospheric value of ^{3}He/H.

3.1.4. π^0 γ-RAYS

The secondary yields of positrons, neutrons and excited nuclear states are most efficient for low energy ($\gtrsim 50$ MeV) incident particles. However, hard spectra of accelerated particles may have a sufficient number of particles above the threshold for π^0 meson production that high energy γ-rays ($>$30 MeV) could be observable ($\sim$135 MeV is the excess energy needed in the centre of mass system). Ramaty *et al.* (1975) have also given estimates of this flux based on the following expression:

$$\phi_{\pi^0} = \phi_{2.2}\,[2\pi^0/n](\bar{f})^{-1} \quad (\mathrm{cm}^{-2}\,\mathrm{s}^{-1}) \tag{III.9}$$

where $[2\pi^0/n]$ is the yield of π^0 mesons per neutron times 2 for given spectrum and target assumptions. $\phi_{2.2}$ and $\bar{f}$ were previously defined. The values of $[2\pi^0/n]$ are given by Figure III-15 for the various assumptions previously discussed, so the predicted value of ϕ_{π^0} can be found from use of the data in Figures III-11 and III-12. It is of interest to note that only low energy γ-ray lines were detected in solar flares (Section V-5.1) and no conclusions could be made concerning π^0 γ-rays. Simultaneous observation of the high

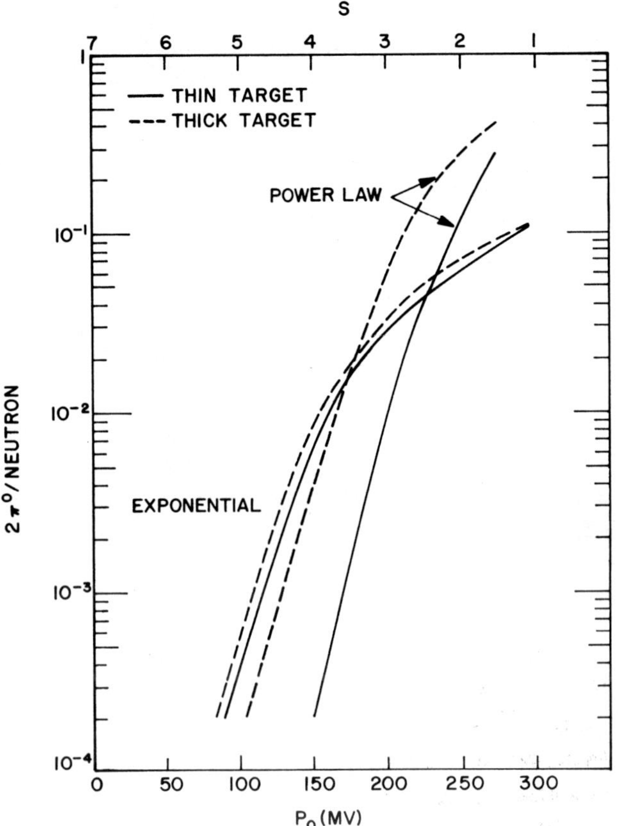

Fig. III-15. Ratio of the γ-ray yield from π^0 decay to the total neutron yield for the thin and thick target models and power law and exponential spectra. (From R. Ramaty *et al.*: 1975, 'Solar Gamma Rays', *Space Sci. Rev.* **18**, 341, D. Reidel Publishing Company, by permission.)

energy γ-ray spectra with the lower energy lines may enable a direct determination of the solar energetic particles' spectral shape and the target conditions and resolve the uncertainty just mentioned.

3.1.5. Solar Flare γ-Ray Continuum

The solar flare γ-ray line predictions discussed above give that component of radiation expected just from proton interactions. However, it is well known that energetic electrons produce a strong bremsstrahlung spectrum which extends into the γ-ray energy region. Therefore, any γ-ray lines must exceed the intensity of this continuum radiation in a given spectral band to be observed. For a discussion of this problem we refer the reader to several theoretical papers (e.g., Korchak, 1967; Cheng, 1972; Ramaty *et al.*, 1975) and to observational results on the γ-ray continuum in relation to γ-ray lines (Suri *et al.*, 1975).

3.2. Cosmic Sources (Point and Localized)

The early work of Morrison (1958) predicted the possibility of γ-ray lines and continuum from the Sun and in particular a strong line at 2.22 MeV during solar flares, a topic we have just discussed in Section III-3.1. In addition, this early work discusses γ-ray lines from the Crab Nebula, extragalactic radio sources such as M87 and Cygnus A, and from the interaction of matter and antimatter. Subsequent reviews have also discussed the evidence for γ-rays >1 MeV from these and similar sources (cf. Garmire and Kraushaar, 1965; Fazio, 1970). However, until the actual experimental observation of γ-rays from cosmic sources many of the theoretical predictions concerning γ-rays must be considered speculative. Our discussion of the theoretical work in this area will therefore be descriptive and no serious attempt is made to evaluate the consistency of the γ-ray predictions with other astrophysical data. A review of the predictions is nonetheless very important from the experimental viewpoint since experiment design should be based on the best prediction of physical parameters such as spectral line width, time history, and flux level, serendipity notwithstanding. The discussion of cosmic point (or localized) sources follows the order given in Table II-2.

3.2.1. Supernova Remnants and Supernovae

An important astrophysical problem in which γ-ray astronomy has particular bearing is the study of γ-ray lines from supernovae (SN) in the process of development or from their remnants (SNR). Several classes of supernovae have been established based on a study of the light curves and spectral data. Several fairly recent papers discuss the general observational data: Minkowski (1964), Zwicky (1965), Shklovsky (1968), and Oke and Searle (1974). Up-to-date reviews of several problems concerning SN or SNR have been covered at two recent conferences for which proceedings have been published (cf. Brancazio and Cameron, 1969; Cosmovici, 1974). An updated list of SN discovered since 1885 and complete to mid-1973 has been published by Sargent *et al.* (1974). The number of known classes may be as large as five and a good general discussion concerning their characteristics has been given by Shklovsky (1968) and Zwicky (1969).

In Table III-4 are shown the general characteristics of the classes I and II most commonly referred to along with events involving much less energy output: novae and recurrent novae. In columns three and four, the association of the events with stellar population type and galaxy type is indicated. The frequency of the events per galaxy, in units of inverse years are given in column five, is a subject of considerable controversy. The estimated absolute photographic magnitudes at different times are shown in the next three columns, and in column seven the corresponding total luminosity, $\int L\, dt$, is given. The last column gives the approximate time for the brightness to decline 3 magnitudes from the maximum. In addition, SN I are usually associated with lower mass stars $\sim 2M_\odot$ and SN II with stars with mass $\geqslant 4M_\odot$. Since the existence of known SNR is an observational fact, most estimates of the possible γ-ray emissions are concerned with the end result of the SN process. Exactly what occurs in the initial phase and how it is

TABLE III-4

Types of common novae and supernovae. (From C.W. Allen, *Astrophysical Quantities*, 3rd Edition, 1973, Athlone Press. Used by permission.)

τ_3 = time for brightness to decline 3 magnitudes from the maximum

IAU desig.	Type	Pop.	Occurrence	Freq.[b] ($gal^{-1}yr^{-1}$)	M_{PG} Pre-nova	M_{PG} Max.	M_{PG} Post-nova	Log energy output (in erg)	τ_3 (day)
SN I	Supernova, type I	II + I	*E* to Sc gal.	0.01		−18.8	+3?	51	30[a]
SN II	Supernova, type II	I	Sb, Sc spiral	0.02		−17	+3?	49	70
N	Nova	II?		40	+5	− 7.7	+4	45	40
Nd	Recurrent nova							43	

[a] After 30 days type I supernovae decline at ~ 1 mag per 70 days.

[b] The counts are per late type galaxy Sb or Sc. The counts are proportional to the mass and luminosity of the parent galaxy.

triggered is subject to considerably more speculation. In our discussion of these objects as γ-ray sources, we will first discuss γ-ray line emission from SNR followed by consideration of γ-ray emission in the initial phase. Estimates of γ-ray continuum emission from these sources will be treated separately at the end of this section.

The Type I SN has a characteristic light decay curve which corresponds to a decrease in brightness of 0.0137 magnitudes per day (cf. Shklovsky, 1968, p. 403) for times longer than about 100 days after the maximum. Thus, during the latter declining phase of the SN, the decay in intensity is exponential, i.e., $L/L_0 = \exp(-t/\tau_I)$, where the mean decay time $\tau_I \sim 70$ days, while the corresponding time for the intensity to decrease by 1/2 is $T_{1/2} \sim 55$ days. Figure III-16 shows the characteristic light decay curves for several Type I SN. It should be noted, though, that the time constant for decay is not always precisely $T_{1/2} \sim 55$ days; however, the total energy emitted under the exponential phase of the light decay curve is $\sim 10^{48}$ erg compared to a total energy release for the outburst of $\geqslant 10^{50}$ erg (Shklovsky, 1968, p. 402). The energy release required, therefore, is $\sim 10^{17}$ erg g^{-1} for an initial mass of $\sim 1 M_\odot$.

a. γ-*Rays from the R-Process*

In view of element building beyond the Fe peak (Burbridge *et al.*, 1957), neutron capture (*r*-process) is required to build the heavy elements through nucleosynthesis. Since this process leads to several transbismuth radioactive nuclides (Baade *et al.*, 1956), Burbidge *et al.* (1956) introduced the ^{254}Cf hypothesis since its spontaneous fission half life is 56 days, in very close agreement with the typical value for the Type I decay curve (see Shklovsky, 1968, pp. 403–410, for an interesting discussion of the compelling reasons for such an idea). This hypothesis was given support by the observed terrestrial synthesis of ^{254}Cf in the 'Mike' thermonuclear device explosion in 1952.

Of course a normal test for a radioactivity hypothesis is to directly observe γ-rays which might accompany the radioactive decay. This idea may be checked either by observing a SN just after its maximum, during its decay phase, or by studying an 'old'

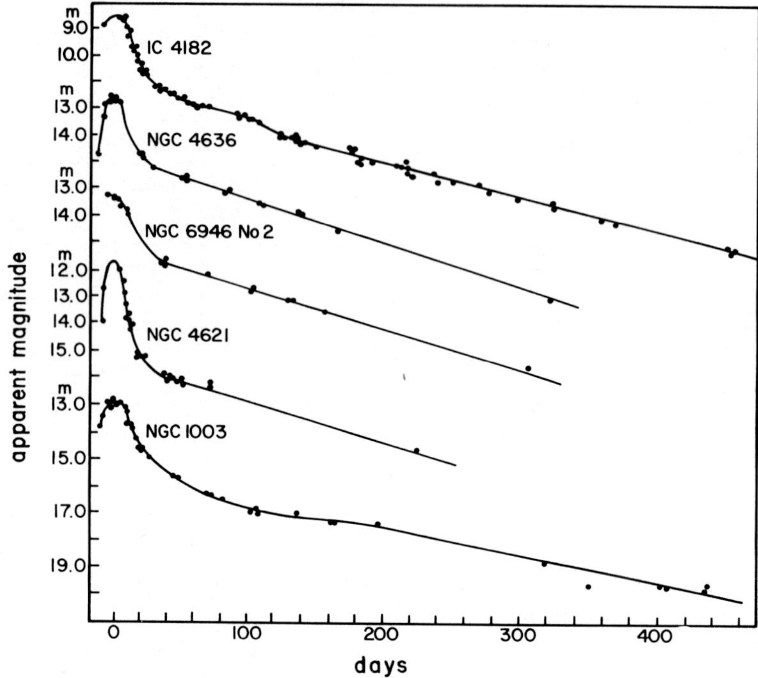

Fig. III-16. Photographic light curves of Type I supernovae. (From R. Minkowski: 1964, *Ann. Rev. Astron. Astrophys.* **2**, 247, Annual Reviews, Inc., by permission.)

SNR. In Morrison's (1958) review of the promise of γ-ray astronomy, he suggested that ^{254}Cf radioactive 'fall-out' probably existed in the Crab Nebula. To be sure, only long half-life activities associated with ^{254}Cf production would be relevant in explaining any Crab γ-flux today, some 900 yr after the explosion. Morrison suggested specifically that nuclear γ-rays from ^{226}Ra should be detectable from the direction of Taurus with an intensity of $\sim 10^{-2}$ photons cm^{-2} s^{-1}. About a year later Savedoff (1959) gave slightly more refined estimates of the expected γ-rays from the ^{254}Cf hypothesis. In particular, he assumed that the maximum total energy flux from the Crab (as a typical SNR) was $6 \times 10^8 L_\odot$ and is due solely to radioactive decay of the ^{254}Cf. This led Savedoff (1959) to an estimate of 5×10^{52} nuclei of ^{254}Cf (or 10^{-2} M$_\odot$) produced in the initial event $\sim$900 yr ago compared to an estimate of 5×10^{-4} M$_\odot$ by Burbidge *et al.* (1956) and used by Morrison (1958) in his γ-ray flux estimates. This high mass estimate for ^{254}Cf is presumably an upper limit. The γ-ray energies and half-lives given were based on nuclear data available at that time and it should be noted that the ^{226}Ra γ-ray lines as proposed by Morrison (1968) are not the strongest expected since only 5% of the decays of ^{226}Ra are to the excited states of its daughter. The strongest lines expected on the basis of the Savedoff (1959) estimates are at 180 keV from ^{251}Cf and at 60 keV from ^{241}Am. These early flux estimates are probably not sufficiently accurate though, since branching ratios and internal conversion were apparently not taken into account; however, the optimistic estimates stimulated experimental activity.

Many γ-ray lines other than those mentioned are probably expected, such as γ-rays from spontaneous fission of ^{250}Cm and its fission products. Hoyle and Fowler (1960) had pointed out, however, that if Type I SN occur at the rate of one every 300 y, and if each event synthesizes $1.5 \times 10^{-4} M_\odot$ of ^{254}Cf, then the abundance of elements such as Pt and U, produced in the *r*-process would be 10^2 larger than that in solar-system material. Hoyle *et al.* (1964) therefore proposed as an alternative to radioactive energy a gravitational mechanism as the direct source of the explosion energy release for a supernova. It is thus left for experiment to decide which of the two possibilities explains the Type I light decay curve (see Shklovsky (1968) for a critique of the gravitational mechanism). Of course, today, following the identification of a pulsar (most likely a rotating neutron star) at the center of the Crab Nebula, the other real possibility for the current energy of the Crab Nebula is dissipation by the rotating neutron star which appears to be 10^{39} erg s^{-1} compared to the present optical emission of $\sim 10^{38}$ erg s^{-1} for the nebula as a whole (Zeldovich and Novikov, 1971).

It can be fairly said, however, that no experiment performed thus far (see a later discussion in Section V-5.2.1) has been capable of verifying the most careful recent calculations of the expected ^{254}Cf induced γ-ray spectrum as carried out by Clayton and Craddock (1965), specifically for the case of the Crab Nebula. There are essentially two questions raised by these authors in evaluating the expected γ-ray line spectrum from the Crab Nebula. The first question concerns the expected γ-ray flux if the ^{254}Cf hypothesis is correct, which requires $1.5 \times 10^{-4} M_\odot$ of ^{254}Cf produced initially. The second question concerns the (lower) γ-ray flux expected if *r*-process nucleosynthesis must occur in the explosion of a Type I SN. For example, during *r*-process nucleosynthesis leading to transbismuth nuclei in a star ($\sim M_\odot$) which develops into a SN, the physical environment required for the element building (via the fission cycle) is $T \sim 10^9$ K and a neutron concentration of $N_n \sim 3 \times 10^{25}$ cm^{-3}. In order to estimate the γ-ray spectrum expected (today) from the Crab Nebula, Clayton and Craddock (1965) have used, for the nuclides originally produced, the results of Burbidge *et al.* (1957) and Seeger *et al.* (1965). In the *r*-process, nuclei with charge up to $Z = 93$ are built, after which neutron induced fission terminates the addition process. The nuclei involved, being highly neutron rich with masses up to $A \simeq 270$, quickly β^- decay and increase the nuclear charge until the nuclei become unstable to spontaneous fission. Apparently, all nuclei produced this way, with $A \simeq 255$, fission rather than undergo α decay. Note that the above β^-decays (at least 5) are needed in order to reach ^{254}Cf ($Z = 98$). The calculations for γ-ray emission from the Crab Nebula today require only knowledge of the activities remaining in sufficient quantity to give detectable γ-rays. The guiding principle used by Clayton and Craddock (1965) in deciding which activities are most important is to consider those nuclides with half-lives near the age of the Crab. (We note, however, that the details of γ-ray emission during the time leading to the buildup of ^{254}Cf may be of very great significance in looking for possible γ-ray emission during the initial phase of a SN I explosion if the ^{254}Cf hypothesis is correct. This question deserves further consideration.) Clayton and Craddock (1965) have also made note of the fission γ-ray spectrum expected from the explosion. The calculations therefore emphasize primarily the production of nuclides produced along with ^{254}Cf which can by β^- or α decay give significant activities remaining after ~910 yr. To clarify the assumptions, we quote from Clayton and Craddock (1965):

> The *r*-process mechanism synthesized (initially) R_A nuclei of atomic weight A in the explosion and all nuclei for $A > 255$ have already undergone spontaneous fission.
>
> The heavy radioactive nuclei characterized by the production curve R_A began radioactive decay in the Crab Nebula 910 yr ago. Species with short half lives (say $T_{1/2} < 100$ yr) will no longer be important in the Crab unless they are daughters of long-lived activities. Thus the radioactivities usually discussed in connection with the light curve (e.g., ^{254}Cf, ^{59}Fe) are not the relevant ones for the present day radiation in the Crab. The long-lived radioactivities like U and Th, although they may be in considerable abundance in the Crab, also have low specific activity due to their very low decay rates. As is well known, the originally produced nuclear species having half lives near the age of the Crab are the ones presently making the largest number of decays per unit time in the remnant.

The number of γ-rays of a given frequency emitted by the Crab per unit time is then

$$N(\nu) = \sum_{A,\,Z} \epsilon(\nu, A, Z)\,\lambda_{A,\,Z} N(A, Z, t) \tag{III.10}$$

where $\epsilon(\nu, A, Z)$ is the fraction of decays of species (A, Z) that is followed by the emission of a prompt γ-ray of frequency ν, $\lambda_{A,\,Z}$ is the decay constant $0.693/T_{1/2}$ for the nuclide (A, Z), and $N(A, Z, t)$ is the total number (abundance) of radioactive nuclides existing at an age t of the nebula. This quantity is calculated from straightforward radioactive decay chain equations and depends directly on the *r*-process production curve, R_A, from which everything starts for this model. This is shown in Table III-5 reproduced from Clayton and Craddock (1965).

TABLE III-5

The *r*-process production curve R_A^*. (From D.D. Clayton and W.L. Craddock, *Astrophys. J.* **142**, 189. Copyright 1965, The American Astronomical Society, by permission of the University of Chicago Press.)

A	R_A(atom $\times 10^{-50}$)	A	R_A(atom $\times 10^{-50}$)
255.	7.1	240.	5.6
254.	7.1	239.	14
253.	9.6	238.	21
252.	5.0	237.	32
251.	12	236.	8.5
250.	7.1	235.	34
249.	7.1	234.	4.3
248.	12	233.	4.3
247.	5.3	232.	33
246.	5.3	231.	6.5
245.	22	230.	6.3
244.	20	229.	1.6
243.	8.0	228.	1.6
242.	13	227.	1.6
241.	11	226.	2.3

*The *r*-process yields of each supernova in units of 10^{50} atoms are taken from Table VIII.1 of Burbridge *et al.* (1957), and normalized to $1.5 \times 10^{-4}\,M_\odot$ of ^{254}Cf.

TABLE III-6

Line γ-fluxes at Earth from Crab Nebula. (From D.D. Clayton and W.L. Craddock, *Astrophys. J.* **142**, 189. Copyright 1965, The American Astronomical Society, by permission of the University of Chicago Press.)[a]

Nucleus	E_γ (MeV)	Branch (%)	Flux ($cm^{-2} s^{-1}$)	Nucleus	E_γ (MeV)	Branch (%)	Flux ($cm^{-2} s^{-1}$)
Cf^{251}	0.18	10 [b]	1.9×10^{-5}	Ra^{226}	0.187	4 [e]	6.4×10^{-7}
Cf^{249}	0.39	75 [c]	9.7×10^{-5}	Ac^{225}	0.099	~10 [c]	3.3×10^{-7}
	0.34	15	2.0×10^{-5}	Ra^{233}	0.338	4.2 [c]	1.3×10^{-7}
	0.26	3	3.9×10^{-6}		0.324	4.9	1.5×10^{-7}
Cm^{245}	0.17	14 [c]	8.1×10^{-6}		0.270	21.4	6.6×10^{-7}
	0.13	5	2.9×10^{-6}		0.154	28	8.7×10^{-7}
Am^{243}	0.075	60 [d]	9.4×10^{-6}		0.144	19	5.9×10^{-7}
Am^{241}	0.06	40 [d]	5.7×10^{-6}		0.122	11	3.4×10^{-7}
Np^{239}	0.28	30 [c]	4.6×10^{-6}	Fr^{221}	0.22	14 [c]	5.0×10^{-7}
	0.23	31	4.8×10^{-6}	Rn^{219}	0.40	4.8 [c]	1.5×10^{-7}
	0.10	24	3.7×10^{-6}		0.27	8.6	2.7×10^{-7}
	0.057	30	4.6×10^{-6}	Pb^{214}	0.352	40 [d]	6.3×10^{-6}
Pa^{231}	0.30	~ 8 [c]	2.5×10^{-7}		0.295	20	3.2×10^{-6}
Th^{229}	0.087	~80 [c]	2.7×10^{-6}		0.24	8	1.3×10^{-6}
Th^{227}	0.334	5.2 [c]	1.6×10^{-7}		0.18	4	6.3×10^{-7}
	0.256	7.1	2.2×10^{-7}	Bi^{214}	0.61	47 [d]	7.4×10^{-6}
	0.236	10.6	3.4×10^{-7}		1.12	20	3.2×10^{-6}
	0.113	4.2	1.3×10^{-7}		1.76	25	4.0×10^{-6}
	0.080	4.6	1.4×10^{-7}	Bi^{213}	0.44	10 [c]	3.2×10^{-7}
	0.060	9	2.9×10^{-7}	Bi^{211}	0.35	~17	5.3×10^{-7}
	0.050	13.6	4.4×10^{-7}				
	0.032	12	3.7×10^{-7}				
	0.030	27	8.4×10^{-7}				

[a] Assuming production-curve of Table III-5 and distance to Crab of 3500 light-yr.
[b] Assumed 10%.
[c] Landolt-Börnstein (Hellwege, 1961).
[d] S.A. Reynolds (Private communication to Clayton and Craddock, 1965).
[e] Adams and Lowder (1964).

The quantities $\epsilon(\nu, A, Z)$ and $\lambda_{A,Z}$ depend only on the basic nuclear properties of nuclide (A, Z) but, even so, are not always well known. Clayton and Craddock (1965) have used primarily the Landolt-Börnstein tables (Hellwege, 1961) for these data. In Table III-6 are shown all nuclear lines with estimated fluxes greater than 10^{-7} photons cm^{-2} s^{-1} at the Earth as given by Clayton and Craddock (1965) for the case of 5×10^{-4} $M_\odot$ of ^{254}Cf produced in *r*-processes nucleosynthesis. Clayton and Craddock (1965) have evaluated what an actual γ-ray spectrum would look like in two hypothetical detectors, if the background in the detector is due solely to a sky background which matches the Ranger 3 measurements of Arnold *et al.* (1962). The mathematical form taken for this assumed isotropic background below 1 MeV is

$$dN/dE = 1.2 \times 10^{-5} E_6^{-7/4} \quad (\text{photons cm}^{-2} \text{ s}^{-1} \text{ keV}^{-1} \text{ sr}^{-1}) \qquad \text{(III.11)}$$

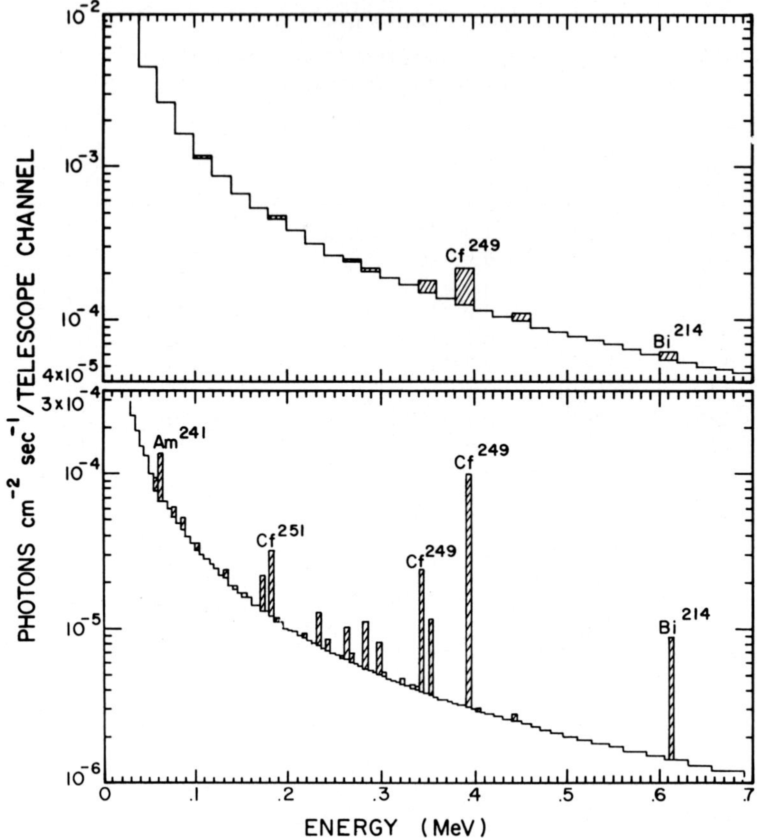

Fig. III-17. The line spectrum from the *r*-process as it would appear atop the continuous extraterrestrial photon flux observed by Arnold *et al.* (1962) when viewed with two idealized γ-ray telescopes having the following properties: (top) channel width of 20 keV and solid angle of 0.1 sr; (bottom) channel width of 5 keV and solid angle of 0.01 sr. (From D.D. Clayton and W.L. Craddock, *Astrophys. J.* **142**, 189. Copyright 1965, The American Astronomical Society, by permission of the University of Chicago Press.)

where E_6 is the photon energy in MeV. This form is in fairly good agreement with the latest experimental results discussed in Section V-5.3.

Figure III-17 gives Clayton and Craddock's (1965) result for telescopes with channel widths of 20 keV and 5 keV and angular apertures of 0.1 and 0.01 sr, respectively. These figures serve to point out the necessity of fine energy and angular resolution in the search for γ-rays from the Crab based on the ^{254}Cf hypothesis.

The γ-ray flux estimates just described are those expected if the light decay curve of a Type I SN after the first 100 days is due to the spontaneous fission energy release from ^{254}Cf (and if the Crab is a Type I SNR). On the other hand, Clayton and Craddock (1965) point out that a test of the hypothesis that Type I SN are the site of *r*-process nucleosynthesis requires a much smaller yield (by a factor of 100) of ^{254}Cf. Thus even if the

^{254}Cf light curve hypothesis is incorrect then the Type I events may still be the site of transbismuth nucleosynthesis, but the γ-ray flux levels would be correspondingly lower by a factor of 10^2 from those shown in Table III-6 and Figure III-17. As to the experimental search for the γ-ray lines expected, we should note here that the hypothetical spectrum shown in Figure III-17 should not be taken too seriously and is optimistic. This is because, first, the background assumed may be greatly underestimated, since other backgrounds will contribute in addition to that assumed, and there will be locally produced lines. Second, the detector's response function must also be folded into the expected source-plus-background spectrum in order to get a realistic measured spectrum. The experimental problems will be discussed in more detail in Chapters V and VI.

Finally, there have been some improvements in the calculations of Clayton and Craddock (1965) by Jacobson (1968) who used newer values of some branching ratios for the radioactive decays in connection with an experiment to detect the *r*-process γ-rays. This is discussed in Section V-5.2.1. Table V-3 gives a summary of the newer flux values expected compared to the earlier values for the lines affected.

Besides γ-ray lines, Clayton and Craddock (1965) have also considered the γ-ray spectrum (essentially a continuum) accompanying spontaneous fission and the γ-ray following the beta decay of neutron-rich fission fragments. Also, X-rays in the energy range 19 keV to 32 keV accompany the fission process, apparently from the internal conversion of prompt fission γ-rays. Their summary of the situation is that each spontaneous fission is accompanied by ~16 MeV in γ-rays. The most likely nucleus contributing now to this fission γ-ray flux would be ^{250}Cm which has a half-life of 2×10^4 yr. From their estimate of the current ^{250}Cm activity, the total fission γ-ray flux would be 8.1×10^{-5} cm^{-2} s^{-1} from the Crab Nebula. It would however be a more or less continuous spectrum peaking at ~ 100 keV and falling as E^{-2} above this energy with a flux even below the so-called diffuse background.

b. *γ-Rays from Si Burning in SN Shells*

More recently a considerable amount of work has been done studying nucleosynthesis in the low mass range $28 \leqslant A \leqslant 57$. This has been carried out by Fowler and Hoyle (1964), Truran *et al.* (1967), Bodansky *et al.* (1968a, b), and Clayton and Woosley (1969). Basically, the synthesis is based on thermonuclear burning of ^{28}Si in SN shells leading eventually to ^{56}Ni which undergoes the decay $^{56}\mathrm{Ni} \xrightarrow[6.4\ \mathrm{days}]{\mathrm{EC}} {}^{56}\mathrm{Co} \xrightarrow[77\ \mathrm{days}]{\beta^+,\ \mathrm{EC}} {}^{56}\mathrm{Fe}$ thus leading to the Fe peak well known in natural solar system abundances. The ^{28}Si is the major nuclide produced after the burning of O. The attractiveness of this scheme is due to the fact that the abundances of different nuclide masses produced in silicon match very well (in addition to ^{56}Fe) the normal abundances in this low mass range (cf. Clayton, 1972). Further, production of ^{56}Ni by this process has led Colgate and McKee (1969) to evaluate the hypothesis that SN light curves (Type I) are due to the $^{56}\mathrm{Ni} \longrightarrow {}^{56}\mathrm{Co} \longrightarrow {}^{56}\mathrm{Fe}$ decay chain, since the half-life involved for the last decay, as noted above, is in approximate agreement with the observed decay times. Since many γ-ray lines may be produced in the radioactivities resulting from silicon nucleosynthesis, Clayton *et al.* (1969) and Clayton and Fowler (1969) have considered the detectability of young SNR through ob-

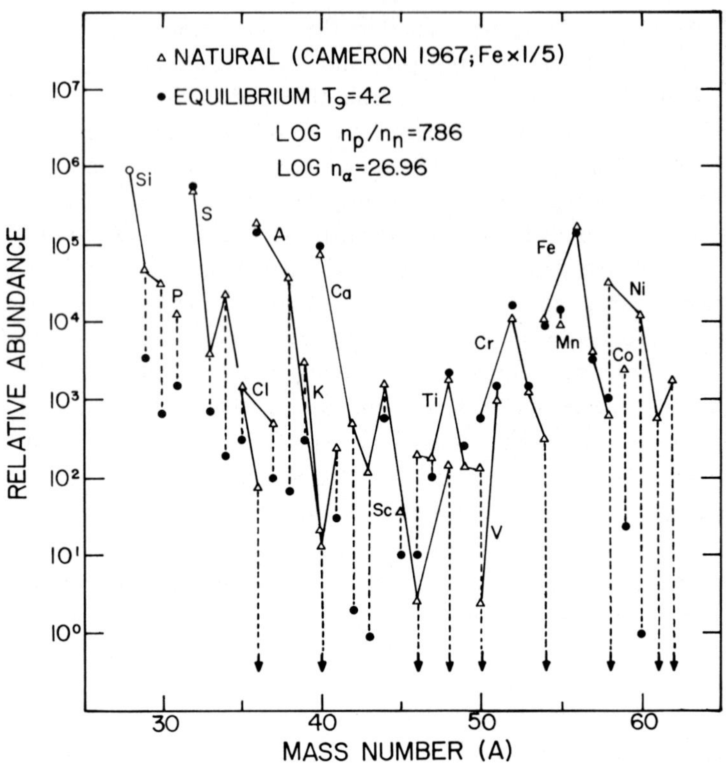

Fig. III-18. Comparison of the natural solar system abundances with the quasi-equilibrium abundances obtained in the thermonuclear burning of ^{28}Si. The alpha particle nuclei and the iron peak nuclei are well characterized by this picture. Many of the species are radioactive; for example, the natural abundance of ^{56}Fe is compared with the calculated abundance of radioactive ^{56}Ni. (From D. Bodansky *et al., Astrophys. J. Suppl.* **16,** 299. Copyright 1968, The American Astronomical Society, by permission of the University of Chicago Press.)

servation of the γ-rays. Detection of the predicted γ-rays from a supernova would therefore be a strong test of the validity of this model of intermediate mass nucleosynthesis. It is valuable to review the basic origins of the γ-ray lines expected.

In Figure III-18, taken from Bodansky *et al.* (1968b), are shown the natural abundances of ^{56}Fe, ^{52}Cr, ^{48}Ti, ^{44}Ca, etc., compared with the calculated abundances from Si burning, giving the radioactive nuclides ^{56}Ni, ^{52}Fe, ^{48}Cr, and ^{44}Ti. The decay chains for the nuclides are:

$$^{44}\text{Ti} \xrightarrow[48\text{ yr}]{\text{EC}} {}^{44}\text{Sc} \xrightarrow[3.92\text{ hr}]{\beta^+,\text{ EC}} {}^{44}\text{Ca}$$

$$^{48}\text{Cr} \xrightarrow[23\text{ hr}]{\text{EC}} {}^{48}\text{V} \xrightarrow[16.0\text{ d}]{\beta^+,\text{ EC}} {}^{48}\text{Ti}$$

$$^{52}\text{Fe} \xrightarrow[8.2\text{ hr}]{\beta^+,\text{ EC}} {}^{52}\text{Mn} \xrightarrow[5.7\text{ d}]{\beta^+,\text{ EC}} {}^{52}\text{Cr}$$

TABLE III-7

Origin of the principal nuclear γ-ray lines in Si burning

Parent	Excited nuclide	Energies (in MeV)
^{44}Ti	^{44}Sc	**0.0678, 0.0784**
^{44}Sc	^{44}Ca	**(0.511), 1.156**
^{48}Cr	^{48}V	**0.116, 0.31**
^{48}V	^{52}Ti	**(0.511)**, 0.945, 0.9833, **1.312**
^{52}Fe	^{52}Mn	**0.165**, 0.383 (weak), **(0.511)**
^{52}Mn	^{56}Cr	**(0.511), 0.7744, 0.935, 1.4336**
^{56}Ni	^{56}Co	**0.163**, 0.276, 0.472, **0.748, 0.812**, 1.56
^{56}Co	^{56}Fe	**(0.511), 0.8469, 1.24**, 1.37, 1.76, 2.02, 2.60, 3.26
^{57}Co	^{57}Fe	0.014, **0.122**, 0.136
^{60}Fe(^{60}Co)	^{60}Ni	**1.17, 1.33**

$e^+ + e^- \rightarrow 0.511$ MeV or 3 γ.

The strongest lines are printed boldface.

$$^{56}\text{Ni} \xrightarrow[6.10\text{ d}]{\text{EC}} {}^{56}\text{Co} \xrightarrow[77\text{ d}]{\beta^+,\,\text{EC}} {}^{56}\text{Fe}\,.$$

The γ-rays which result from the chains are shown in Table III-7, listed according to the excited nuclide giving the γ-ray following decay of the parent. See Lederer *et al.* (1967) for the decay schemes and relative intensities of the γ-rays.

Clayton *et al.* (1969) have assumed a model Type I SN at 10^6 parsec (1 Mpc) which has ejected 0.5 $M_\odot$ of silicon-burning debris containing 0.14 $M_\odot$ of ^{56}Ni. Based upon the conventional model of the rate of expansion of the shell (Colgate *et al.*, 1961) and the assumption that there is no modification of the line spectrum due to scattering in the shell, the time history of the nuclear line flux was derived for the strongest lines and is shown in Figure III-19. The fluxes increase with time after the event due to the buildup of the particular radioactive species involved and the thinning of the expanding shell which allows more γ-rays to escape. The strongest γ-ray lines expected at 0.511 MeV ($e^+ + e^-$), 0.812 MeV (^{56}Ni) and 0.847 MeV (^{56}Co) could be easily observable at the intensities indicated ($\sim 10^{-4}$ photons $cm^{-2}\ s^{-1}$) in balloon or satellite instruments. The frequency of Type 1 SN within 1 Mpc of the Solar System is however quite low, probably $\lesssim 10^{-1}\ yr^{-1}$, which applies essentially to our local group of galaxies. The most recent nearby SN 1972e occurred in early 1972 in NGC 5253 (Kowal, 1972), a dwarf elliptical galaxy at a distance of 2.3 to 4.6 Mpc. Apparently, then, one would have expected γ-ray lines from this event diminished by about at most a factor of 20. In Section V-5.2.1, we discuss the experimental efforts to detect γ-ray lines from this event.

Clayton (1971) has also extended calculations of explosive nucleosynthesis to include the production of ^{60}Fe. The ^{60}Fe undergoes β^- decay to ^{60}Co which in turn undergoes β^- decay to ^{60}Ni, giving the characteristic cascade pair of γ-rays at 1.17 MeV and 1.33 MeV. He estimates that SNR toward the Galactic Center may be sufficiently intense in γ-ray lines of ^{60}Co (1.17 MeV, 1.33 MeV) and ^{44}Sc (1.156 MeV) to give fluxes at the Earth of

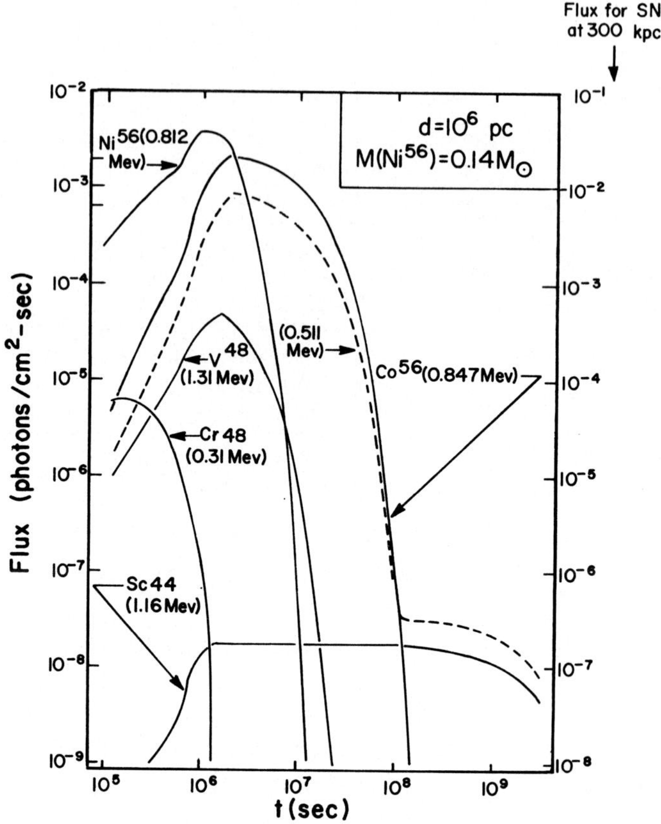

Fig. III-19. Expected γ-ray line emission from a SN at 1-Mpc according to the Silicon Burning Hypothesis. (From D.D. Clayton *et al.*, *Astrophys. J.* **155**, 75. Copyright 1969, The American Astronomical Society, by permission of the University of Chicago Press.)

$$F_{60}(\text{galactic}) = 1.7 \times 10^{-5} \rightarrow 1.7 \times 10^{-3}\ \gamma\,\text{cm}^{-2}\,\text{s}^{-1}$$

$$F_{44}(\text{galactic}) = 4.5 \times 10^{-4}\ \gamma\,\text{cm}^{-2}\,\text{s}^{-1}\,.$$

Clayton (1974) has pointed out recently that ^{57}Co is also produced in Si burning in significant quantities and γ-ray lines from its K capture decay to ^{57}Fe will give lines at 136 and 122 keV. The stronger line at 122 keV may be the most significant line feature ~2 yr after a galactic SN since the half-life of ^{57}Co is 270 days. In this regard it should be noted that a line at 14.4 keV would also result from the decay of ^{57}Co since the 14.4 keV level in ^{57}Fe is fed by γ-ray decay from the 136.3 keV level which is populated by the K capture.

This bright prospect of using γ-ray line astronomy to resolve the basic problem of the site of nucleosynthesis may, however, be impossible to achieve. Brown (1973) has carried out Monte Carlo calculations on the transport of γ-ray lines out of the presumed silicon-burning shell of a Type I SN. Since the shell thickness can be an appreciable fraction of the mean free path of an MeV γ-ray (~ 10 g cm^{-2}) and the Compton scattering process is the primary phenomena at this energy, an initial γ-ray line spectrum can be strongly distorted. The calculations of Brown (1973) assume isotropic emission of γ-ray lines at a depth of 18.6 g cm^{-2} from the surface of a Fe sphere of radius 37.2 g cm^{-2}. This corresponds to the thickness for a Type I shell in the model of Colgate and McKee (1969) intermediate in its time history. The γ-ray transport calculations take into account coherent scattering and incoherent (Compton) scattering and photoelectric and pair production absorption processes. The Monte Carlo program does not include fluorescent emissions or the bremsstrahlung of Compton electrons. The ejected shell in the model has the properties of neutral Fe as far as γ-ray interactions are concerned, but otherwise has the silicon-burning debris abundance given by Clayton *et al.* (1969).

The effect of scattering on the emitted γ-ray spectrum in such a situation was carried out by Monte Carlo calculations (Brown, 1973) for hypothetical lines at 1.25 MeV and 3.5 MeV for the geometry given above and with a total of 30 000 photons emitted in each case. The line spectra were strongly modified at low energies due to the Compton scattering, yet one might expect that with a detector of sufficiently high energy resolution, lines would still be easily resolvable. The results of extending these calculations to a more practical case are shown in Figure III-20. This shows the γ-ray spectrum emitted from a

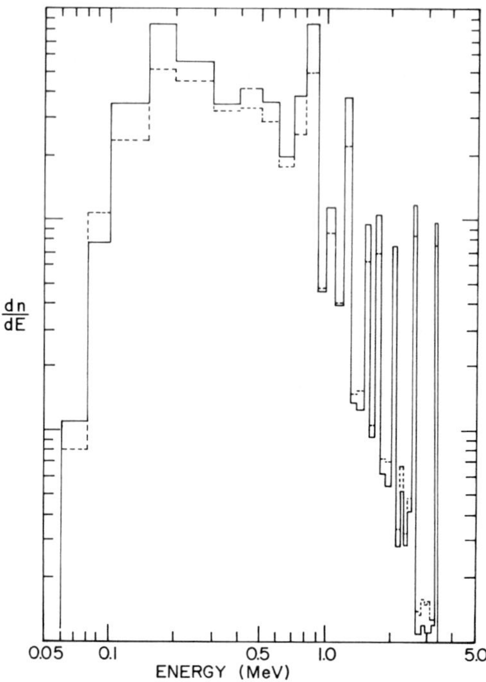

Fig. III-20. γ-ray line spectra (arbitrary units) from decay of ^{56}Ni and ^{56}Co emitted uniformly and isotropically within Fe spheres of radius 37.2 g cm^{-2} (solid histogram) and 750 g cm^{-2} (dashed histogram). (From R.T. Brown, *Astrophys. J.* **179**, 607. Copyright 1973, The American Astronomical Society, by permission of the University of Chicago Press.)

^{56}Ni and ^{56}Co source at depths of 37.2 g cm^{-2} and 750 g cm^{-2} inside a shell as described above. The thick shell corresponds to a very early time in the history of a Type I event. Since the early-time thickness is ~ 40 mean free paths for the γ-ray at 1.25 MeV, the total emitted flux from the shell will be very different in the two cases. The solid spectrum is for the smaller thickness and the dotted spectrum is for the greater thickness. Only the relative shapes of the spectrum should be compared. In each case it is seen that Compton scattering has strongly modified the spectrum with all but the highest energy lines superposed on a relatively strong continuum. As Brown (1973) points out, the 0.51 MeV line has essentially disappeared. The strongest lines clearly separated from the continuum are from levels in ^{56}Fe giving lines at 3.26 MeV, 2.60 MeV, 2.02 MeV, 1.76 MeV, 1.24 MeV, 1.03 MeV, and 0.847 MeV, and from levels in ^{56}Co giving lines at 1.56 MeV, 0.812 MeV, and 0.748 MeV (see Table III-7). Therefore, these lines should be the easiest to detect from a few days to several days following a sufficiently nearby Type I event.

From the experimental viewpoint one cannot stop here, however, since Doppler broadening may be sufficient to appreciably broaden the emitted lines as in the case of optical emission. This broadening would make line detection more difficult and even possibly negate (for this application) the advantages gained by using the high resolution detectors planned for HEAO-C (cf. Section VI-6.3.1). In addition, the Doppler-shifted, Compton-distorted spectrum arriving at the Earth will be further distorted by the response function of the detector itself, and must be detectable above the background measured in the detector. In spite of these frightening prospects, the detection of (or failure to detect) γ-rays from SNR will give valuable knowledge on the conditions in SN shells or put strong constraints on theoretical models. Clayton and Hoyle (1974) have also pointed out the importance of searching for γ-ray line emission from galactic novae.

c. *Gum Nebula*

A recent Goddard Space Flight Center conference on the Gum Nebula (Maran *et al.*, 1971) has brought to light a possible additional source for γ-ray lines from SNR. If the Gum Nebula is a result of the hypothetical Vela X SN (PSR 0833-45) and cosmic rays are produced in the process, then γ-ray lines may be emitted as a result of energetic particle interactions in the nebular gas (Ramaty and Boldt, 1971). The authors predict fluxes of several γ-ray lines from different processes. For the reaction ^{28}Si (p, p$'\gamma$) giving a line at 1.78 MeV, a flux of ~8×10^{-3} γ cm^{-2} s^{-1} is predicted on the basis that 2/3 of the mass of the SN ejecta is silicon. Similarly, estimates are made for line emission fluxes at 4.43 MeV and 6.13 MeV from ^{12}C (p, p$'\gamma$) and ^{16}O (p, p$'\gamma$), respectively. The results are:

$$F(4.43 \text{ MeV}) = 4 \times 10^{-2}\ M_C/M_\odot \quad (\gamma \text{ cm}^{-2}\ \text{s}^{-1}) \tag{III.12}$$

and

$$F(6.13 \text{ MeV}) = 1.5 \times 10^{-2}\ M_O/M_\odot \quad (\gamma \text{ cm}^{-2}\ \text{s}^{-1}) \tag{III.13}$$

where the ratios give C and O masses in units of the solar mass. The above examples suggest which γ lines may be expected; however, flux estimates may be uncertain by several orders of magnitude either way.

TABLE III-8

Parameters of three γ-ray sources identified with SNR. (From J.A. De Freitas Pacheco: 1973, 'Supernova Remnants and Gamma-Ray Sources', *Astrophys. Letters* **13**, 97, by permission.)

Object	Event	Type	L_γ (meas)[a] (photons s^{-1})	L_γ (theo) (photons s^{-1})
Cygnus loop		II?	2.3×10^{39}	1.6×10^{37}
Cas A	(−270 yr)	II?	7.1×10^{39}	1.0×10^{39}
Tyco	1572	I	4.3×10^{40}	5.0×10^{37}

[a] Based on fluxes of Browning *et al.* (1972) and converted to luminosity for comparison with theory by Pacheco (1973).

d. *γ-Ray Sources* (>100 MeV)

Supernova remnants could provide a possible source of the cosmic γ-rays observed from the galactic plane (see Section V-5.2.2 for a more complete discussion). For example, Hayakawa and Tanaka (1970) had proposed that the OSO-3 γ-ray flux $\gtrsim$ 100 MeV (Clark *et al.*, 1968) observed in the galactic plane could be explained by π^0 decay γ-rays emanating from many discrete SNR. Pacheco (1973) has carried out an additional analysis of this model taking into account the time evolution of the SNR and finds the γ-ray luminosity (~100 MeV) versus the radius of an expanding SN shell. The basic γ-ray yields from the reaction $p + p \longrightarrow p + p + \pi^0$ were taken from Cavallo and Gould (1971). These calculations were compared with the measured fluxes of high energy γ-rays associated with specific SNR given by Browning *et al.* (1972). The measured γ-ray luminosities and those calculated are compared in Table III-8. It is therefore concluded that π^0 γ-rays cannot explain the high energy γ-ray emission (~100 MeV) from these specific sources and, further, that the π^0 γ-rays from several discrete SNR sources cannot explain the galactic plane γ-ray source unless current ideas about SN energetics are drastically changed (Pacheco, 1973).

e. *Supernovae (Prompt Emission)*

Considerable speculation exists concerning the expected emission of γ-rays during the initial phase of a SN. This is largely due to the fact that no γ-ray measurements have been carried out during the 'instant' of occurrence of an SN and through its development to maximum light intensity. The experimental problem is therefore to know exactly what to look for in terms of γ-ray energies and time structure. The recently reported Vela γ-ray bursts (Klebesadel *et al.*, 1973; see Section V-5.4) have given impetus to the idea that γ-ray bursts may occur in the initial SN phase. It is of value, then, to review in historical order some of the theoretical predictions of prompt emissions. Burbidge (1966) had suggested that the energy released in burning 10^{-2} solar masses of H could be emitted in MeV γ-rays in 10^3 s during the initial phase of a SN outburst. In this case, specific γ-ray lines at 0.51 MeV, 2.2 MeV, and 5.5 MeV might be produced, and the flux for a SN at 4 Mpc might be a total of 30 photons cm^{-2} s^{-1}.

Colgate (1968) had predicted the occurrence of a γ-ray burst when a shockwave breaks through the expanding shell of an exploding star. According to this work, the explosion process always gives rise to a strong shockwave (presumably independent of the energy generating mechanism causing the rapid energy release). This shock has accelerated to relativistic velocities by the time it proceeds to a radius where the outer mass fraction of the star is 10^{-4} of the total star mass. Immediately behind such a strong shockwave, the internal energy per g equals the kinetic energy per g or $\gamma_s c^2$, where $\gamma_s = (1 - \beta^2)^{-1/2}$ and β is the (fluid velocity)/c immediately behind the shockwave. We are interested in the temperature of the matter when the shockwave breaks out of the surface of the star. Colgate (1968) has estimated this temperature, T, from $aT^4 = \gamma_s \rho_s c^2$ where ρ_s (g cm^{-3}) is the rest mass density behind the shock and $a = 4\sigma/c = 7.56 \times 10^{-15}$ erg cm^{-3} deg^{-4}. For his values of $\gamma_s = 1.5 \times 10^3$ and $\rho_s = 0.2$ g cm^{-3} one finds $T \simeq 2.4 \times 10^9$ K or 200 keV. Since the black body photon average energy is ~3 kT, the Doppler shifted photon energy along the observer's line of sight will be

$$E_\gamma = 3\,kT \sqrt{\frac{1+\beta}{1-\beta}} \simeq 6\,kT\,\gamma_s \simeq 2\ \text{GeV}\,. \qquad \text{(III.14)}$$

This would be the upper limit energy for the earliest photons, and the photons diffusing from lower depths in the shell could be Compton scattered to lower energies. Colgate (1968) estimated the total γ-ray energy in the initial pulse as $W_{ts} = MF_s\,\gamma_s{}^2 c^2 \simeq 5 \times 10^{47}$ erg, where $MF_s \simeq 5 \times 10^{20}$ g is the mass in the surface layer. Such a burst could occur over a time as short as 1.5×10^{-5} s giving an initial luminosity of $L \sim 3 \times 10^{52}$ erg s^{-1}. If the photons have an energy of ~1 GeV (1.6×10^{-3} erg), then for a SN at 1 Mpc the peak intensity of the γ-ray burst could be quite high, 2×10^5 photons cm^{-2} s^{-1}. This instantaneous flux could be considerably higher or lower depending on what one uses for the average shock photon energy. As another means of detecting such an event, Colgate (1968) suggested that a SN in this galaxy could give rise to an optical fluorescent pulse about 1/10 that of moonlight. Later, however, Colgate (1970) revised these initial ideas, since in cases of very high shock temperatures ($kT \sim m_0 c^2 \sim 0.5$ MeV) the density of electron-positron pairs would be much higher than the original electron density and the opacity much greater than originally estimated. Thus the current view is that for a SN giving rise to a relativistic shockwave, the emitted prompt γ-rays would be Doppler-shifted positron-electron annihilation γ-rays, still giving GeV photons. These remarks presumably refer to what is expected for Type I SN.

In connection with the origin of the Vela γ-ray bursts of duration 0.1 to 1 s (Klebesdal *et al.*, 1973; Section V-5.4), Colgate (1974a) has extended his shockwave ideas to Type II SN. These have characteristic H emission spectra and occur in young stellar populations from probably massive stars. Presumably the Type I SN are from less massive ($<4M_\odot$) stars, which have evolved to a compact structure, while the Type II SN is a large radius, large mass system. In both cases it is estimated that the energy given off in γ-rays is the shockwave energy imparted to the surface layer of the star, but the nature of the γ-ray burst is quite different in the two cases. In the Type II case, Colgate (1973) has argued that a high temperature precursor to the shockwave contains ion energies as high as 10 MeV per nucleon with electron energies of about 0.5 MeV. Nucleons in the high energy tail of a plasma this hot are sufficient to produce neutrons from the breakup of ^{4}He (~25 MeV) and from (p, n) reactions on ^{14}N and ^{13}C ($\leqslant$6 MeV) as well. This leads to the production of D through the capture of neutrons produced in the spallation (Colgate, 1973, 1974b). In this precursor, electrons are also heated to a temperature of $\sim mc^2$(0.5 MeV) and give rise to bremsstrahlung photons which are heated by inverse Compton scattering up to $1\to 2\ mc^2$(0.5 to 1 MeV). This nonthermal photon distribution relaxes to a Planck distribution of $kT \simeq 1$ keV as the ions in the precursor cool. As the precursor and the shock break the surface of the star, the photons emitted are expected to have a mean energy of $\sim mc^2$(0.5 MeV) with total energy $\sim 10^{48}$ erg (Colgate, 1974a). The time duration of this γ-ray burst in the frame of the expanding shock (velocity $U_{\mathbf{exp}}$) is estimated to be $\sim t \simeq r/U_{\mathbf{exp}}$ or 100 s for a cloud of radius $r \sim 2 \times 10^{12}$ cm (~0.1 AU). Following Rees (1966) and Petschek (1967), the time of the pulse for an observer watching the relativistically expanding surface must be contracted by a factor $(1-\beta) \simeq 1/(2\gamma^2)$. For an expansion velocity giving $\gamma \sim 10$, then, the pulse is shortened by a factor of 5×10^{-3} to ~0.5 s in approximate agreement with the time scale of the Vela bursts.

If this phenomenon explains the observed bursts, then Colgate (1974a) expects a second pulse due to the capture of neutrons produced in the spallation reactions mentioned above. The neutrons are produced at ~1 GeV relative to the expanding fluid and are slowed down and captured in an outer layer of ~100 g cm^{-2} over a time scale of $\lesssim$1000 s. The observer time contraction is again $\sim 1/(2\gamma^2)$, giving a burst of 3 s mean duration. The energy of these photons would be 2.22 MeV in the frame of the expanding debris. Colgate (1974a) suggests this burst could explain the second pulse seen in the Vela measurements. Possibly a sharp line emission would be seen.

In terms of the frequency of these Type II SB, Colgate (1974a) concludes that to explain γ-ray bursts ($\sim 10^{-5}$ to 10^{-4} erg cm^{-2}) a source of 10^{48} erg would be placed at about 10 to 30 Mpc, which would include a rich cluster such as in Virgo. Thus with 2000 to 5000 galaxies and with a Type II SN rate of 10^{-2} yr^{-1} the expected event frequency would be 20 to 50 yr^{-1}, which is 4 to 12 times larger than the observed Vela rate. Colgate (1974a) therefore expects that there might be many more slower bursts lasting ~100 s, which would not have been detected by the Vela recording system.

In conclusion we note that the variety of theoretical predictions for the energy and time characteristics of γ-ray bursts from SN is manifold. Only sufficient observations will enable us to adequately test the theories.

3.2.2. NEUTRON STARS

The interpretation of the pulsar NP 0532 as a rotating neutron star gives strong evidence (if not proof) for the belief in the existence of neutron stars. Because of their small size (~10 km), the gravitational potential energy of a particle of matter of nucleon mass can be quite large (~100 MeV per nucleon). This realization has led several authors to calculate the γ-ray spectrum resulting when the accelerated infalling matter (accretion) interacts with the neutron star surface (cf. Shvartsman, 1970; Zeldovich and Novikov, 1971; Ramaty *et al.*, 1973; Reina *et al.*, 1974). Most of the calculations have considered only the gravitational effect on ions and no effect of magnetic fields. In such a case, the energy of free fall of an ion at the surface of the star is:

$$E_\mathrm{i} \simeq 1350 \left(\frac{M}{M_\odot}\right) / R\ (\mathrm{km}) \qquad (\text{MeV per nucleon}). \tag{III.15}$$

With ion energies this large, several reactions are possible depending on the composition of the star's surface. Thus meson production gives π^+ and π^0 from which positron-electron annihilation γ-rays and π^0 decay γ-rays may be produced. Also, neutron production from He breakup or (α, n) reactions, as discussed in Section III-3.1, can lead to emission of the 2.2 MeV neutron-proton capture γ-ray.

In particular, Ramaty *et al.* (1973) have calculated the yield of positrons from accretion onto neutron stars. Normally it is assumed that the neutron star surface is ^{56}Fe; however, if the accretion rate is sufficiently large, the surface composition for gamma ray production may be significantly modified and reflect that of the accreted matter. Therefore, assuming that the composition of CNO is in the ratio 0.5:0.1:1 (Cameron, 1967), they have calculated the yield, Q^+, of positrons (0.51 MeV γ's), the yield, $Q^{(4.43)}$, of the ^{12}C line at 4.43 MeV and the yield, $Q^{(6.13)}$, of the ^{16}O line at 6.13 MeV for a range of neutron star masses, as well as the surface gravitational red shifts of the γ-ray lines. Table III-9 shows the result of these calculations for a neutron star radius of $R = 10$ km. The first column gives the neutron star mass in solar masses, the second the corresponding red shift according to Equation (II.75), and the third the proton kinetic energy from Equation (III.15). Since the slowing down and annihilation times are very short in the neutron star's surface, $2Q^+$ gives the number of 0.51 MeV γ-rays per proton with corresponding units for the other lines.

Ramaty *et al.* (1973) have used the results of these calculations to attempt to explain the 473 keV line reported from the Galactic Center (Johnson *et al.*, 1972; Section V-

TABLE III-9

Surface redshifts, incident proton energies, and γ-ray yields as functions of neutron-star mass. (From R. Ramaty *et al.*, *Astrophys. J.* **181**, 891. Copyright 1973, The American Astronomical Society. Used by permission of the University of Chicago Press.)

$M/M_\odot$	Z_s	E_p (MeV)	Q^+	$Q^{(4.43)}$	$Q^{(6.14)}$
0.37	0.053	39	0.0062	0.003	0.002
0.55	0.085	57	0.015	0.005	0.003
0.77	0.13	77	0.028	0.006	0.004
1.00	0.18	103	0.048	0.007	0.005
1.24	0.24	130	0.070	0.008	0.006
1.44	0.31	160	0.095	0.009	0.007
1.56	0.38	184	0.12	0.01	0.008
1.68	0.45	208	0.14	0.011	0.008
1.72	0.51	230	0.18	0.012	0.009

NOTE – Per proton of given energy, Q^+, $Q^{(4.43)}$, and $Q^{(6.14)}$ are the yields for positron annihilation radiation, 4.43-MeV C line, and 6.14-MeV O line, respectively.

5.2.3) as a red-shifted annihilation line from many neutron stars in the Galactic Center. If this interpretation is correct, then from Table III-9 it can be deduced that the neutron stars responsible should have masses of $\sim 0.8 M_{\odot}$, since the red shift must be $\lesssim 0.1$. In order to explain the intensity of the reported line of 1.8×10^{-3} photons cm^{-2} s^{-1}, Ramaty *et al.* (1973) have estimated that 10^5 to 10^{10} neutron stars would be necessary in the Galactic Center region depending, respectively, on whether the accretion occurs on neutron stars in binary systems or by accretion from the interstellar medium on single neutron stars. In the former case it is suggested that the accretion rate can be as high as $\sim 10^{16}$ g s^{-1} and in the latter $\sim 10^{11}$ g s^{-1}. As pointed out by these authors, the primary observational consequence of this theory would be the emission of several other γ-ray lines, all shifted by the same amount as the positron annihilation line. The specific lines may be innumerable depending on the composition; however, in the case of the ^{12}C and ^{16}O lines, the predicted fluxes are $\sim 5 \times 10^{-4}$ and 3×10^{-4} photons cm^{-2} s^{-1} at 4.43 and 6.13 MeV, respectively, which are near or below instrument sensitivities thus far reported.

Reina *et al.* (1974) have also carried out calculations and have estimated the luminosity in the positron annihilation line to be 4×10^{-7} of the total energy input due to accretion. This would correspond to $\sim 4 \times 10^{32}$ erg s^{-1} in the line based on a total X-ray luminosity of 10^{39} erg s^{-1} from the Galactic Center region as inferred from Uhuru data (Giacconi *et al.*, 1972). The measured luminosity of the Galactic Center line is $\sim 10^{37}$ erg s^{-1}, several orders of magnitude larger than estimated by Reina *et al.* (1974). If the model of Ramaty *et al.* (1973) is correct, it would imply that the γ-ray luminosity should be $\sim 10^{-2}$ of the total X-ray luminosity. Reina *et al.* (1974) have also estimated the γ-ray line flux from several known X-ray sources assuming the known X-ray luminosity is due to accretion. For example, a red-shifted 0.511 MeV annihilation line from SCO X-1 would be expected to have a flux of 1.4×10^{-7} photons cm^{-2} s^{-1}, which is several orders of magnitude below present instrument capabilities. Reina *et al.* (1974) have also suggested that a γ-ray line at 22.5 MeV would be expected from inelastic scattering of protons on ^{4}He; however, there is no known excited state of ^{4}He which decays by γ-ray emission.

Shvartsman (1970) has calculated for single neutron stars the γ spectrum resulting from π^0 meson decay and π^+ meson decay. The mesons originate in (α, p) and (α, α) reactions. In the π^0 case the γ-ray spectrum peaks at ~ 70 MeV, and in the π^+ case Shvartsman has deduced that the positrons from the decay chain $\pi^+ \rightarrow \mu^+ \rightarrow e^+$ annihilate in flight, so one of the photons carries off essentially all kinetic energy of the positron (cf. Section II-2.4.4). This gives another maximum in the spectrum below ~ 52.5 MeV, the maximum energy of the μ^+ decay positron. This latter spectrum is strongly distorted to lower energies due to energy losses of the positrons prior to annihilation and due to the gravitational potential. Shvartsman (1970) has also pointed out that a portion of the π^- mesons produced can recombine with H prior to decay (in a dense medium) according to the reaction $\pi^- + p \rightarrow n + \gamma$ giving a γ-ray line at 129.4 MeV. In addition, Shvartsman (1970) has suggested that the temperatures under the accreting layers can be sufficiently high (10^7 K to 10^9 K) that several thermonuclear reactions can contribute to the γ-ray luminosity above 5 MeV.

The γ-ray luminosity estimated for a strong X-ray source such as SCO X-1 has been estimated by Shvartsman (1970) as $\sim 3 \times 10^{-6}$ photons cm^{-2} s^{-1} for $\bar{E}_\gamma = 30$ MeV assuming the ratio of γ-ray luminosity, L_γ, to the X-ray luminosity, L_x, is $\sim 3 \times 10^{-4}$. Generally the γ-ray luminosity for a specific nuclear line from these calculations ranges from 10^{-4} to 10^{-7} L_{tot}, where L_{tot} is the total energy input due to accretion.

It is generally agreed that the energy required to explain the full range of electromagnetic emissions from the Crab Nebula is derived from the slowing down of a rotating neutron star near the center of the nebula. Indeed, all pulsars may be rotating neutron stars with a strong magnetic field of a dipole nature as advocated by Gold (1968). In the case of the Crab Nebula, experimental observations by Kurfess (1971) and Orwig *et al.* (1971) have demonstrated that pulsed γ-rays in the MeV energy region are emitted with the same period as the radio, optical and X-ray emissions. Pulsed γ-rays from the Crab Nebula have also been observed at 100 MeV by Parlier *et al.* (1973) and at $\geqslant 10^5$ MeV by Grindlay (1972).

In order to explain pulsar observations in detail, Sturrock (1971) has developed a specific pulsar model comprised of a rotating neutron star with radius $R \sim 10^6$ cm and a strong surface dipole magnetic field with $B \sim 10^{12}$ G. The axis of the magnetic dipole is, in general, not parallel to the rotational axis of the star. In such a case, Sturrock (1971) proposes that those magnetic field lines which extend to the light cylinder, where the rotational speed is equal to the velocity of light, are open. The radius of this cylinder is given simply by $R_L = 10^{9.7}\ T$ where T is the pulsar period in seconds. If the field lines were closed beyond R_L, then a particle tied to the field lines would be moving faster than the speed of light. Sturrock (1971) has shown that electrical currents must flow along the open magnetic field lines from the polar caps of rotating neutron stars. Two current streams are produced, one of electrons and one of protons, and the regions where these particles flow are termed the electron polar zone (EPZ) and the proton polar zone (PPZ). Electrons and protons accelerated away from the surface of the star by electric fields along the open field lines will have a negligible energy perpendicular to the magnetic field, and, therefore, synchrotron losses will be negligible. Since the field lines are curved and the particles are also moving along a curved path, there will be an associated energy loss from the accleration which is termed 'curvature radiation'. Sturrock (1971) has shown that this radiation can fall in the γ-ray energy region.

Since we are not interested here in discussing the details of pulsar radiation mechanisms, we will give only the relation showing how the energy of the curvature radiation is related to pulsar parameters. For the case of curvature radiation from protons, which seems to be applicable to the Crab Nebula, Sturrock (1971) gives the energy of the γ-rays produced in the proton polar zone as

$$E_{\gamma P \eta} = 10^{-87.4}\ B^3\ R^{17/2}\ T^{-13/2}\ \eta^{-1/2}\ (\text{eV}) \tag{III.16}$$

where the radius where the emission occurs is given by ηR, and the other quantities were defined above. The quantity η is greater than one, but is limited by the velocity-of-light radius. Using values for the Crab Nebula, $T = 10^{-1.5}$ s, $R = 10^6$ cm, and $B = 10^{12}$ G, it follows that photons with energies as high as 10^9 eV can be produced by this process. Reference to Sturrock (1971) should be made for a similar expression for electron curvature radiation, as well as several other aspects of pulsar emission in general.

3.2.3. FLARE STARS

There are two types of flare stars which might be considered as sources of γ-rays or hard X-rays (>100 keV). The first class are the *observed* flare stars which are UV Ceti-type variable stars. *Hypothetical* stars which produce flares similar to super solar flares may be considered a second class. The former are cool dwarf stars of spectral class dM3e-dM6e which are known to produce optical flares producing magnitude changes from ~0.5 to 6 mag. The prototype of this class of stars was discovered by Luyten (1949) who observed a 2 magnitude variation in the brightness of the binary star L726-8AB (UV Ceti) in late 1948. This star often undergoes magnitude changes of $\Delta m \sim 0.5$ every few hours, $\Delta m =$ 1 to 2 every few days and occasionally, $\Delta m = 6$. The rise time of some flares is less than a minute and their duration is ~10 min. Lovell (1971) has given an interesting account of how these observations led him to search for (and successfully observe) radio bursts at 240 MHz from UV Ceti. About 26 optical flare stars are known at distances less than 20 pc. Based on the optical observations for large flares, the total energy output is $\sim 10^{34}$ erg with an average emission rate of $\sim 10^{30}$ erg s^{-1}. This instantaneous emission rate is comparable with the total quiescent optical emission of $\sim 10^{30}$ erg s^{-1} for such a cool dwarf star.

This is in considerable contrast to the case of solar flares where the optical emission rate of a large flare is 10^{-6} that of the solar luminosity of 4×10^{33} erg s^{-1}. Also only on rare occasions are white (continuum) light flares seen against the solar disk. In the case of the radio bursts from flare stars, Lovell (1971) points out that the relative energy in radio emission to the optical emission can be several orders of magnitude larger than the same ratio for large solar outbursts, which is $\sim 10^{-5}$. It is this circumstance that has facilitated the detection of flares by radio telescopes, and Lovell (1971) cites this as an example why "one should never be deterred from making observations by theoretical calculations that effects are unobservable." Thus far efforts to observe X-ray bursts from flare stars have been unsuccessful (e.g. Hudson and Tsikoudi, 1973).

By assuming that the radio emission is due to synchrotron radiation of a power law electron spectrum, Grindlay (1970) has calculated the X-ray emission from the nonthermal bremsstrahlung of the same spectrum of electrons. His calculation gives an estimated flux of ~3 photons cm^{-2} s^{-1} at the Earth for photon energies >10 keV for a model UV Ceti flare of $\Delta m_V = 1$. Crannell *et al.* (1974) have estimated the X-ray flux for a similar flare by using the ratio of the solar X-ray flux to the solar radio flux at 210 MHz for a typical solar flare and scaling to a large UV Ceti type radio burst to predict a flux of ~1 photon cm^{-2} s^{-1} at the Earth $\gtrsim$4 keV. If the flare star phenomenon is really like a typical large solar flare, then extrapolating these fluxes to the γ-ray region (~0.5 MeV) using a typical integral power law spectrum of E_γ^{-3} would give an estimated flux of $\sim 10^{-6}$

photons cm^{-2} s^{-1} which is at least 3 orders of magnitude below what can be detected experimentally.

Stecker and Frost (1973), inspired by the Vela γ-ray burst observations, have predicted γ-rays from super solar flares on as yet hypothetical objects. The Vela burst phenomenon is discussed in Sections III-3.2.6 and V-5.4.

3.2.4. GALACTIC CORE AND DISK

There have been several theoretical estimates of the intensity of characteristically nuclear γ-rays from the disk and center of our galaxy. The expected distribution of galactic cosmic rays and matter throughout the galaxy along with the known γ-ray production mechanisms have provided a reasonably solid basis for estimating γ-ray fluxes. For our purposes the most important estimates concern the explanation of the observed high energy (>30 MeV) galactic core and disk γ-rays (cf. Section V-5.2.2), the prediction of a flux of galactic 0.51 MeV photons, and γ-ray lines from excited nuclear states. We will discuss briefly the recent theoretical results for these three cases.

a. *γ-Rays* $\gtrsim$30 MeV

A flux of high energy γ-rays ($\gtrsim$ 30 MeV) from the Galactic Center and disk is now established (see Section V-5.2.2 for a discussion of the experimental observations). There are three possibilities seriously discussed for the origin of this radiation: 1) bremsstrahlung interactions, 2) inverse Compton scattering on starlight, and 3) cosmic ray produced π^0 mesons. The spectrum, as measured with instruments of relatively wide angular resolution $\lesssim 2^\circ$, indicates that the source is diffuse (although there are disagreements on this) and, due to its shape, possibly made up of two components. There is still the possibility that several point sources at the Galactic Center could produce the observed spectrum but a prevailing theoretical view is that the spectrum results from cosmic ray-interstellar matter interactions giving a π^0 decay γ-ray spectrum with a possible contribution from inverse Compton scattering of cosmic ray electrons due to an increased photon density from starlight toward the Galactic Center. Stecker (1970, 1973a) has investigated extensively the differential and integral π^0 γ-ray spectrum resulting from cosmic ray interactions in the interstellar gas. The current results are based on the latest available experimental cross section data for the production of π^0 mesons in p-p interactions (see Figure II-9). It should be noted that the resulting γ-ray spectrum depends directly on the energy spectrum and angular distribution of the π^0 mesons in center of mass system of the collision. Stecker (1970) has used a model which assumes there are two production modes for the π^0's observed at accelerator energies which extend to over 10 GeV. The first mode gives a pion component, known as the isobar component, which dominates pion production in collisions where the cosmic ray protons have energies of a few GeV, and the second component is known as the fireball component, which supplies the majority of pions for protons of energies greater than 5 GeV. To be more specific, in the first mode the pions are produced through an isobaric state $\Delta^*(1.238)$ associated with a broad resonance at 1.238 GeV which decays as follows:

$$\text{p} + \text{p} \rightarrow \text{p} + \Delta^* \,(1.238)$$

(*i* process) $\qquad \hookrightarrow \text{p} + \pi^0$

$$\hookrightarrow 2\gamma.$$

The isobaric state is assumed to decay isotropically in its own rest system.

In the second mode it is assumed that after the interaction, there is energy available in the *center of mass system* which creates a thermal pion gas with an energy distribution similar to a Maxwell-Boltzman distribution. This mode is defined by

$$\text{p} + \text{p} \rightarrow \text{p} + \text{p} + \text{f}$$

(*f* process) $\qquad \hookrightarrow \zeta_{\pm}\pi^{\pm} + \zeta_0\pi^0$

$$\hookrightarrow 2\zeta_0\gamma$$

where *f* symbolizes the fireball or pion gas and is not a separate particle. The quantity ζ represents the meson multiplicities. In general, the energy distributions of the π^0 mesons in the observer's system will be different in the two cases leading to different γ-ray spectra. The resulting γ-ray spectra are shown in Figure III-21 for both the *i* component and the *f* component (see Section II-2.4.2). The sum of the two spectra is also shown as the solid curve, and an additional component from π^0 mesons produced in pα, αp, and $\alpha\alpha$ interactions increases the spectrum giving the upper dashed curve. The integral γ-ray spectrum derived (Stecker, 1970) from the differential spectrum of Figure III-21 is in good agreement with a similar spectrum derived by Cavallo and Gould (1971).

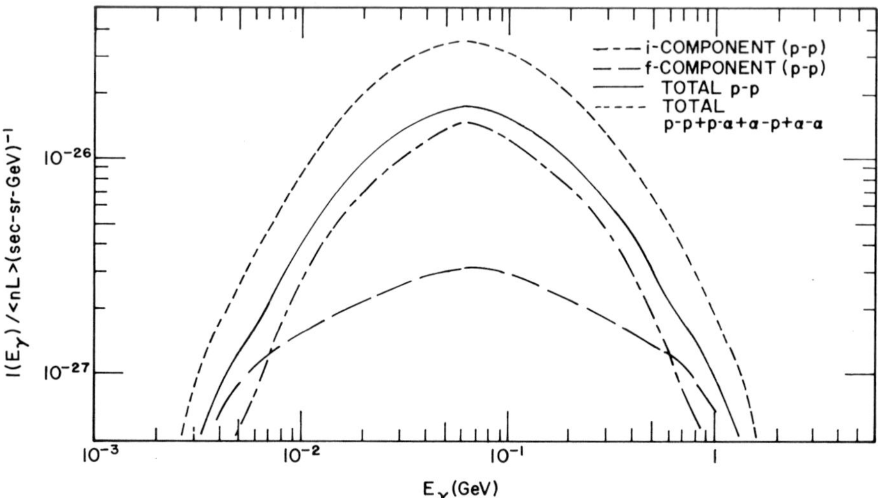

Fig. III-21. The calculated differential production spectrum of γ-rays produced in cosmic ray interactions based on the isobar-plus-fireball model. (From F.W. Stecker: 1970, *Astrophys. Space Sci.* **6**, 377, D. Reidel Publishing Company, by permission.)

TABLE III-10

The γ-ray production rate from the decay of π^0 mesons produced in interstellar pp, pα, and αp cosmic ray interactions as calculated by various workers. (From F.W. Stecker: 1973, *NASA SP–339*, 211.)

Reference	Energy Range	Rate/H Atom ($\times 10^{25}$) s^{-1}
Pollack and Fazio (1963)	> 0 MeV	1.2
Dilworth *et al.* (1968)	> 50 MeV	1.1
Stecker (1970)	> 100 MeV	1.3 ± 0.2
Cavallo and Gould (1971)	> 100 MeV	1.8 ± 0.54
Levy and Goldsmith (1972)	> 100 MeV	3.2
Kraushaar *et al.* (1972)[a]	> 100 MeV	< 1.6
Stecker (1973a)	> 100 MeV	<(1.51 ± 0.23)

[a] Inferred from observations.

Stecker (1973c) has reviewed the results on the total γ-ray production rate, Q_{γ,π^0}, deduced by several authors. This is related to the integral γ-ray flux $I_\gamma\,(>E)$ by:

$$Q_{\gamma,\pi^0}\,(>E) \equiv 4\pi I_\gamma(>E)/\langle nL\rangle \quad (s^{-1}) \tag{III.17}$$

where $\langle nL\rangle$ is the average value for the integrated number of H nuclei cm^{-2} for galactic H gas along the line of sight. The results are shown in Table III-10. Pollack and Fazio (1963) used the observed cosmic ray spectrum near the Earth to calculate Q_{γ,π^0} and Stecker (1970) used a demodulated cosmic ray spectrum, while the value used by Kraushaar *et al.* (1972) is based on their observations on OSO-3 and a neutral H density derived from 21 cm radio observations. The latest value by Stecker (1973a) assumed a maximum demodulated cosmic ray spectrum from Comstock *et al.* (1972). Only the production rate of Levy and Goldsmith (1972) is greatly discordant with the other values. Stecker (1973a) has investigated this discrepancy and finds that the latter value is high because it is based on a theoretical multiplicity law rather than the experimental cross section used by Stecker (1973a). The calculations of Levy and Goldsmith (1972) also lead to a significantly wider differential π^0 γ-ray spectrum than shown in Figure III-21. Presumably, then, it is safe to consider this galactic differential γ-ray spectrum as correct and appropriate throughout the galaxy if the demodulated cosmic ray energy spectrum is correct.

The absolute γ-ray (line) spectrum measured by Kniffen *et al.* (1973) toward the Galactic Center is, however, too high and of the wrong shape to be fully accounted for by π^0 decay alone. Cowsik (1973) has suggested that inverse Compton scattering on starlight photons of mean energy 3 eV can account for 70% of the integral γ-ray flux from the Galactic Center. On the other hand, Stecker *et al.* (1974) have suggested that a model with 70% of the γ-ray flux from π^0 decay with the remainder from a Compton spectrum of slope E_γ^{-1} (for the integral spectrum) fits the observations of Kniffen *et al.* (1973). One of the problems in deciding experimentally which of these alternatives is correct is the fact that the inverse Compton γ-ray spectrum is a power law which is produced by a power law distribution of electrons, whereas the π^0 γ-ray differential spectrum has a broad peak centered at ~70 MeV. An instrument of good energy resolution (≤20%) will be necessary to determine the spectral shape carefully, but it could turn out that, if these two components are of comparable magnitude, experimental resolution of the problem will be difficult.

The γ-ray spectrum for nonthermal electron bremsstrahlung has also been considered as a source for the galactic γ-rays. The form of the γ-ray spectrum due to relativistic bremsstrahlung is nearly the same as the primary electron spectrum interacting with the galactic matter (see Section II-2.4.1). Stecker (1975) gives an approximate form of the γ-ray production spectrum resulting from bremsstrahlung as

$$q_b(E_\gamma) = 4.3 \times 10^{-25}\, n \frac{I_e(>E_\gamma)}{E_\gamma} \quad (\mathrm{cm^{-2}\ s^{-1}\ MeV^{-1}}) \tag{III.18}$$

where n is the number density of nuclei in the interstellar medium which consists of pure H and He, and $I_e(>E_\gamma)$ is the integral electron spectrum (in the medium) in appropriate units.

If the differential electron spectrum is a power law of the form $E_e^{-\Gamma_e}$, then it follows that the spectral shape for the photon energy is a power law with the exponent $\Gamma_\gamma = \Gamma_e$. Cowsik (1973) has described the electron spectrum that should be appropriate for determining the bremsstrahlung contribution to the galactic γ-ray spectrum. Below a few GeV, Cowsik (1973) gives the index for the differential electron spectrum as $\Gamma_e \simeq 1.6$ and above a few GeV as $\Gamma_e \simeq 2.6$. In general then, the shape of the γ-ray spectrum from bremsstrahlung is steeper than the spectrum from the inverse Compton interaction, since the latter gives a power law spectrum with exponent $\Gamma_\gamma = (\Gamma_e + 1)/2$ (see Section II-2.3.1).

As mentioned above, a complete understanding of the high energy galactic γ-ray emission must await new measurements with the best possible energy resolution along with the best calculations of the contribution of the three production mechanisms just discussed.

b. 0.51 MeV *Line Radiation*

Observation of the positron-electron annihilation line from the galaxy would be of great interest since its presence would allow a direct determination of the positron production rate from cosmic rays, which can also be related to the observed flux of cosmic ray positrons and the matter density in the galaxy. It is therefore a good probe for estimating the intensity of galactic cosmic rays, assuming the matter distribution can be determined by other means (e.g. the 21 cm radio emission). Fazio (1967) has reviewed the earlier estimates of the intensity of this line. The flux values ranged from $\sim 10^{-8}$ photons $\mathrm{cm^{-2}\ s^{-1}\ sr^{-1}}$ for the galactic pole (Ginzburg and Syrovatskii, 1964a) to $\sim 5 \times 10^{-4}$ photons $\mathrm{cm^{-2}\ s^{-1}\ sr^{-1}}$ for the Galactic Center (Pollack and Fazio, 1963). These early estimates had not considered the 3-photon positronium annihilation mode discussed in Section II-2.4.4.

Stecker (1971) has reviewed the positronium question in detail and gives the most up-to-date estimate of the two photon annihilation (0.51 MeV) intensity observed along the line of sight. This is expressed as

$$I_{0.51}(l^{\mathrm{II}}, b^{\mathrm{II}}) = \frac{Q_{\mathrm{T,rest}} M(l^{\mathrm{II}}, b^{\mathrm{II}})}{8\pi} \quad (\mathrm{cm}^{-2}\ \mathrm{s}^{-1}\ \mathrm{sr}^{-1}) \tag{III.19}$$

where $Q_{\mathrm{T,rest}}(\mathrm{g}^{-1}\ \mathrm{s}^{-1})$ is the total positron annihilation rate at rest (*Ps* at ~35 keV) and $M(l^{\mathrm{II}}, b^{\mathrm{II}})$ (g cm^{-2}) is the amount of interstellar gas in the direction of the observation. The factor 8π results from multiplying the solid angle by two, since through *Ps* formation and singlet annihilation only one 0.51 MeV photon is produced for every 2 positrons which annihilate. The evaluation of $Q_{\mathrm{T,rest}}$ was carried out by Ramaty *et al.* (1970) using Stecker's (1969) analysis for positron production and annihilation. Positron production from radioactive CNO emitters and π^+ decay mesons was considered for different assumed galactic cosmic ray spectra. Power law and modulated primary electron spectra were also assumed. (The original work of Ramaty *et al.* (1970) should be consulted for details.) In addition, $Q_{\mathrm{T,rest}}$ was calculated for the case of a solar minimum cosmic ray spectrum and also for the assumption that the upper limit positron flux of Cline and Hones (1968) is a true interstellar positron flux. The interstellar mass function $M(l^{\mathrm{II}}, b^{\mathrm{II}})$ was taken from Ginzburg and Syrovatskii (1964b) based on 21 cm observations, which measure only the neutral H contribution. The resulting 0.51 MeV intensity at the Earth is given in Table III-11 (cf. Ramaty *et al.*, 1970; Stecker, 1971). Reference should be made to Stecker (1971) and Ramaty *et al.* (1970) for an explanation of the values of M used for the lower two rows in the table (designated missing mass hypothesis). The results indicate a considerable range for the values of the estimated flux of 0.51 MeV photons.

The question is "what is the lowest intensity capable of being measured?" At the present time there is the report of a line from the Galactic Center with energy just under 0.5 MeV (476 keV) at an intensity of 1.8×10^{-3} photons cm^{-2} s^{-1} (see Section V-5.2.3 for a discussion of this observation). Since the solid angle of the telescope used for this measurement was ~0.38 sr, the above flux value corresponds to an intensity of ~4.7×10^{-3} photon cm^{-2} s^{-1} sr^{-1}. As can be seen from Table III-11, the largest predicted flux is 2×10^{-3} in the same units, so in terms of numbers, positron annihilation could explain this observed line if the most optimistic theoretical estimates are valid. (See Leventhal, 1973a and Section V-5.2.3 for further discussion as to how the energy difference between 511 and 476 keV can be reconciled.)

Stecker (1971) points out, however, that these calculated γ-ray intensities were obtained assuming that the cosmic ray primaries are uniformly spread along the line of sight path where annihilation takes place. Since this may not be true, these estimates may be too high; however, the question is left for experiment to resolve. Certainly the intensity of this line should be highest toward the Galactic Center and a search there with a detector with sufficient energy and spatial resolution could reveal a measurable flux.

c. *γ-Ray Lines from Galactic Cosmic Rays*

The calculations of the expected intensities of nuclear gamma ray lines from cosmic ray interactions in galactic matter were first made by Ginzburg and Syrovatskii (1964a), and Hayakawa *et al.* (1964). For example, Hayakawa *et al.* (1964) had predicted an average

TABLE III-11

Resultant 0.51 MeV γ-ray intensities ($cm^{-2}\ s^{-1}\ sr^{-1}$) for various directions of observation. (From F.W. Stecker: 1971, *NASA SP-249* and Monobook Company, Baltimore, Maryland; also R. Ramaty *et al.*, *J. Geophys. Res.* **75**, 1141, 1970, copyrighted by American Geophysical Union.)

	Solar minimum	$R_r = 350$ MV		Power law		$I_+ = 2 \times 10^{-2}\ cm^{-2}\ s^{-1}\ sr^{-1}$	
		$R_0 = 500$ MV	$R_0 = 200$ MV	$T_c = 100$ MeV/Nucleon	$T_c = 5$ MeV/Nucleon	Mean	Maximum
Average (including halo) (21-cm) $\frac{1}{4\pi}\int dl^{II} \sin b^{II}\, db^{II}\, M(l^{II}, b^{II}) \simeq 1.6 \times 10^{-3}$	7.0×10^{-8}	1.1×10^{-7}	2.8×10^{-6}	2.1×10^{-6}	8.3×10^{-6}	$\frac{1.4 \times 10^{-6}}{n^{5/2}}$	3.2×10^{-5}
Anticenter (21-cm)$M(\pi, 0) \simeq 1.2 \times 10^{-2}$	5.6×10^{-7}	8.6×10^{-7}	2.1×10^{-5}	1.6×10^{-5}	6.2×10^{-5}	$\frac{1.0 \times 10^{-5}}{n^{5/2}}$	2.4×10^{-4}
Galactic center (21-cm)$M(0, 0) \simeq 6 \times 10^{-2}$	2.7×10^{-6}	4.3×10^{-6}	1.0×10^{-4}	7.9×10^{-5}	3.1×10^{-4}	$\frac{5.3 \times 10^{-5}}{n^{5/2}}$	1.2×10^{-3}
Disk average (missing-mass hypothesis) $\frac{1}{2\pi}\int dl^{II}, M(l^{II}, 0) \simeq 3 \times 10^{-2}$	1.4×10^{-6}	2.2×10^{-6}	5.0×10^{-5}	4.0×10^{-5}	1.6×10^{-4}	$\frac{2.7 \times 10^{-5}}{n^{5/2}}$	6.0×10^{-4}
Galactic center (missing-mass hypothesis) $M(0, 0) \simeq 0.1$	4.4×10^{-6}	7.0×10^{-6}	1.7×10^{-4}	1.3×10^{-4}	5.0×10^{-4}	$\frac{8.5 \times 10^{-5}}{n^{5/2}}$	2×10^{-3}

galactic intensity of $\sim 10^{-5}$ cm^{-2} s^{-1} sr^{-1} for the 4.4 and 6.1 MeV lines from $^{12}C^*$ and $^{16}O^*$, respectively, and $\sim 6 \times 10^{-5}$ cm^{-2} s^{-1} sr^{-1} for the 0.5 MeV line intensity.

More recently it has been realized that the heating of H I regions might be a result of higher than expected fluxes of low energy cosmic rays (cf. Dalgarno and McCray, 1972). Also, an excessive flux of low energy cosmic rays was postulated to account for the stellar abandance of Li, Be, and B by Reeves *et al.* (1970). Fowler *et al.* (1970) evaluated the nuclear γ-ray spectrum resulting from cosmic ray matter interactions. This work has now been extended by Meneguzzi and Reeves (1973, 1975), who have considered in detail the kinematic effects on the shape of the γ-ray spectrum. Since these calculations are most illuminating for galactic γ-ray line astronomy, it is of value to review them here briefly. In general, the deexcitation γ-rays can result from two processes: 1) energetic heavy nuclei which have been excited in collisions with an interstellar gas of H and He and lose little energy in the interaction, and 2) heavy nuclei at rest which are excited by cosmic ray protons and α-particles. In the first case the γ-ray lines are strongly Doppler-broadened, and in the second case sharp nuclear lines can be expected, as we have discussed above, except that even in this case the recoil of heavy nuclei is sufficient to produce a measurably broadened line. Meneguzzi and Reeves (1975) have calculated the production rate, $q_1(E_\gamma)$ and $q_2(E_\gamma)$ of γ-ray lines, by a cosmic ray spectrum, ϕ, in essentially the same manner as Ramaty *et al.* (1975) (cf. Section III-3.1 above). $q_1(E_\gamma)$ corresponds to the heavy ion excited while moving, and $q_2(E_\gamma)$ corresponds to the heavy ion excited at rest. They have taken account of the Doppler broadening of the emitted γ-rays by use of a modified cross section for line production. The line width they calculate for the case of C lines is $\sim$80 keV (FWHM) and for Fe lines $\sim$8 keV (FWHM) (Meneguzzi and Reeves, 1975). The γ-ray flux at the Earth is given by $\phi(E_\gamma) = q(E)N_H$ (cm^{-2} s^{-1} sr^{-1} MeV^{-1}) where $q(E_\gamma) = q_1(E_\gamma) + q_2(E_\gamma)$ is the total contribution at E_γ from both processes described above. N_H is the integrated number of H atoms cm^{-2} in the line of sight.

In order to calculate the source functions, $q_2(E)$ and $q_1(E)$, for γ-ray lines from the galaxy, Meneguzzi and Reeves (1975) have used a differential cosmic ray injection spectrum of the form

$$\phi_i \propto k_i E_G^{-2.7} \quad (\text{cm}^{-2}\ \text{s}^{-1}\ (\text{GeV per nucleon})^{-1}) \tag{III.20}$$

where k_i is the cosmic ray abundance of the nuclide i and E_G is its total energy in GeV per nucleon. This spectrum was modified by propagation in the interstellar gas as given by Meneguzzi and Reeves (1975). The γ-ray line production spectrum was determined using cross sections for excitation of ^{12}C, ^{14}N, ^{16}O, ^{20}Ne, ^{24}Mg, ^{28}Si, and ^{56}Fe.

Figure III-22 shows the results of Meneguzzi and Reeves (1975) for the total source function spectrum $q(E_\gamma)$ on the left ordinate. The spectra from both types of collision processes discussed above are clearly shown. The γ-ray flux expected at the Earth may be estimated using an integrated 'equivalent' H density of $N_H = 10^{23}$ atoms cm^{-2} in the direction of the Galactic Center (Meneguzzi and Reeves, 1975). The strongest line flux is $\sim$ 100 times weaker than the minimum apparent background flux as measured on Apollo 15 (Trombka *et al.*, 1973).

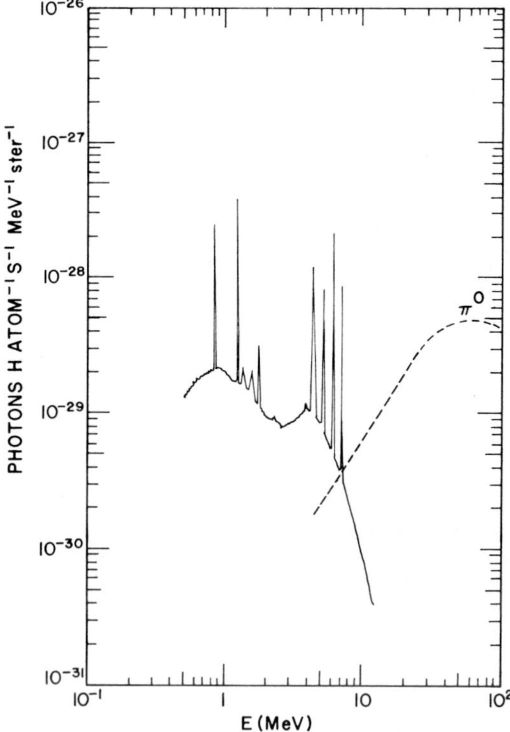

Fig. III-22. Calculated γ-ray source function for the interstellar gas interacting with cosmic rays. (From M. Meneguzzi and H. Reeves: 1975, *Astron. Astrophys.* **40**, 91. Used by permission.)

Meneguzzi and Reeves (1975) have also considered the possibility that the cosmic ray flux for $E_G < 50$ MeV per nucleon is much higher than assumed for Figure III-22. The magnitude of this flux is constrained by the ionization rate of H I regions and the ^{7}Li production rate. Taking these factors into account, the intensity of the line fluxes could be $\gtrsim 10$ times higher than shown in Figure III-22 and, in this case, the lines could be detectable with γ-ray telescopes with adequate energy and spatial resolution.

Fishman and Clayton (1972) have also considered the production of γ-rays from ^{7}Li by low energy cosmic rays in order to explain the ~480 keV emission line from the Galactic Center region discussed in Section V-5.2.3. It was proposed by Fishman and Clayton (1972) that cosmic ray ^{7}Li nuclei, in bombarding the Galactic Center matter, are excited by the process $^7\text{Li}\,(\text{p}, \text{p}')\,^7\text{Li}^{*(478)}$. In the present case, the predicted γ-ray line flux at the Earth is given by

$$F_\gamma \simeq \frac{1}{4\pi} \int_V \int_2^{10} r^{-2}\, n_\text{p} \phi(^7\text{Li}, E) \sigma_{\text{p},\,\text{p}'}(E)\; \text{d}V\; \text{d}E \quad (\text{cm}^{-2}\ \text{s}^{-1}) \tag{III.21}$$

where r is the distance to the Galactic Center, n_p is the Galactic Center proton number density, $\phi(^7\text{Li}, E)$ is the cosmic ray flux of ^{7}Li in the energy region 2 to 10 MeV per nucleon, and $\sigma_{\text{p},\,\text{p}'}(E)$ is the cross section for inelastic scattering in the energy region 2 to 10 MeV per nucleon, which has an average value of $\bar{\sigma} \simeq 150$ mb. The integration is taken over the Galactic Center volume V and the ^{7}Li effective energy range.

In order to explain the intensity of the 473 keV line, reported by Johnson *et al.* (1972) as $F_\gamma \simeq 1.8 \times 10^{-3}$ cm^{-2} s^{-1}, it is found that the average flux of cosmic ray ^{7}Li in the energy band 2 to 10 MeV per nucleon must be ~ 48 cm^{-2} s^{-1}. This flux was found by using a value $N_p = n_p V = 3 \times 10^{66}$ nuclei for the total number of H nuclei in the Galactic Center region. The average flux of cosmic ray protons implied by the above ^{7}Li flux is $\bar{\phi} = 6 \times 10^4$ cm^{-2} s^{-1} in the energy range 2 to 10 MeV per nucleon (Fishman and Clayton, 1972). It has been pointed out by these authors that this appears to be an excessive cosmic ray flux, but that little is known for certain about the flux in this range from cosmic ray observations near the Earth. In fact, they point out that such an apparently high flux of low energy cosmic rays must exist if their interpretation is correct. As a test for their model, they point out that a γ-ray line should also be present at 432 keV from an excited state of ^{7}Be, which would be produced by the reaction ^{7}Li(p, n) ^{7}Be*. The cross section for producing this reaction was assumed to be about one third of the cross section for the reaction ^{7}Li(p, p′) ^{7}Li* in the energy region 2 to 50 MeV, so the 432 line should be of comparable intensity to the 478 keV line. The experiment of Johnson *et al.* (1972) did not have the ability to resolve these two lines; however, the line width due to reaction kinematics, which was not considered by Fishman and Clayton (1972), may blend the two lines together.

Fishman and Clayton (1972) have also pointed out that the low energy cosmic rays should produce positrons from reactions of protons in CNO, which would in turn produce an annihilation line at 511 keV. Such a line is not seen, but the threshold for the positron-producing reactions is higher ($>$ 10 MeV) than for ^{7}Li excitation and the cross sections are also somewhat smaller. Finally, if this model is correct, then other nuclei excitation lines should be seen from ^{12}C and ^{16}O at 4.43 and 6.13 MeV, respectively. These lines also are not seen in the Galactic Center γ-ray spectrum; however, the flux limits are rather high $\sim 10^{-2}$ photons cm^{-2} s^{-1} (see Section V-5.2.3 and Rygg and Fishman, 1973 for more details).

3.2.5. γ-RAYS FROM BLACK HOLES

The possible existence of Black Holes is most strongly supported by the observational data on the time variations of the X-ray source Cygnus X-1 and the characteristics of its optical counterpart (cf. Boldt *et al.*, 1974). It has been remarked that isolated Black Holes may exist and their detection may be impossible; however, the local gravitational effects present in a binary system in which one of the counterparts is a Black Hole may indeed provide the explanation for many time-varying objects in the Universe. It is natural, therefore, to consider what to expect for the γ-ray emission from a Black Hole, since the accretion process expected to be operative at a neutron star (cf. Section III-3.2.2) would be even more dramatic near a Black Hole. The γ-ray luminosity of Black Holes has been considered by Shvartsman (1971), Shapiro (1973), and Dahlbacka *et al.* (1974).

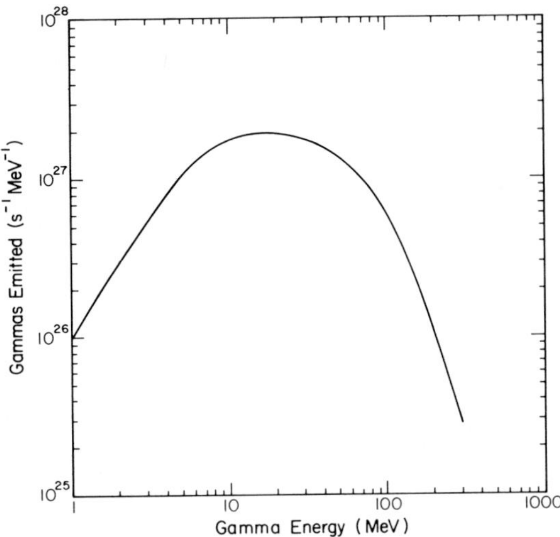

Fig. III-23. Differential γ-ray spectrum from a $10 M_\odot$ Black Hole in a region of density $1\,\mathrm{cm}^{-3}$ and $\theta_0 = 1$ eV. (From G.H. Dahlbacka *et al.*: 1974, *Nature* **250**, 36. Used by permission.)

In the model of Dahlbacka *et al.* (1974), the authors consider Black Holes with mass $M \leqslant 10^4\, M_\odot$ and study the temperature of the infalling matter. If the accreted gas is *not* ionized, then according to Shapiro (1973) the gas will not become hotter than 10^9K (~ 100 keV); however, if the gas is ionized at ~ 1 eV, then Dahlbacka *et al.* (1974) estimate that the temperature for the ions can be > 100 MeV. This is sufficiently hot that meson production can occur, so $\pi^0 \to 2\gamma$ is possible. Dahlbacka *et al.* (1974) suggest that this mechanism provides the main contribution to the γ-ray luminosity of a Black Hole rather than nonthermal electron bremsstrahlung as proposed earlier by Shapiro (1973).

The π^0 γ-ray spectrum expected from a single Black Hole of mass $10\, M_\odot$ has been determined by Dahlbacka *et al.* (1973) and is shown in Figure III-23. π^0 production through the baryon intermediate state Δ^* (1.238) is the predominant production mode for protons of kinetic energy < 4 GeV, so the resulting γ-ray spectrum from the π^0 energy spectrum can be determined by transforming from the Δ^* rest frame to the observer's frame. This requires taking into account the thermal distribution of Δ^* energies and the Doppler shift of infalling gas, as well as the gravitational red shift. The curve shown in Figure III-23 corresponds to an initial density and temperature $n_0 = 1\,\mathrm{cm}^{-3}$ and $\theta_0 \sim 1$ eV, respectively. The shape of the spectrum is independent of the Black Hole mass and the interstellar density n_0. Thus the characteristic broad peak at ~ 18 MeV gives an indentifiable feature for a large number of Black Holes. The width of the broad peak is ~ 80 MeV and falls off as $\sim E^{-3}$ above 100 MeV. The total γ-ray emission is given by

$$\dot{N}_\gamma \simeq 3 \times 10^{26} (M/M_\odot)^3\ (1\,\mathrm{eV}/\theta_0)^3\ n_0^2 \quad (\mathrm{s}^{-1}). \tag{III.22}$$

Dahlbacka *et al.* (1974) have suggested that the apparent diffuse γ-ray background which has a hump at ~ 10 MeV could be due to Black Holes in external galaxies with red-shifted spectra. If the gas density $n_0 \sim 10^2\,\text{cm}^{-3}$, then $\sim 10^{10}$ Black Holes per galaxy of mass $\sim 30\,M_\odot$ would be required to explain the flux at 10 MeV of $\sim 10^{-4}\,\text{cm}^{-2}\,\text{s}^{-1}\,\text{sr}^{-1}\,\text{MeV}^{-1}$ observed by Trombka *et al.* (1973). If our Galactic Center contains Black Holes, then the γ-ray spectrum should peak at ~ 18 MeV as shown in Figure III-23, but apparently only 10^7 30-$M_\odot$ Black Holes would be required to explain the Galactic Center γ-ray flux.

3.2.6. COSMIC γ-RAY BURSTS (VELA CLASS)

In June 1973 came the formal announcement of the discovery of the existence of bursts of cosmic γ-rays. They were measured in the energy region from ~ 100 keV to several hundred keV and lasted for time intervals of 1 to 30 s (Klebesadel *et al.*, 1973). The first observations were made from the nuclear-burst-monitoring satellites called Vela, which formed part of the surveillance network for the nuclear test ban treaty. The discovery, which was completely unexpected, is reviewed in some detail in Section V-5.4. It was readily determined by independent observations on other satellites that the phenomena were genuine and likely to be of cosmic origin, since no known objects in the solar system coincided with the possible source locations.

Unfortunately, all that can be said about the Vela-class bursts at the present time is that they exist and to give a description of some of their signficant properties. Understandably, this situation has given rise to a number of theoretical models which, in the best Aristotelian tradition, exploit every conceivable physical process that could produce γ-rays.

The important properties of the Vela-class events known at the present time (early 1975) are

(a) *Rate of Occurrence* – An average of about 7 to 8 events are observed per year and the energy content of the larger events is $\simeq 10^{-4}\,\text{erg cm}^{-2}$ for photons with energy above 100 keV. By the end of 1974, 30 events had been recorded in 4 yr.

(b) *Spectral Shape* – In cases where it has been determined, the differential photon energy spectrum, dN/dE, is approximately $\propto E^{-1.4}$ for photons below 200 keV and $\propto E^{-2.4}$ above 200 keV. Bursts have been recorded with photon energies as low as 2 keV and above 1 MeV.

(c) *Time Characteristics* – Some bursts are as short as 0.1 s. Six have been recorded with durations of ~ 0.5 s, and one of the largest events, which occurred on 1972, May 14, lasted over 80 s. Substructure in the time profile of some bursts is as short as 10 ms, so the maximum size of the source must be $\lesssim 5000$ km or smaller than the Earth.

(d) *Spatial Distribution* – With the number of events available, there is no clear preference for one source direction over an isotropic spatial distribution. About five events may lie close to the direction of Cygnus X-1.

(e) *Correlated Phenomena* – At the present time, no Vela-class bursts have occurred in coincidence with flare star optical bursts, radio bursts, gravity wave events, or any type of optical events, nor were there any supernovae lying in the possible directions allowed by the observations.

If the location of the burst sources in space is truly isotropic and there is no preference for source directions lying in the galactic plane, either the sources are relatively local, within 100 pc of the Sun, or they are extragalactic at distances $\gtrsim$ 10 Mpc. The energy requirement for a source at a distance of r (pc) giving an integrated flux at the Earth of 10^{-5} erg cm^{-2} is $E_T \sim 10^{33} r^2$ (erg). Thus for local sources at distances of 1 to 10 pc, the energy requirements range from 10^{33} to 10^{35} erg, and for extragalactic sources at distances of 10 to 1000 Mpc, the energy requirements range from 10^{47} to 10^{51} erg.

A test of the possibility that the burst sources are distributed isotropically in space and with a constant density per unit volume can be made by determining the integral size spectrum of bursts. If S is the size of a burst measured in erg cm^{-2} and $N(>S)$ is the number of bursts observed whose size is greater than S, then $N(>S) \propto S^{-3/2}$ (see Harwit, 1973). Thus far there is not sufficient experimental data to test this relation.

Besides the lack of more sensitive instruments required for the above test, the most serious experimental deficiency in this field is probably the lack of precise source positions (to $\lesssim 1'$). Only when such information is available, and hopefully correlated with emissions at other wavelengths, will it be possible to restrict the possible theoretical models available to explain the phenomenon or to lead to a new, correct model. At the present time, it does not seem worthwhile to describe the existing theoretical proposals but a mention of some of the suggested sources is of interest. They include supernovae, stellar-superflares on magnetic white dwarf stars, 'directed' flares on dwarf stars, nuclear flashes on novae, comets impacting on neutron stars, accretion of material onto a neutron star from a giant binary companion, radiation from relativistic dust grains interacting with sunlight, decay of a hypothetical super-massive nucleus or 'goblin', the collapse of a dwarf star to a neutron star or Black Hole following the accumulation of sufficient material by accretion, matter-antimatter collisions, radiation from starquakes (or glitches) on old neutron starts, and so on. Several of these models are discussed in the Proceedings of the Conference on "Transient Cosmic Gamma and X-ray Sources" (Strong, 1974). The supernova model of Colgate was described in Section III-3.2.1 where a discussion of the prompt emission from supernovae was considered.

3.3. Cosmic Diffuse Sources

It should be emphasized again that diffuse γ-ray source, in the present context, refers to an extragalactic radiation which may be isotropic in space and, therefore, similar to the 3 K isotropic microwave radiation associated with the big bang origin of the Universe. The γ-radiation associated with our galactic disk is also sometimes referred to as a component of a general class of diffuse γ-ray sources (see, for example, Cowsik, 1973; Stecker, 1973c, 1975). We have considered separately the experimental evidence for such a galactic γ-ray flux in Section V-5.2 and theoretical models in Section III-3.2.4.

The experimental evidence for an observable extragalactic γ-ray flux in the energy region 100 keV to 100 MeV is discussed in Section V-5.3. It is clear from our discussion that there is considerable uncertainty as to the shape of the γ-ray energy spectrum in this energy region, and therefore it is difficult, if not impossible, to construct theoretical explanations. The current best estimate of the diffuse (extragalactic) γ-ray flux is shown in Figure V-12. Particularly in the energy region from a few hundred keV to 100 MeV, the spectral shape shown should be considered an upper limit. It is clear from this figure and the discussion in Section V-5.3 that, especially between 1 and 50 MeV, there is a significant excess flux over a simple extrapolation of the power law valid at energies below 100 keV (see Section V-5.3). It is the existence of this excess, which gives structure to the spectrum, that has stimulated theoretical interest and which also suggests a cosmological origin. Several of the difficulties that have been encountered in obtaining the experimental data were reviewed by Pal (1973), and a critical review of the complete problem of the diffuse X-and γ-radiation has been given by Silk (1973). The latter review covers the photon energy range from ~ 0.25 keV to 300 GeV, the highest energy where evidence for cosmic γ-rays exists, and should be consulted for the broad astrophysical perspective.

The energy region from 100 keV to 100 MeV is of primary interest here, and because of the uncertainty of the experimental situation mentioned above, we will only make a brief mention of some of the models advanced to explain the data shown in Figure V-12. The discussion by Silk (1973) and the original references should be consulted for details. Brecher and Morrison (1969) have proposed the inverse Compton scattering of relativistic electrons on the microwave photons in intergalactic space as the source. In this model, the electron injection spectrum is taken as that given from the observed synchrotron spectra of radio galaxies. A similar model uses a hypothetical spectrum of electrons produced at rotating white dwarfs (Cowsik, 1971). Clayton and Silk (1969) have proposed that the MeV excess is due to γ-rays from the decay of ^{56}Co and ^{56}Ni produced in silicon-burning nucleosynthesis. Stecker *et al.* (1971a, b) have suggested that π^0 production, either by cosmic rays or matter-antimatter annihilation at an early cosmological epoch ($z \sim 70$), could produce the excess MeV γ-ray flux. The nonthermal bremsstrahlung of electrons in intergalactic space was considered by Arons *et al.* (1971) and thermal bremsstrahlung by Cruddace *et al.* (1972). Leventhal (1973b) has proposed that the apparent excess near 1 MeV in the diffuse γ-ray spectrum (Section V-5.3) is the result of cosmologically red-shifted neutron-proton capture γ-rays originally of energy 2.2 MeV.

It is of value to note, however, that Silk (1973) has pleaded "for a moratorium on any further theories until the observational situation becomes clarified."

CHAPTER IV

INTERACTION OF γ-RAYS WITH MATTER

The previous chapters of this book have dealt with the production mechanisms for γ-rays in the nuclear transition region and the expected intensity of γ-rays (particularly of discrete energy) from various astrophysical sources. In order to discuss the experimental side of this field, it is necessary to review the fundamentals of the interaction of γ-rays with matter. This question is treated in this chapter. Appendix A gives data on gamma ray absorption coefficients in various materials.

4.1. γ-Ray Properties

Classification of the electromagnetic radiation of interest here can be conveniently referred to the ultimate 'source' of the radiation, whether atomic or nuclear. Thus, *X-rays* result from transitions in atoms excited by some process and *γ-rays* result from excited nuclei. Such radiation is usually of discrete energy; however, there are electromagnetic processes, such as bremsstrahlung, in any Coulomb field which give photons of continuous energy as is the case for the usual classical X-rays. In γ-ray and X-ray astronomy the emphasis has been on the apparent continuous nature of the spectra and, therefore, an energy convention has somewhat arbitrarily divided the two fields, reserving the former for photon energies $\lesssim 100 \rightarrow 200$ keV. In the following discussion the 'source' definition will be used, although it is not possible in astronomy to determine, *a priori*, the origin of the detected electromagnetic radiation, unless identifiable line features are evident in the measured spectra. Table II-1 gave a summary of the production mechanisms of EM radiation in the energy range of interest and Table II-2 listed possible astrophysical sites where the production mechanisms can operate. For the broadest appraoch to instrument design, it is necessary to review all possible interaction mechanisms of the cosmic γ-rays.

In considering the detection of γ-rays, we are concerned with the energy spectrum between the X-ray (<10 keV) energies and the more efficiently detected γ-rays of energies above 50 MeV. For our purpose, in most instances, it is sufficient to use the photon picture for the detection of electromagnetic radiation. Here the massless particle, which obeys Bose statistics, is endowed with four properties:

(a) Energy – E_γ

(b) Momentum – E_γ/c

(c) Integral spin angular momentum – $l\hbar$, and

(d) Polarization which corresponds to the orientation of the 'electric vector' of the incident photon.

In any detection scheme which would involve interference phenomena such as Bragg reflection techniques, it is customary to treat the incident radiation as waves with the usual properties of frequency, wavelength, propagation vector, **E** and **B** vectors and polarization, although we will use the photon concept throughout (see Heitler, 1954, pp. 54 and 401).

Fano (1953a, b) has summarized the basic photon interactions, considering the myriad possibilities for the interaction of γ-rays with matter. This scheme, shown in Table IV-1, is based on that given by Davisson (1966). The interactions are grouped into five broad categories, but the detailed description of the kind of interaction involved is shown in column three of the table, while the fourth column gives the historical name of the interaction

4.2. Interaction Processes

In practice it is well known that the bulk of the γ-rays in our energy range of interest, 50 keV → 50 MeV, interacts through only three of the thirteen processes listed in Table IV-1; these are:

(a) the photoelectric effect,
(b) Compton scattering on free electrons, and
(c) pair production in the field of the nucleus.

In experimental design, two other processes are also of interest: Rayleigh scattering (IIa – Table IV-1) and the nuclear photoeffect (III, – Table IV-1). The energy range of principal importance of the interaction is shown in column 4, and the approximate Z dependence of the cross section is given in column 5.

Textbooks and literature abound with the basic theory and applications of the three basic processes, and the most useful are given in Heitler (1954), Evans (1955), and Davisson (1966). Excellent tables and graphs have also been prepared at the National Bureau of Standards by Nelms (1953), Hubbell (1969), and Hayward (1970) which are valuable for experimental work.

4.2.1. PHOTOELECTRIC EFFECT

The basic photoelectric process involves the complete transfer of the incident photon energy to an atomic electron in process I. The resulting electron then has a kinetic energy $E_k = h\nu - \phi$ where ϕ is the binding energy of the particle. A free electron, however, cannot become a photoelectron since a third particle is necessary to conserve momentum. The kinetic energy of the recoiling residual atom or nucleus is usually negligible because of its large mass. Only some of the important features of the atomic photoeffect will be discussed here.

In principle, photoelectrons can be ejected from any of the K, L, M . . . shells of an atom, but the tighter the electron is bound the greater the probability for the process. Therefore, most of the available theoretical and experimental work concerns the K electron. As a rule, one must also take into account the fact that the recoil atom has a vacancy, thus a characteristic X-ray of the atom is emitted along with the photoelectron, or the atom relaxes to a lower state by emitting an Auger electron. This is of particular importance at energies <100 keV; and, in a detector of small size (e.g., a small scintillator), where the K X-ray can escape, the total photon energy is not lost in the detector, except for the amount of energy transferred to the electron. In spectrometers this gives rise to the so-called 'K electron escape peak' shifted down in energy from the full energy peak, for which all the incident photon's energy is contained in the detector.

Important experimental results that have been confirmed generally by theory have been reviewed by Davisson (1966) and Roy and Reed (1968). For detector design, of significance is the strong dependence of the photoelectric cross section on atomic number ($\propto Z^5$), energy dependence of the cross section, and the angular distribution of the photoelectrons.

According to Heitler (1954), the K electron nonrelativistic photoelectric absorption cross section for an incident photon energy $h\nu \ll m_0c^2$ may be expressed as:

$$ {}_a\tau_K = \phi_0 Z^5 \alpha^4 \, 2^{5/2} (m_0c^2/h\nu)^{7/2} \quad (\text{cm}^2 \text{ per atom}) \qquad \text{(IV.1)}$$

where:

$\phi_0 = \frac{8}{3}\pi r_0^2$ is the Thomson cross section in which

$r_0 = e^2/m_0c^2$ is classical electron radius, and

$\alpha = \dfrac{2\pi e^2}{hc} = (137)^{-1}$ is the fine structure constant.

The theoretical treatment for the relativistic case leads to a dependence on photon energy of $(h\nu)^{-1}$ as discussed by Davisson (1966) and Roy and Reed (1968). These basic results are illustrated in Figure IV-1 (Davisson, 1966) where ${}_a\tau/Z^5$ is plotted vs. the reciprocal photon energy. Note that the units on the ordinate are 10^{-32} cm^2 per atom for H to $\sim 4 \times 10^{-23}$ cm^2 per atom for Pb! The curves on the figure all apply to the case of photon energies above the K electron binding energy. Of course, below the K edge only photoelectric absorption by L, M, or higher shells is possible. The above brief description is intended to illustrate the basic ideas; for applications the reader is referred to the most recent tables of Hubbell (1969), which cover the full energy range of interest as well as many materials used in actual detectors, and to Pratt *et al.* (1973) for a comprehensive review.

The distribution in solid angle of the direction of the photoelectron is expected to behave as

$$\sin^2\theta \cos^2\phi = \cos^2\psi \qquad \text{(IV.2)}$$

as given by Heitler (1954) for the nonrelativistic region (see also Roy and Reed, 1968). Here ψ is the angle between the direction of ejection of the electron and the *E vector* of the incident photon, θ is the angle between the emitted electron and the incident photon direction, and ϕ is the angle *between the plane* containing the incident photon and the emitted electron and the *plane* defined by the incident photon direction and its polarization direction.

This behaviour is in good agreement with experiments only for low energy photoelectrons according to Roy and Reed (1968). They have reviewed the agreement between modern theory and experiment on the angular distribution of the photoelectrons and

TABLE IV-1.

Classification of γ-ray interaction mechanisms. (From C.M. Davisson: 1966, in K. Siegbahn (ed.), *Alpha-, Beta-, and Gamma-Ray Spectroscopy* **1.** Used by permission North-Holland Publishing Company, Amsterdam.)

Process		Kind of interaction	Other name	Approximate energy range of maximum importance	Approximate interaction with Z	Present notation	Remarks
I Photoelectric effect	a	With bound atomic electron (all energy given to electron)		Dominates at low E (1 keV to 500 keV) Decreases as E increases	Z^5	τ	
II Scattering from electrons							
Coherent	a	With bound atomic electrons	Rayleigh scattering, Electron resonance scattering	< 1 MeV and greatest at small scattering angles	Z^2 (small angles) Z^3 (large angles)	σ or $\sigma^{(R)}$	Combines coherently with nuclear resonance, nuclear Thomson, and Delbrück scattering
	b	With free electrons	Thomson scattering†	Independent of energy	Z	σ	
Incoherent	c	With bound atomic electrons		< 1 MeV; least at small scattering angles	Z	σ	
	d	With free electrons	Compton effect	Dominates in region of 1 MeV, Decreases as E increases	Z	σ	
III Photonuclear absorption							
Nuclear photoeffect	a	With nucleus as a whole (emits γ or particles)	(γ,γ) (γ,n) (γ,p), etc. processes, Particle production, Nuclear photo-disintegration	Above threshold has broad maximum in range of 10-30 MeV			

TABLE IV-1 (continued)

IV	Nuclear scattering Coherent	a	With material as a whole, Dependent on nuclear energy levels	Mössbauer effect, Nuclear resonance scattering	Important only in very narrow resonance range			Combines coherently with Rayleigh, nuclear Thomson, and Delbrück scattering
		b	With nucleus as a whole, Dependent on nuclear energy levels	Nuclear resonance scattering	Narrow resonance maxima at low energies, Broad maxima in range of 10-30 MeV	Z^2/A^2	σ or $\sigma^{(NR)}$	Combines coherently with Rayleigh, nuclear Thomson, and Delbrück scattering
	Incoherent	c	With nucleus as whole, Independent of nuclear energy levels	Nuclear Thomson scattering	$\lambda \gg$ nuclear radius independent of energy	Z^4/A^2	σ or $\sigma^{(T)}$	Combines coherently with Rayleigh, nuclear Thomson, and Delbrück scattering
		d	With individual nucleons	Nuclear Compton scattering	$\lambda \leqslant$ nuclear radius, i.e. >100 MeV			
V	Interaction with a Coulomb field	a	In Coulomb field of nucleus	Elastic pair production	Threshold about 1 MeV, Dominates at high E (i.e. $E > 5$ or 10 MeV), Increases as E increases	Z^2	κ or κ_{pair}	
	Pair production	b	In Coulomb field of electron	Triplet production, Inelastic pair production	Threshold at 2 MeV, Increases as E increases	Z	$_e\kappa$ or $_e\kappa_{triplet}$	
	Delbrück	c	In Coulomb field of nucleus	Nuclear potential scattering	Real part > imaginary below 3 MeV, < imaginary above 15 MeV, Real and imaginary both increase as E increases	Z^4	σ or $\sigma^{(D)}$	Combines coherently with Rayleigh, nuclear resonance, and nuclear Thomson scattering

†Thomson scattering is the low frequency limit of the Compton effect

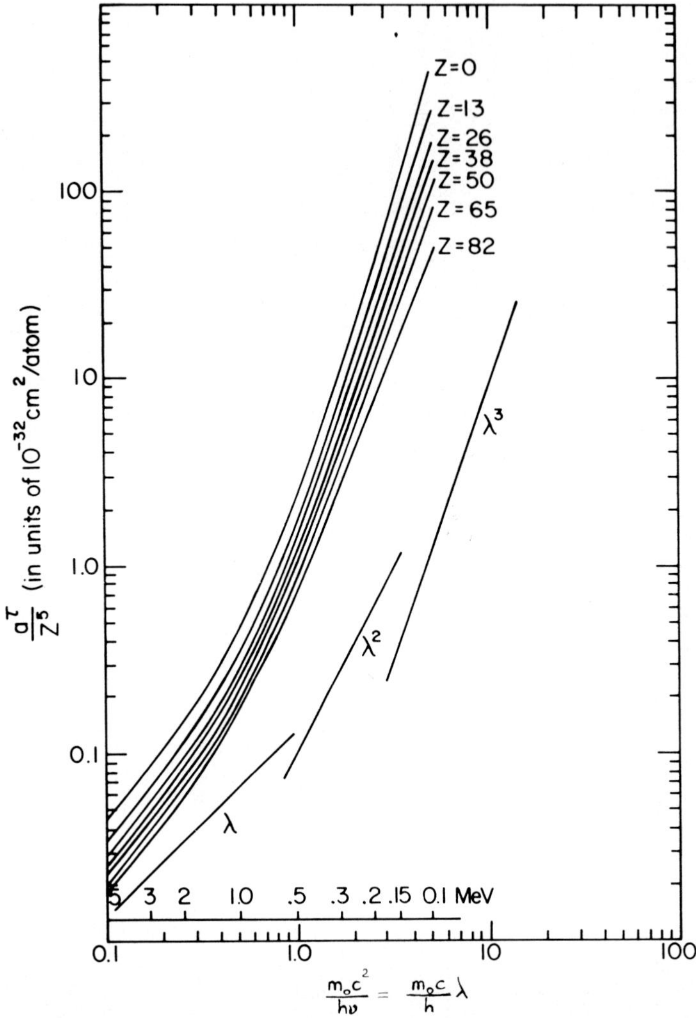

Fig. IV-1. The (photoelectric absorption cross section)/Z^5 is given for several values of Z vs. $m_0c^2/h\nu$, the reciprocal of the initial photon energy. (From C.M. Davisson: 1966, in K. Siegbahn (ed.), *Alpha-, Beta-, and Gamma-Ray Spectroscopy*, **1**. Used by permission of North Holland Publishing Company.)

found it quite incomplete. The basic situation seems to be that, for low energy photons, the photoelectrons seem to be ejected in the direction of the electric vector of the incident photon. At higher energies ($>m_0c^2$) the emission becomes predominantly perpendicular to the electric vector of the incident photon. The main reason for noting this feature of the photoelectric effect here is its possible application in determining the polarization of an incident photon beam by use of a high Z gas target. To study the polarization of photons, another method employed is Compton scattering, which is discussed in the next section.

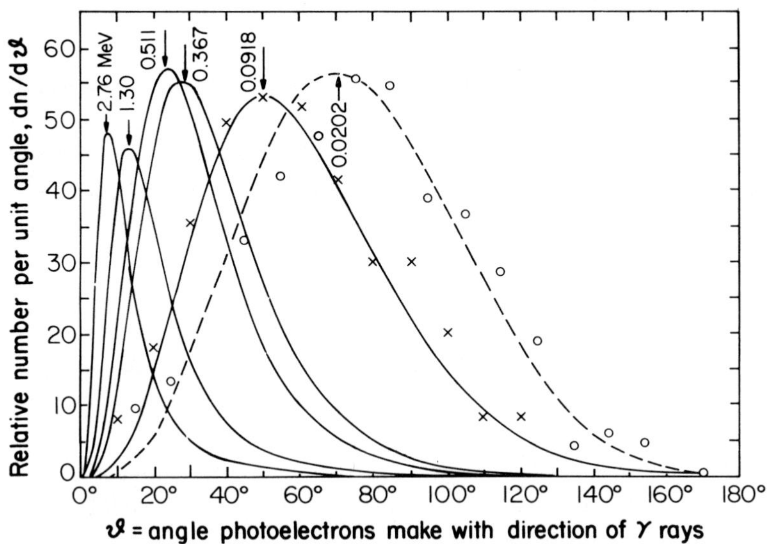

Fig. IV-2. The angular distribution of the photoelectrons emitted for different discrete energy γ-rays vs. the angle of emission of the electron. The measurements were made with a cloud chamber or magnetic lens spectrometer. (From C.M. Davisson and R.D. Evans: 1952, *Revs. Mod. Phys.* **24**, 79. Used by permission.)

In Figure IV-2 the experimentally determined angular distribution of photoelectrons is presented and compared with theory for unpolarized photons. This is taken from Davisson and Evans (1952). The ordinate gives the relative number of photoelectrons between two cones having half-angles of θ and $\theta + d\theta$. The solid curves are the theoretical predictions and the symbols are early experimental results using cloud chambers or magnetic lens spectrometers. At high energies the electrons tend to be emitted at small angles to the incident photon's direction. Later experiments by Roy *et al.* (1955) using 1.17 MeV and 1.33 MeV ^{60}Co γ-rays have confirmed the forward emission at higher photon energies as shown in Figure IV-2; however, the experiments indicate emission at $\theta = 0$ contrary to theory. The angular distribution at high Z does not agree with that shown in Figure IV-2 (see Pratt *et al.*, 1964; Brysk and Zerby, 1968).

4.2.2. THE COMPTON EFFECT

Compton was led to the discovery of incoherent scattering off free electrons while studying the energy of characteristic X-rays (e.g., Mo $K\alpha$) scattered in different directions from various targets (Compton and Allison, 1935). The basic observations gave an intensity peak of scattered radiation in the forward direction, which had the same energy as the incident radiation and was termed coherent radiation. The strongly scattered (incoherent) radiation at different angles, θ, had a wavelength shifted by an amount

$$\lambda_{\text{scatt}} - \lambda_0 = \frac{h}{m_0 c}(1 - \cos\theta). \tag{IV.3}$$

The observations could not be explained by the classical scattering of an electromagnetic wave from free electrons (Thomson scattering), but required a full quantum mechanical treatment. Today we know that the coherent radiation, seen by Compton as mentioned above, was actually Rayleigh scattering from the bound electrons, a collective effect (process IIa – Table IV-1). Thomson scattering is actually the low energy limit of Compton scattering (process IId – Table IV-1).

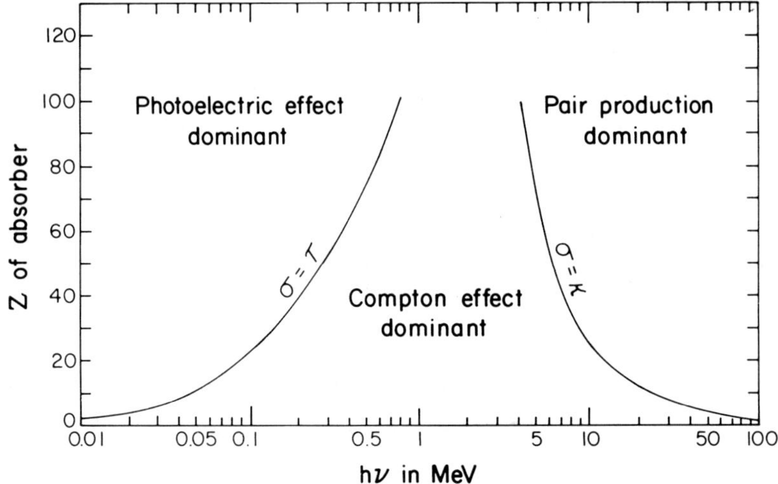

Fig. IV-3. The Z-dependent boundary regions are shown where the photoelectric and Compton cross sections are equal ($\sigma = \tau$) and where the pair production and Compton cross sections are equal ($\sigma = \kappa$) vs. the initial photon energy. (From R.D. Evans, *The Atomic Nucleus,* Copyright 1955, McGraw-Hill Book Company, by permission.)

For our purpose, we need to summarize the salient features of this process, both because its unique properties, which are important in detector design, and because it is the dominant process for absorbing photons in the energy range of interest. Figure IV-3 shows the energy region where the three major types of γ-ray interaction processes dominate for different atomic numbers. It is clear that, for all Z below $Z = 60$, which covers all practical detector materials up to CsI, the Compton effect is most important. Note however that if very high Z materials could be used as scintillators or solid state devices, then the photoelectric and pair production effects would compete strongly with the Compton process. Figures IV-4 and IV-5, taken from Evans (1955), define all the parameters necessary to discuss all aspects of the Compton scattering process needed in our work. In Figure IV-4 a photon of energy $E_{\gamma_0} = h\nu_0$ with an electric

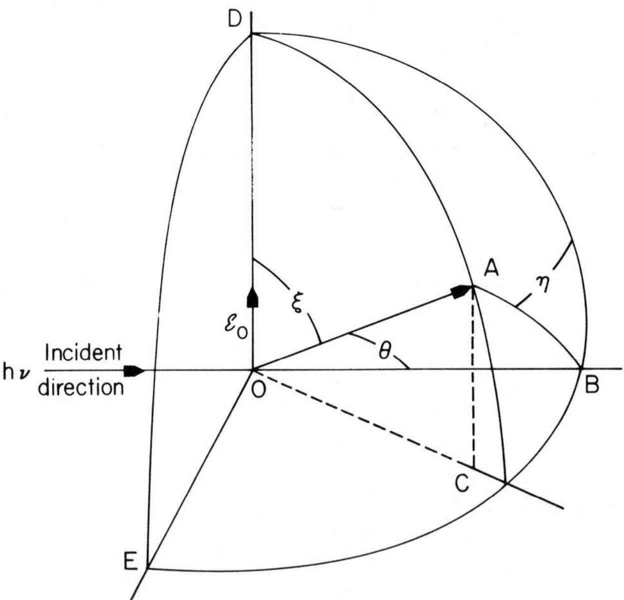

Fig. IV-4. The scattering geometry is shown for a Compton interaction at point O. $\mathcal{E}_0$ is the electric vector of the incident photon, and OA is the direction of the scattered photon at angle θ. (From R.D. Evans, *The Atomic Nucleus,* Copyright 1955, McGraw-Hill Book Company, by permission.)

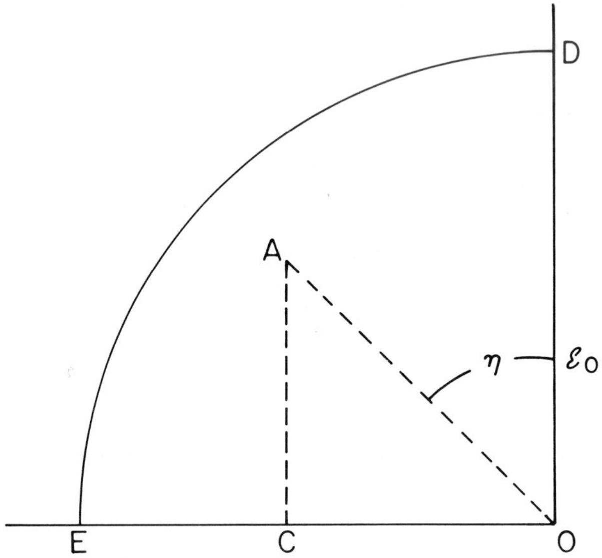

Fig. IV-5. The projection of A and C onto the ODE plane defines the angle η shown in Figure IV-4. (From R.D. Evans, *The Atomic Nucleus*. Copyright 1955, McGraw-Hill Book Company. Used by permission.)

vector $\mathcal{E}_0$ in the direction OD is incident on a free electron at rest. The important quantities in the diagram are defined as follows:

OA = direction of scattered photon energy $h\nu$,
OAB = scattering plane,
θ = scattering angle of photon,
$\mathcal{E}_0$ = electric vector of incident photon,
ξ = angle between $\mathcal{E}_0$ and OA,
ODB = plane of polarization of incident photon,
η = angle between plane of incident polarization ODB and scattering plane,
$\mathcal{E}'$ = electric vector of scattered photon,
β = angle between $\mathcal{E}'$ and ODAC plane,
$\mathcal{E}'_{\parallel} = \mathcal{E}' \cos\beta$ = component of $\mathcal{E}'$ in ODAC plane,
$\mathcal{E}'_{\perp} = \mathcal{E}' \sin\beta$ = component of $\mathcal{E}'$ perpendicular to ODAC plane, and
Θ = angle between $\mathcal{E}_0$ and $\mathcal{E}'$.

The energy of the scattered photon is

$$E_{\gamma'} = h\nu' = \frac{E_{\gamma_0}}{1 + (E_{\gamma_0}/m_0c^2)(1 - \cos\theta)} \tag{IV.4}$$

where θ is angle through which the photon is scattered. The kinetic energy of the scattered electron is

$$K_e = E_{\gamma_0} - E_{\gamma'} = E_{\gamma_0} \frac{2\alpha \cos^2\phi}{(1+\alpha)^2 - \alpha^2 \cos^2\phi} \tag{IV.5}$$

where ϕ is the angle the target electron's momentum vector makes with the direction of the incident photon and $\alpha = h\nu_0/m_0c^2$. The kinetic energy of the electron may also be conveniently expressed as

$$K_e = \frac{E_{\gamma_0}(1 - \cos\theta)}{m_0c^2[1 + (E_{\gamma_0}/m_0c^2)(1 - \cos\theta)]}. \tag{IV.6}$$

The electron's direction lies in the plane containing the directions of the incident and scattered photon or the OBA plane shown in Figure IV-4; however, ϕ is not shown. The relation between the scattering angles for the secondary photon and the electron is

$$\cot\phi = (1+\alpha)\tan\theta/2.$$

In most detection schemes using the Compton process, the angle ϕ cannot be determined, so it is customary to use conservation of energy to obtain the energy of the incident photon, if the energies of the scattered electron and photon are separately measured. It is conceivable, though, that a position-sensitive detector made of a high-Z gas or thin plates could make use of knowledge of the angle ϕ. In Figure IV-5 the angle η is shown in the ODE plane. Tabulations are available giving the kinematical relationships for this process, and the most useful are the NBS tables prepared by Nelms (1953). The maximum energy that can be transferred to an electron by this process is also of considerable importance in detector applications. This, of course, occurs when the photon is scattered through 180° so

$$K_e\,(\text{Max}) \;=\; \frac{E_{\gamma_0}}{1+\frac{1}{2}\alpha}. \tag{IV.7}$$

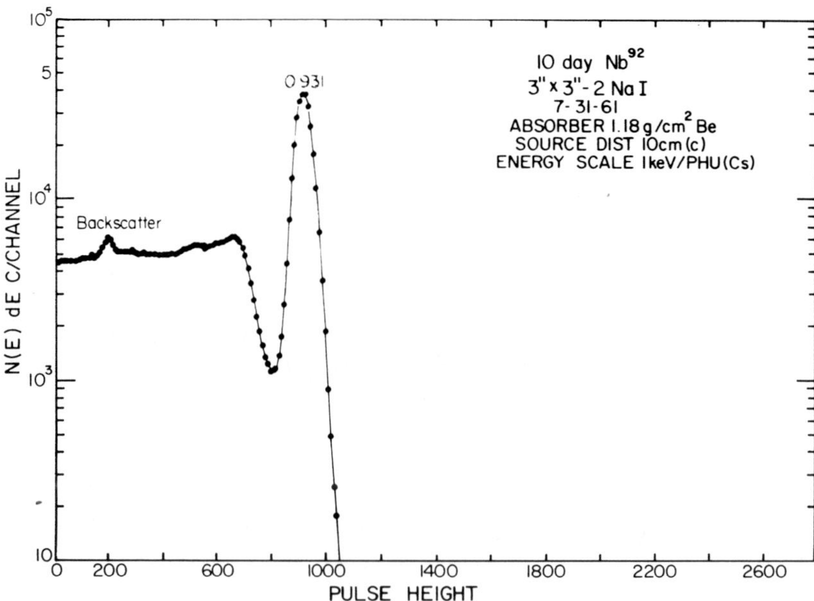

Fig. IV-6. ^{92}Nb (0.931 MeV) γ-rays incident on a standard NaI(Tl) crystal produce a full energy peak at 0.931 MeV, a Compton edge at 0.73 MeV, and a backscatter peak at 0.20 MeV. (From R.L. Heath: 1964, *Scintillation Spectrometry – Gamma-Ray Spectrum Catalogue* **2**, 2nd Edition, AEC Research and Dev. Report TID-4500, Phillips Petroleum Company, Atomic Energy Division, Idaho Operations Office, USAEC.)

In scintillation crystals, for example, this relation explains two effects; namely, the Compton edge and the backscatter peak.

In Figure IV-6 is shown the relative energy loss spectrum for a beam of 0.931 MeV photons incident on a NaI(Tl) scintillation crystal. The so-called 'full energy peak' is due to the photoelectric effect, which occurs when the kinetic energy of the photoelectron, as well as the K X-ray or other deexcitation energy of the excited atom, is fully absorbed by the crystal. The 'Compton edge' produced by the maximum energy transferred in the Compton process occurs at ~0.73 MeV for $E_{\gamma_0} = 0.931$ MeV. The 'backscatter peak' at ~0.20 MeV is due to the primary photons which pass through the crystal with no interaction and scatter through 180° from material behind the crystal returning to the crystal to undergo photoelectric, or full energy, absorption at energy $E_{\gamma_0} - K_e\,(\text{Max})$. Sharper features are obtained with cooled Ge(Li) detectors. For the

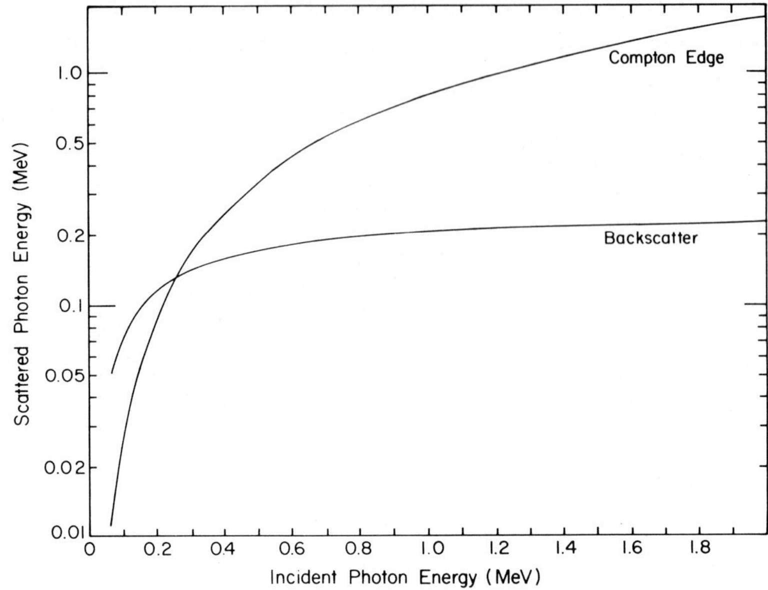

Fig. IV-7. The position of the Compton edge and the minimum backscatter peak is shown as a function of the energy of the incident photon.

reader's convenience we present Figure IV-7, giving the location of the Compton edge and the backscatter peak versus incident photon energy.

a. *Polarized Incident Radiation*

There is a strong possibility that cosmic γ-ray production often leads to polarized photons; therefore, it is important to consider first the probability of Compton scattering of a polarized beam of photons. Of interest, then, is a differential collision cross section for plane-polarized photons incident on free electrons. Figure IV-8 shows the geometry and the definition of the important angles and photon properties for the process following the discussion by Evans (1955). The most fundamental expression which describes the process, is given by Heitler (1954) as *

$$d({}_e\sigma) = \frac{r_0^2}{4}\, d\Omega \left(\frac{\nu'}{\nu_0}\right)^2 \left(\frac{\nu_0}{\nu'} + \frac{\nu'}{\nu_0} - 2 + 4\cos^2\Theta\right) \quad (\text{cm}^2 \text{ per electron}) \tag{IV.8}$$

where the quantities have all been previously defined. The quantity Θ is of particular interest, since it is the angle between the electric vectors of the incident and scattered photons. Physically, then, $d({}_e\sigma)$ is the effective area which a single free electron presents to a photon of initial energy $h\nu_0$, such that the photon is scattered into a solid angle $d\Omega = 2\pi \sin\theta \; d\theta$ with a new energy $h\nu'$ and with the electric vector rotated by an angle Θ. From this basic relation all of the useful 'practical' cross sections and

* This is known as the *Klein-Nishina* formula when expressed as $d({}_e\sigma)/d\Omega$ cm^{-2} sr^{-1} per electron (Klein and Nishina, 1929).

absorption coefficients have been derived and are summarized by Heitler (1954), Evans (1955), Davisson (1966), Marmier and Sheldon (1969) and others. For detector applications, this is clearly the most fundamental Compton cross section of interest, since it directly determines the number of photons scattered into dΩ with specific properties $h\nu'$, Θ, and polarization, as long as each electron in a scatterer acts independently of the others.

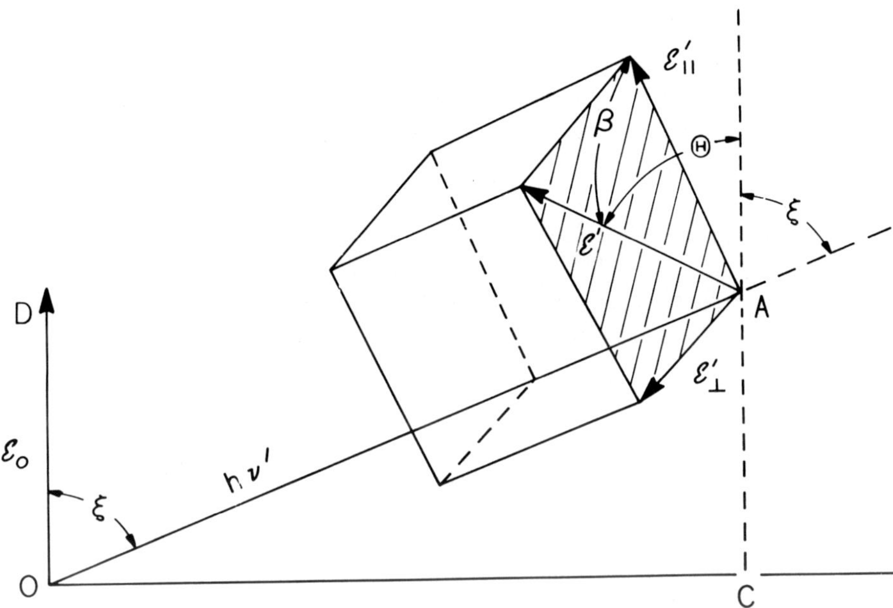

Fig. IV-8. The Compton interaction is shown in the ODAC plane (see Figure IV-4), giving the polarization of the scattered photon as shown by the electric vector $\mathcal{E}'$, which makes an angle Θ with the direction of $\mathcal{E}_0$ and an angle β with the ODAC plane. (From R.D. Evans, *The Atomic Nucleus.* Copyright 1955, McGraw-Hill Book Company. Used by permission.)

In practical situations where one detects the scattered electron and the scattered photon, the polarization of the latter is not significant as a rule. Therefore, it is customary to sum the above cross section over all possible directions of polarization of the scattered photon, while dependence on the polarization of the incident photon is retained. The result (expressed in a form of most interest here) for Compton scattering at an angle θ is:

$$d({}_{e}\sigma) = \frac{r_0^2}{2}\, d\Omega \left(\frac{\nu'}{\nu_0}\right)^2 \left(\frac{\nu_0}{\nu'} + \frac{\nu'}{\nu_0} - 2\sin^2\theta\cos^2\eta\right) \quad (\text{cm}^2 \text{ per electron}) \tag{IV.9}$$

where η is the projection of the angle ξ in Figure IV-4 onto the plane normal to the incident photon's direction and is shown also in Figure IV-5.

This relation demonstrates that, for a completely polarized incident photon beam, the scattering probability will be maximum when $\eta = 90°$. Therefore, the photon (and electron) tend to be scattered into a plane normal to the incident electric vector.

This expression is applied in the design of polarimeters as described by Metzger and Deutsch (1950) for laboratory studies of polarized γ-rays emitted from radioactive nuclei. For X-ray astronomy, Novick *et al.* (1972) have used the properties of this equation to determine the polarization of the X-rays from the Crab Nebula with a passive scatterer in the Thomson limit when $\nu' = \nu_0$.

In constructing a polarimeter, a scattering angle θ must be specified, and the variation of scattered intensity with the azimuthal angle η investigated. For a given value of θ, Equation (IV-9) shows that for $\eta = 0$ the scattered intensity will be at a minimum, and for $\eta = 90°$ it will be at a maximum. The asymmetry ratio between maximum and minimum scattered intensities is:

$$R(\text{Max/Min}) = \frac{(\nu_0/\nu' + \nu'/\nu_0)}{(\nu_0/\nu' + \nu'/\nu_0 - \sin^2\theta)} \tag{IV.10}$$

and depends on both ν_0 and θ. In order to maximize R, it is found that θ should be $\lesssim 90°$, but the exact value is strictly energy dependent. Figure IV-9 shows a plot of R vs. θ for a range of values of incident photon energies.

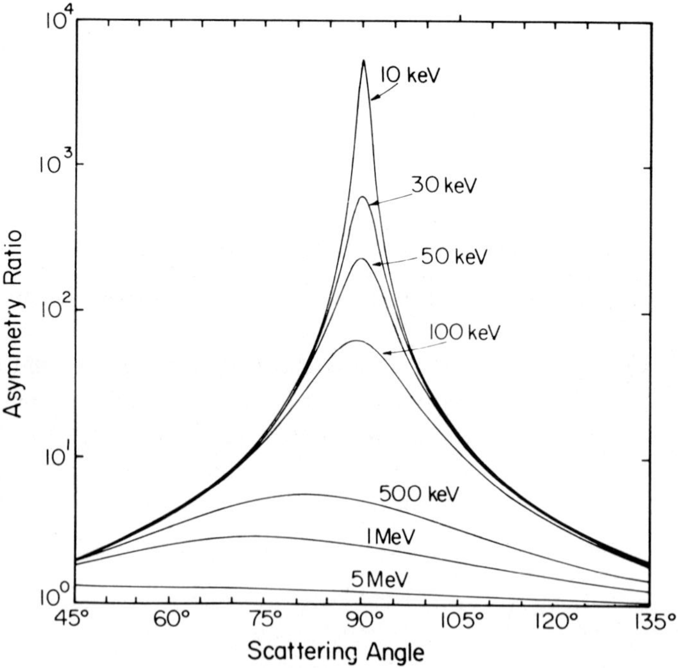

Fig. IV-9. The ratio of Compton scattering of 100% polarized photons into a plane normal to the incident electric vector $\mathcal{E}_0$ to those scattering into a plane parallel to $\mathcal{E}_0$.

An important application of this property to astronomy would be the measurement of polarization of γ-rays from the Sun or from other cosmic sources. Leaving aside for the moment the fact that the flux of such photons is very low (only $\sim 10^{-1}$ photons $cm^{-2} s^{-1}$ were detected from the Sun during the August 1972 flares), we determine from Figure IV-8 that the asymmetry ratio is $R \sim 5$ for a perfectly polarized incident 500 keV photon beam, if the polar scattering angle θ is set at 82°. Of course, this large a value of R would not be achieved in practice because of the finite size of detectors, which span a range of θ and η. As yet, no workable Compton polarimeter has been developed for the γ-ray energy range in astronomy, although such a device is clearly feasible.

b. *Unpolarized Incident Radiation*

The case of unpolarized radiation has received more theoretical attention than the polarized case. The differential collision cross section for unpolarized radiation (Heitler, 1954; Evans, 1955) is

$$d({}_e\sigma)/d\Omega = r_0^2/2(\nu'/\nu_0)^2 (\nu_0/\nu' + \nu'/\nu_0 - \sin^2\theta) \quad (\text{cm}^2\ \text{sr}^{-1}\ \text{per electron}) \qquad \text{(IV.11)}$$

which has the same physical interpretation as Equation (IV-9) except here the incident radiation is unpolarized. This form is the most useful in detector design unless polarization is important. A polar plot of $d({}_e\sigma)/d\Omega$ in Figure IV-10, taken from Davisson and Evans (1952), gives the cross section (probability) per electron in the target for a single unpolarized photon to scatter into a solid angle $d\Omega$ at θ. This is given for different values of α, which is the incident photon energy in units of electron rest energy. Two significant aspects of the Compton process as illustrated in Figure IV-10 are:

(a) The increasing possibility that the scattered photon is emitted in the forward direction as the incident photon energy, $\alpha = h\nu_0/m_0c^2$, is increased; and

(b) The fact that at very low energies the cross section actually approaches the classical Thomson cross section, which has the angular distribution $\propto (1 + \cos^2\theta)$.

This latter behavior is practically attained when the photon energy exceeds the electron binding energy, but is small compared to $m_0c^2 = 511$ keV, for example ~ 25 keV. It should be noted that the units in Figure IV-10 are cm^2 per (electron-solid angle), so in a practical application one must multiply Equation (IV-11) by $d\Omega = 2\pi \sin\theta\ d\theta$ and integrate over the appropriate limits in θ.

The number distribution for scattering into an angle $d\theta$ at θ may also be easily found from Equation (IV-11).

$$d({}_e\sigma)/d\theta = d({}_e\sigma)/d\Omega \cdot d\Omega/d\theta = 2\pi \sin\theta\,[d({}_e\sigma)/d\Omega]. \qquad \text{(IV.12)}$$

Therefore the number of photons scattered into $d\theta$ at $\theta = 0°$ is zero.

It is appropriate to make a distinction between the differential collision cross section $d({}_e\sigma)$, so far discussed, and the differential (energy) *scattering* cross section $d({}_e\sigma_s)$. The former is often tabulated or published in the form of graphs (e.g. Nelms, 1953) as the Klein-Nishina cross section for unpolarized radiation. The latter quantity, $d({}_e\sigma_s)$, is actually proportional to the energy scattered into $d\Omega$ at θ and is related to the former by

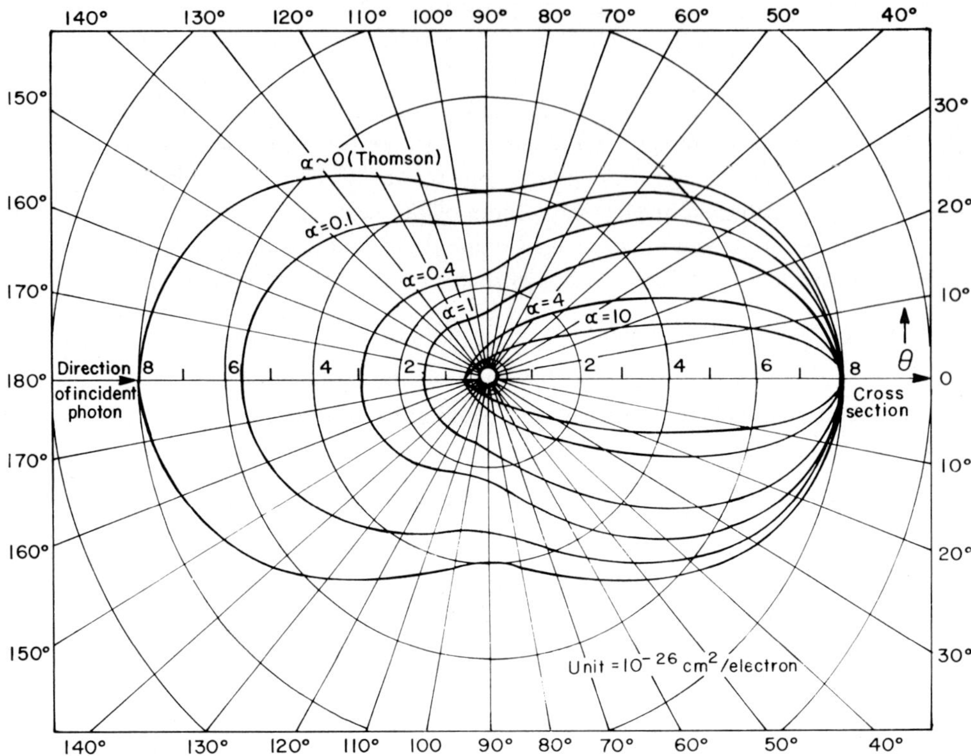

Fig. IV-10. The probability (in cm^2/unit solid angle) of scattering an unpolarized photon into a solid angle $d\Omega$ at θ for different incident photon energies (α) in units of electron rest energy. (From C.M. Davisson and R.D. Evans: 1952, *Revs. Mod. Phys.* **24**, 79. Used by permission.)

$$d({}_{e}\sigma_{s}) = \nu'/\nu_0 \; d({}_{e}\sigma).$$

Which one is used depends on the application, but caution should be exercised with published tables and graphs. See Davisson (1966) for a further discussion.

In Figure IV-11 the differential energy distribution is shown for the Compton electrons produced for several intial photon energies from 0.5 MeV to 3.5 MeV. The units are expressed as a probability (10^{-27} cm^2 keV^{-1} per electron). This curve should be compared with Figure IV-6. In the real world, of course, the electrons involved in Compton scattering are seldom free or at rest, a circumstance which introduces many complications into a situation which seems naively simple. If the photon energy is sufficiently high compared to the binding energy of electrons, then the 'atomic' cross section $d({}_{a}\sigma) = Zd({}_{e}\sigma)$.

More realistically, however, there are two cases when there are many electrons present in the atom. If the atomic scattering is 'coherent' then $h\nu' = h\nu_0$ (unmodified radiation) and scattering *amplitudes* are added from each electron before finding the scattered *intensity*. If $h\nu' < h\nu_0$ (modified radiation), then the scattering is 'incoherent' and scattered intensities from each electron are combined linearly. The total atomic differential Compton cross section is then

$$d({}_a\sigma) = d({}_a\sigma_{\mathbf{coh}}) + d({}_a\sigma_{\mathbf{incoh}}). \tag{IV.13}$$

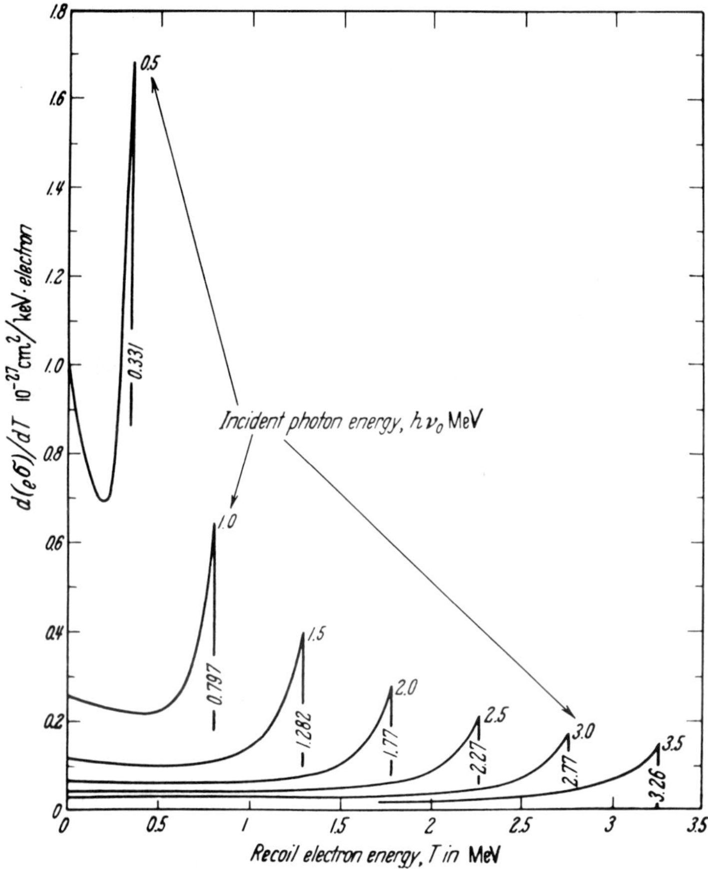

Fig. IV-11. Number-vs.-energy distribution of Compton recoil electrons, for 7 values of the incident photon energy $h\nu_0$, in 10^{-27} cm^2 (millibarn)/keV, per free electron. The energy spectrum of scattered photons is obtained by transforming the energy scale from T to $h\nu_0 - T$ for each curve. (From R. Evans: 1958, in S. Flugge (ed.), *Handbuch der Physik Vol. XXXIV*, p. 268, Springer-Verlag, Berlin. Used by permission.)

The magnitude of some of these effects can be seen in Figure IV-12 from Motz and Missoni (1961) for the case of 662 keV photons on Au. The curve (a), labeled K electron-coherent (dot-dash curve), represents the case when the momentum transfer to the inner shell electrons is sufficiently small that the electron reemits the incident photon. The separate amplitudes in this case are first added and then the differential scattering cross section per electron is found. In the curve (b), labeled K electron-incoherent, the

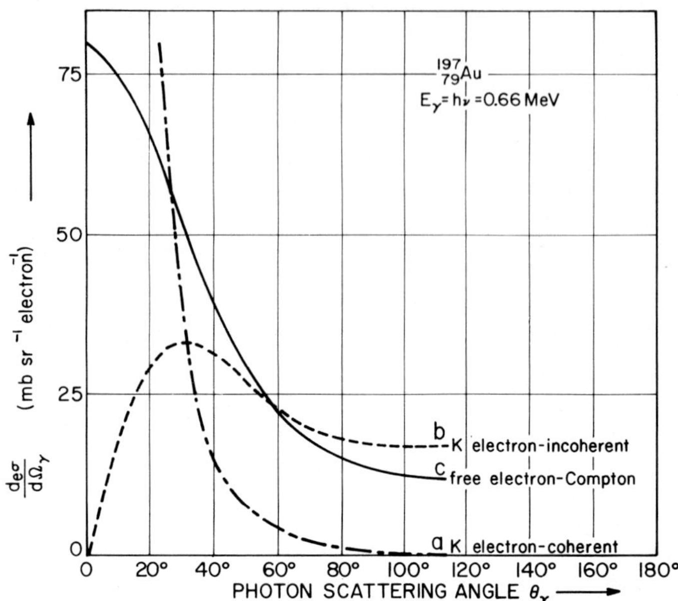

Fig. IV-12. Comparison of the angular distribution for scattering of 0.66 MeV photons on Au: (a) Coherent scattering from electrons in the K-shell; (b) The experimental results for incoherent K-electron scattering; (c) Compton scattering on free electrons using the Klein-Nishina formula for unpolarized radiation. (From J. W. Motz and G. Missoni: 1961, *Phys. Rev.* **124**, 1458. Used by permission.)

momentum transfer to the bond is too small to consider the electron free for angles less than or $\sim 40°$. This dashed curve is from experimental data and shows the actual radiation in the expected true Compton scattering. The effect is dramatic as can be seen by comparing the dotted curve with the solid curve (c), which is the theoretical free-Compton differential scattering cross section as given by equations for unpolarized radiation. This reduction in the true Compton scattering is more than compensated at small angles by the coherent or Rayleigh scattering from all the electrons of the atom. Detailed discussions of many corrections to Compton scattering, important in applications, are given by Nelms (1953) and Hubbell (1969).

4.2.3. PAIR PRODUCTION

The third major interaction process, which applies only to photons with energy exceeding 1.022 MeV ($2m_0c^2$), is the creation of electron pairs in the Coulomb field of a nucleus with the disappearance of the photon. The threshold for pair production in the field of an electron is 2.044 MeV ($4m_0c^2$) and is always less probable than the former. In general, the threshold kinetic energy for pair production is given by

$$E_{\text{th}}(\text{Pair}) = 2m_0c^2(1 + m_0/M) \qquad \text{(IV.14)}$$

where M is the mass of the Coulomb charge whose presence is required for the interaction, as well as for taking up the excess momentum, and m_0 is the electron mass.

Normally the electron and positron do not emerge with equal energies, so it is necessary to know the probability distribution of the electron and positron energies. This was first calculated by Bethe and Heitler (1934) using the first order Born approximation and later by Maximon and Bethe (1952) and Bethe and Maximon (1954) without using the Born approximation. The latter theory gives a result for the differential cross section $d\sigma_p/dE_+$ for producing a positron with total energy E_+ to $E_+ + dE_+$ in a single pair production process in the nuclear Coulomb field, neglecting the screening effect of atomic electrons. The exact expression is very complex, so we refer the reader to a recent review by Roy and Reed (1968) or the original papers mentioned above. We include here, however, the graphical results given by Bethe and Ashkin (1953). The basic cross section may be expressed in the form (Marmier and Sheldon, 1969)

$$\sigma_p(E_+) \equiv d\sigma_p/dE_+ = \frac{\bar{\sigma}}{h\nu - 2m_0c^2} f(\epsilon, Z) \quad (\text{cm}^2\ \text{MeV}^{-1}\ \text{per atom}) \tag{IV.15}$$

where

$$\bar{\sigma} \equiv \sigma_0 = \alpha r_0^2 = 5.794 \times 10^{-28}\ (\text{cm}^2)$$

and

$$\epsilon = k/\mu \equiv h\nu/m_0c^2\ .$$

Figures IV-13a and IV-13b taken from Bethe and Ashkin (1952) give the quantity

$$[\sigma_p(E_+)/\bar{\sigma}] \times (h\nu - 2m_0c^2) \equiv f(\epsilon, Z) \equiv [\phi(E_+)/\bar{\phi}]\ (k - 2\mu) \tag{IV.16}$$

plotted vs.

$$(E_+ - m_0c^2)/(h\nu - 2m_0c^2) \equiv (E_+ - \mu)/(k - 2\mu)$$

which is the fraction of the total kinetic energy that the positron receives, for several values of the incident photon energy. The electron, of course, has a total energy $E_- = h\nu - E_+$ neglecting the recoil energy of the nucleus. This cross section is symmetrical with respect to the interchange of the positron and electron. This is only approximately true, however, since the electron is attracted to the nuclear charge and the positron is repelled. This results in a non-symmetric distribution with a lower net energy being received by the electron. The effect is most pronounced when the kinetic energy of the pair is of the order of the K-shell electron binding energy (Bethe and Ashkin, 1952). The total atomic pair production cross section to this approximation is simply the area under a given curve or

$${}_a\sigma_p = \int \sigma_p(E_+)\, dE_+\ .$$

Marmier and Sheldon (1969) summarize useful forms for this total cross section as

$${}_a\sigma_p = \bar{\sigma} Z^2 \left(\frac{28}{9} \log \frac{2h\nu}{m_0c^2} - \frac{218}{27} \right) \quad (\text{cm}^2\ \text{per atom}) \tag{IV.17}$$

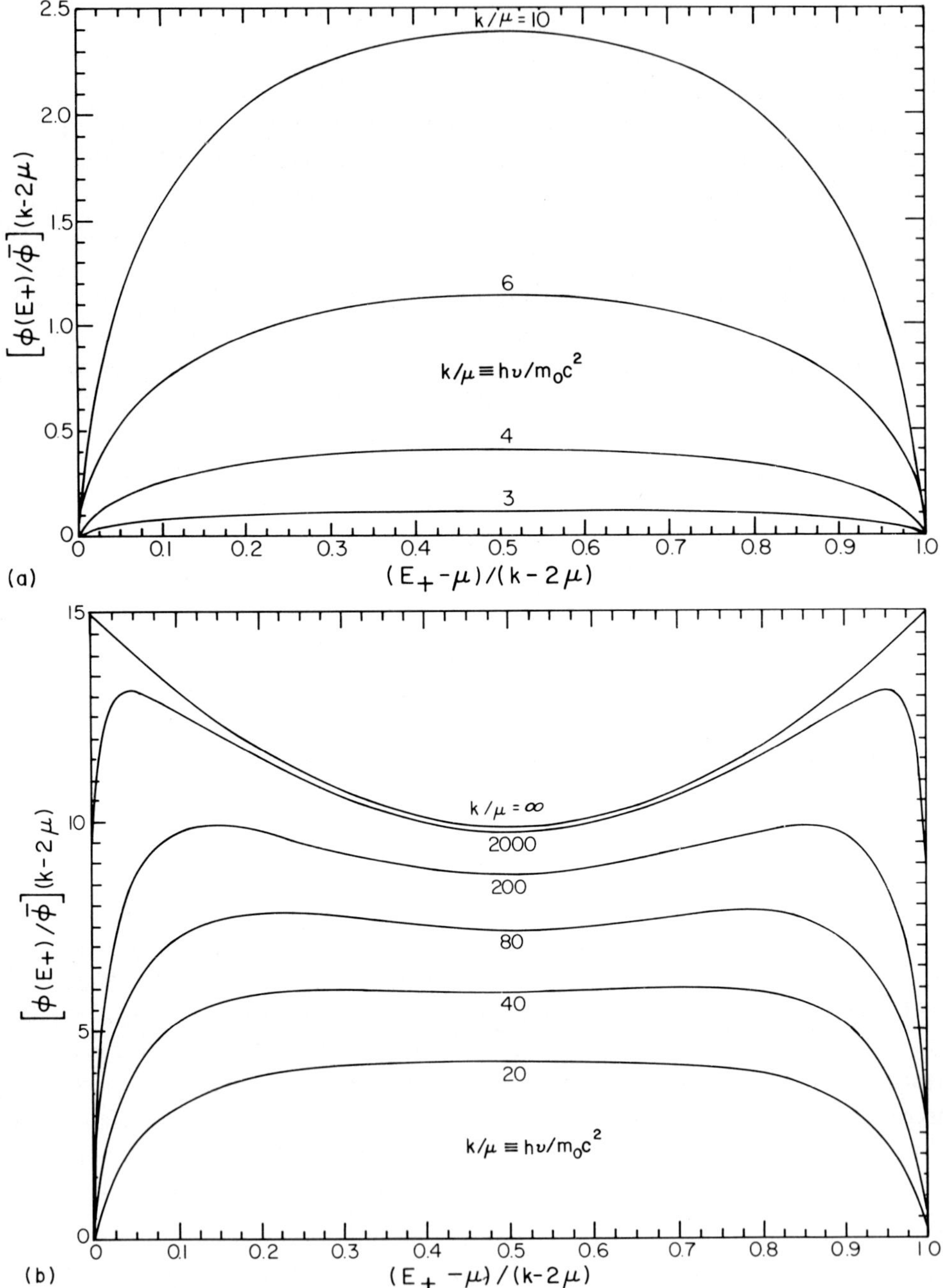

Fig. IV-13a, b. Energy distribution of the electron in an electron pair. Abcissa: kinetic energy of one electron, divided by available energy $h\nu - 2m_0c^2$. Ordinate: cross section. The scale of the ordinate is so chosen that the area under each curve gives the total cross section in units of $\phi = Z(Z+1) r_0^2/137$. The number on each curve indicates the γ-ray energy in units of $m_0c^2 \equiv \mu$. (From H.A. Bethe and J. Ashkin: 1953, in E. Segré (ed.), *Experimental Nuclear Physics I,* John Wiley & Sons, by permission.)

for no screening when $1 \ll h\nu/m_0c^2 \ll 1/\alpha Z^{1/3}$; for complete screening when $h\nu/m_0c^2 \gg 1/\alpha Z^{1/3}$,

$$_a\sigma_p = \bar{\sigma} Z^2 \left(\frac{28}{9} \log \frac{183}{Z^{1/3}} - \frac{2}{27} \right) \quad (\text{cm}^2 \text{ per atom}). \tag{IV.18}$$

Note that this expression is independent of the photon energy. In intermediate energy regions, an approximate form for the total cross section is

$$_a\sigma_p = \bar{\sigma} Z^2 \left(28/9 \log(2h\nu/m_0c^2) - 218/27 - 1.027\right) \quad (\text{cm}^2 \text{ per atom}).$$

Figure IV-14 shows an example of the agreement between the Bethe-Maximon theory and experimental results for pair production on Pb (cf. Davies *et al.*, 1954). The earlier Born approximation results are somewhat higher at all energies.

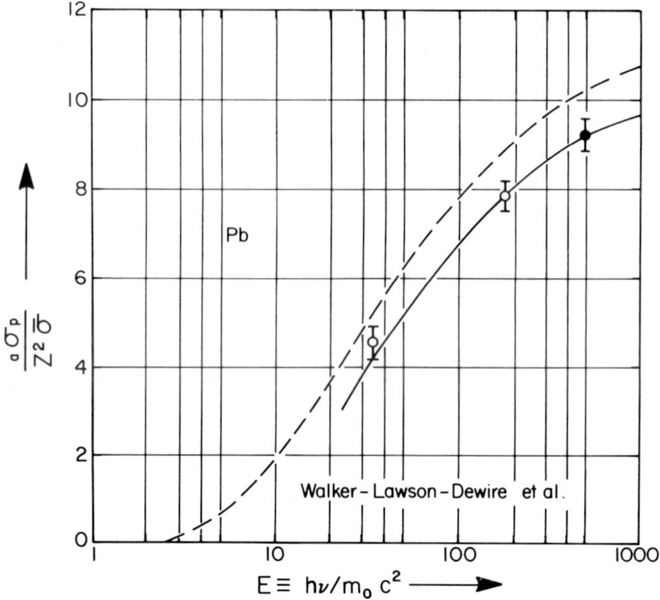

Fig. IV-14. Comparison of experimental results with theoretical curves derived from the Born approximation (dashed curve) and the Bethe-Maximon theory (solid curve) for the pair production cross section on Pb. (From H. Davies, H.A. Bethe, and L.C. Maximon: 1954, *Phys. Rev.* **93**, 788 and as modified by P. Marmier and E. Sheldon, *Physics of Nuclei and Particles*, Vol. 1, 1969. Used by permission of Academic Press Inc., New York.)

In detectors such as spark chambers where both of the pair electrons are observed, their angle of separation measured from the point of materialization can be used to obtain a measure of the incident photon's energy. Sandhu *et al.* (1962) have compared their experimental observations of the distribution in angle of divergence between the pairs,

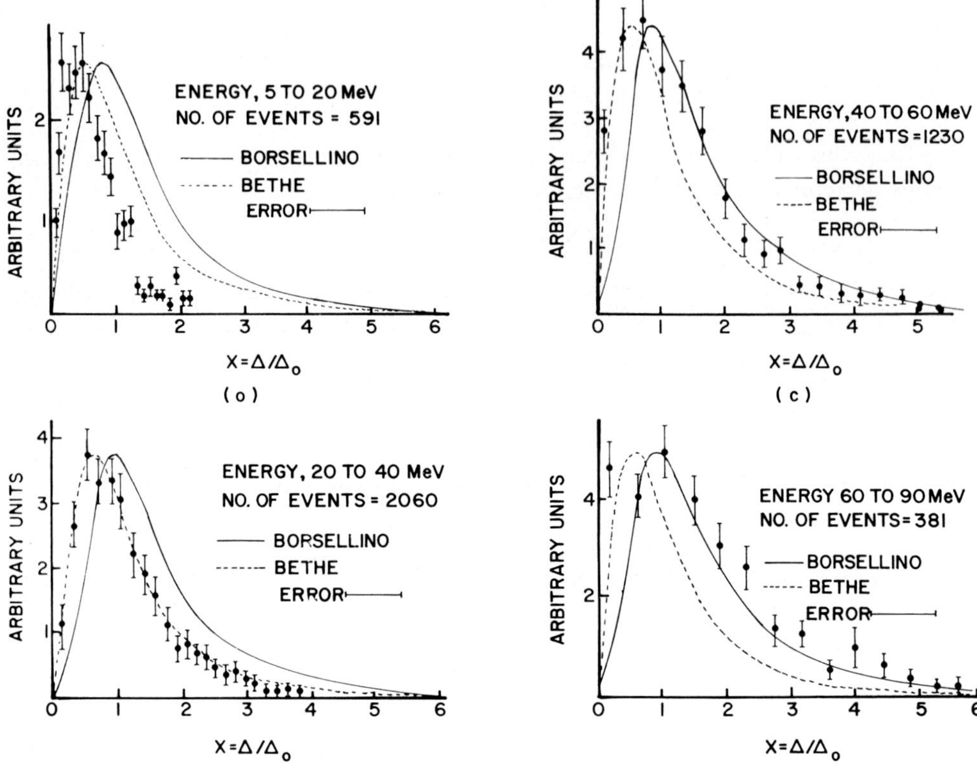

Fig. IV-15. Distribution of the opening angle Δ between the two partners of the pairs at different energy intervals. The abscissa represents the ratio $X = \Delta/\Delta_0$ (From H.S. Sandhu, E.H. Webb, R.C. Mohanty, and R.R. Roy: 1962, *Phys. Rev.* **125**, 1017. Used by permission.)

Δ, with the theoretical calculations of Bethe (1934) and Borsellino (1953). The results are shown in Figure IV-15 plotted vs. Δ/Δ_0, where $\Delta_0 = 4E_0/h\nu$ is the angle of divergence of the pair if they share the photon energy equally, and E_0 is the rest energy of the electron. At the lower energies, 5 to 20 MeV and 20 to 40 MeV, the experimental data seems to agree better with that of the Bethe (1934) distribution. This shows a very significant deviation from unity for the expected value of Δ/Δ_0 at these low energies, thus a smaller angle is expected than is given by Δ_0. Hence, one would estimate the photon energy as too high, if the pair angle was used for a measure of energy. This is important in determining the spectrum of γ-rays in the energy range below 50 MeV, since the effect is to give too many high energy photons and make the spectrum flatter.

In practice, with scintillation counters or solid state detectors, the pair production interaction can give rise to three peaks in a γ-ray spectrum. Figure IV-16 is an energy loss spectrum in a NaI(T1) crystal resulting from an incident 2.41 MeV γ-ray. The full energy peak at 2.41 MeV corresponds to absorption of the total kinetic energy of both electrons of the pair as well as the energy of the two 0.511 MeV annihilation photons

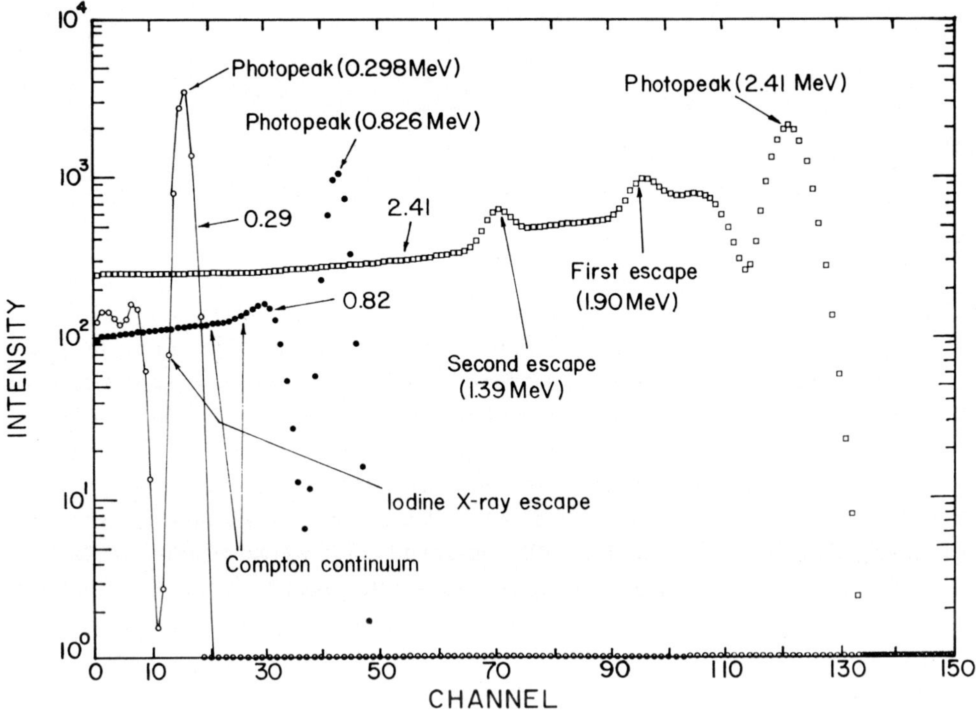

Fig. IV-16. The full energy and escape peaks are shown for 2.41 MeV photons on NaI (Tl). Spectral features produced by incident 0.826 MeV and 0.298 MeV photons are also shown.

resulting when the positron of the pair annihilates releasing the rest energy of two electrons. In case one or both of these annihilation photons escapes the crystal, a first or second escape peak is produced, reduced in energy loss by 0.511 MeV or 1.022 MeV, respectively. This effect is more pronounced the smaller the crystal. Also, for comparison, the spectral features produced are shown when photons below the pair production theshold impinge on the scintillator crystals. Similar but sharper features are seen when using Ge(Li) detectors (see also Figure IV-6).

CHAPTER V

γ-RAY FLUX OBSERVATIONS

As mentioned in the Introduction, there are now five pieces of evidence for positive γ-ray observations in the nuclear transition region, but only two reports of γ-ray lines: the Galactic Center feature and the solar flare lines. Even though our interest here is the observation of γ-ray lines, there is an intimate relation between line and continuum fluxes. For example, the galactic γ-rays > 30 MeV may be a mixture of π^0 γ's and Compton scattered photons. We therefore review briefly the current significant results in γ-ray observations above ~ 50 keV that we feel bear most significantly on γ-ray line astronomy. (It should be pointed out that current terminology would refer to the lower part of this energy range as hard X-rays.) We do not, however, review here the great amount of experimental work that has been done in studying the so-called hard X-ray sources, for which the reader should refer to reviews by Peterson (1973), Laros (1973), Gruber (1974), Peterson (1975), and the current literature. The prime reason for this omission is that the experiments did not emphasize the measurement of discrete γ-ray lines since the detectors have inherently low energy resolution. We therefore will describe, in order, the results of experiments to detect γ-rays from the Sun (Section V-5.1) and point or localized cosmic sources (Section V-5.2), including the galactic disk. Evidence for a celestial diffuse source of γ-rays from beyond our galaxy is discussed in Section V-5.3 and the apparently isotropic Vela γ-ray bursts are described in Section V-5.4.

5.1. Solar Observations

The experimental investigations to detect solar neutrons and γ-rays which were reported in the literature by 1970 were reviewed previously by Chupp (1971). Up to that time there was no conclusive evidence for either solar neutron or γ-ray fluxes. On the other hand, there were at least three highly disputed claims of observations of both solar neutrons and γ-rays, all in times of modest or low solar activity. None of these 'possible' events occurred in coincidence with the optical phase of any flare. Nonetheless, since they are published as positive fluxes, we should keep the reports in mind and the conditions of solar activity under which they were observed. The Tata result of Apparao *et al.* (1966) was obtained under very quiet solar conditions; that of Daniel *et al.* (1967) was made several hours before a subflare. The latter result was seriously questioned by Holt (1967), since no neutron decay protons were seen by the OGO-A satellite which was in orbit at the time and should have seen them if the neutron flux was 10^{-1} neutrons cm^{-2} s^{-1} as reported. This criticism has been countered by Daniel *et al.* (1971) who have revised their result downward nearly an order of magnitude to 1.5×10^{-2} neutrons $cm^{-2} s^{-1}$ based on a new measurement of the atmospheric neutron flux, which allowed them to convert the measured solar neutron counting rates to an absolute flux. In the case of γ-rays, Kondo and Nagase (1969) reported an extremely large (800%) increase in the γ-ray flux (3 to 10 MeV) 10 min after a 1N flare and associated radio burst. The last positive report of a solar γ-ray increase was given by Hirasima *et al.* (1969), who reported a γ-ray line flux coincident with a 1000 MHz radio burst. Very recently, Koga *et al.* (1974) have reported evidence for a γ-ray burst on 1972, May 14 at 2012.47 UT which had similar characteristics to the two just mentioned. This event did not coincide with any major solar event nor with the cosmic γ-ray burst on the same date (Section V-5.4).

As satellite experiments in the future continue to search for γ-ray and neutron events, it will be interesting to see if any enhancements are found under similar activity conditions as in the cases just discussed. Then we can decide if indeed these peculiar observations are truly of solar origin. Table V-1 summarizes the results just discussed on solar neutrons and γ-rays. Recent work and several other experiments are treated in more detail by Chupp *et al.* (1973a).

5.1.1. OSO-7 γ-RAY OBSERVATIONS IN AUGUST 1972

The current evidence for γ-rays associated with solar flares was obtained during the August 4 and 7 events with the University of New Hampshire (UNH) γ-ray spectrometer on the OSO-7 satellite (Chupp *et al.*, 1973a, b, 1974a). Only a brief description of the OSO-7 instrument will be given here since this has been described thoroughly elsewhere (e.g. Higbie *et al.*, 1972; Forrest *et al.*, 1972; Chupp *et al.*, 1973b). Figure V-1 shows a schematic of the detector and its arrangement in the rotating wheel compartment of the OSO-7 spacecraft. The γ-ray monitor (Higbie *et al.*, 1972), designed to measure the γ-ray spectrum from 0.3 to 10 MeV, consists of a 3″ × 3″ NaI(Tl) crystal surrounded by a cup-shaped anticoincidence shield of CsI(Na). This arrangement also enhances the relative sensitivity for lines by suppressing the Compton continuum.

The instrument is automatically calibrated each orbit at day/night and night/day transitions with a ^{60}Co source giving photopeaks at 1.17, 1.33 and 2.50 MeV. The calibration source (Forrest *et al.*, 1972) consists of a neutron-activated Co compound in a plastic scintillator button which is viewed by a photomultiplier. The beta decay of the ^{60}Co provides a gating pulse which is used to prevent the coincident γ-rays from ^{60}Ni from being recorded during a normal γ-ray spectrum accumulation. At each day/night or night/day transition of the spacecraft, a signal automatically puts the ^{60}Co beta pulse in coincidence with a signal in the main γ-ray detector, which is due in this case primarily to an absorption of one of the γ-rays at 1.17 MeV or 1.33 MeV. Occasionally, both γ-rays are simultaneously absorbed, giving a sum peak at 2.50 MeV. During normal operation, when the calibration lines are gated out of the spectrum, there is still a small leakage of ^{60}Co lines, ~5%, into their respective channels. This scheme allows the energy

TABLE V-1

Summary of recent solar neutron and γ-ray observations before August 1972. (From E. L. Chupp *et al.*: 1973a, *NASA SP-342*, 285.)

Reference	Instrument	Radiation and energy range	Neutron or γ-ray flux ($cm^{-2}\ s^{-1}$)		
			Continuous	Flare associated	
Joseph thesis (1970)	Plastic scintillator telescope with antishield	n 15 to 150 MeV γ 5 to 30 MeV		2B, 1N $n < 1.2 \times 10^{-2}$ $\gamma < 10^{-2}$	1969, February 26
Cortellessa *et al.* (1971)	Plastic scintillator with antishield	n 10 to 200 MeV	$< 5.5 \times 10^{-3}$	1N Protons $< 6 \times 10^{32}$ above 30 MeV $P_0 = 60$ MV	1970, June 30
Eyles *et al.* (1972)	Scintillator recoil telescope with antishield	(i) n 50 to 350 MeV γ > 80 MeV	$n < 3 \times 10^{-3}$ $\gamma < 4 \times 10^{-4}$	(i) 1B $n < 23\ cm^{-2}$ $\gamma < 6\ cm^{-2}$	1969, May 29
		(ii) Test of Elliot model		(ii) 2B $n < 4.2 \times 10^{-3}$ Theo. $\sim 5 \times 10^{-2}$ for Class 4	1965, March 21
Sood (1972)	Cerenkov pair telescope with Pb converter	γ > 20 MeV		1N $\gamma < 2.6 \times 10^{-4}$ (10 min rise) $< 1.1 \times 10^{-5}$ (active disc)	1968, October 17
Lockwood *et al.* (1973) OGO-6	He^3 proportional counter with antishield	n 1 to 20 MeV	$< 1.8 \times 10^{-3}$	Null results	
Leavitt *et al.* (1972) OSO-6	Scintillator recoil telescope	n > 40 MeV	$< 4 \times 10^{-4}$	Null results	

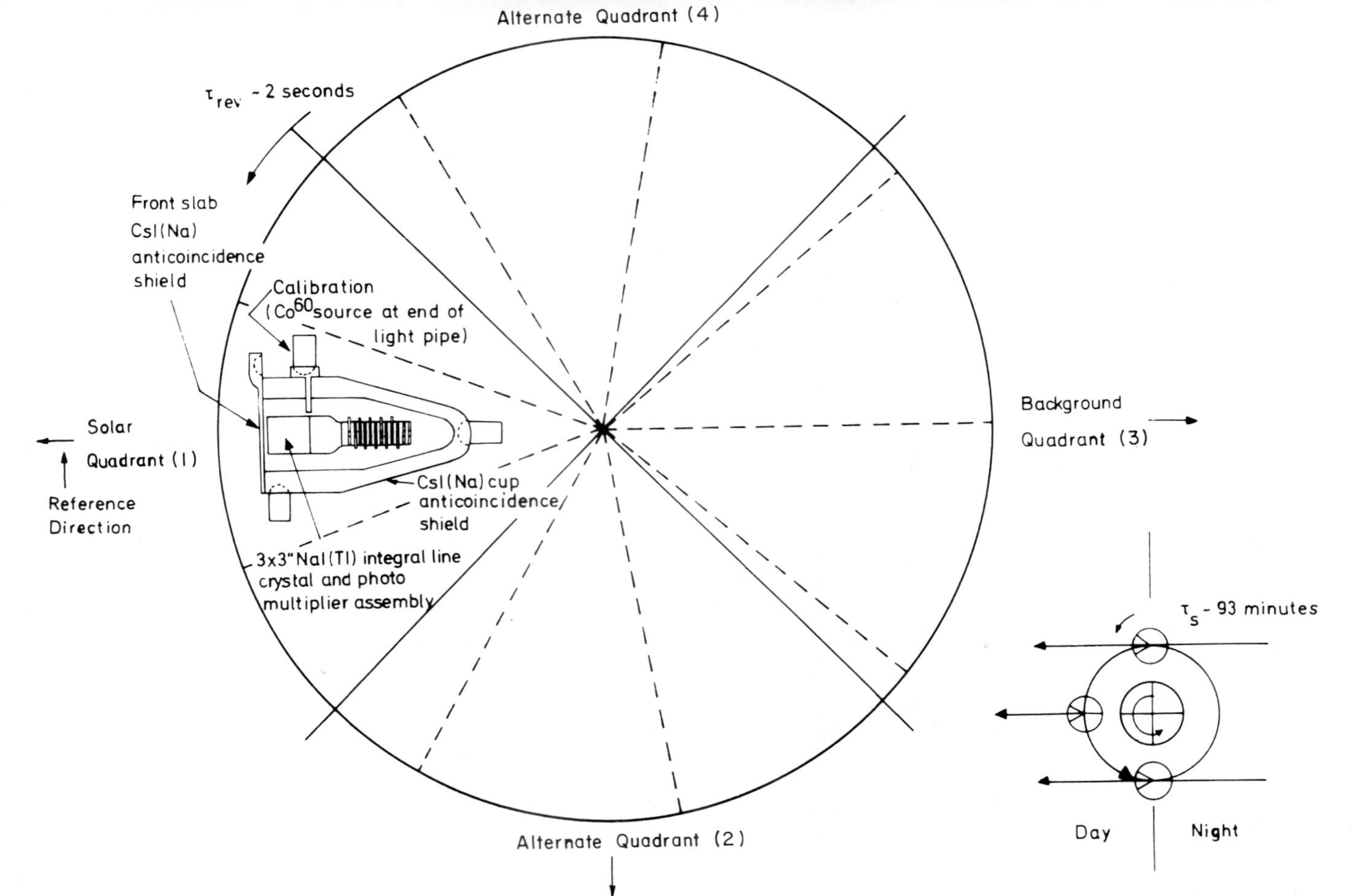

Fig. V-1. A schematic diagram showing the OSO-7 γ-ray spectrometer and its operation in orbit. The inset at the right shows the relative aspect of solar and background quadrants at satellite sunrise, noon, and sunset. (From E.L. Chupp *et al.*: 1973, *Nature* **241**, 333. Used by permission.)

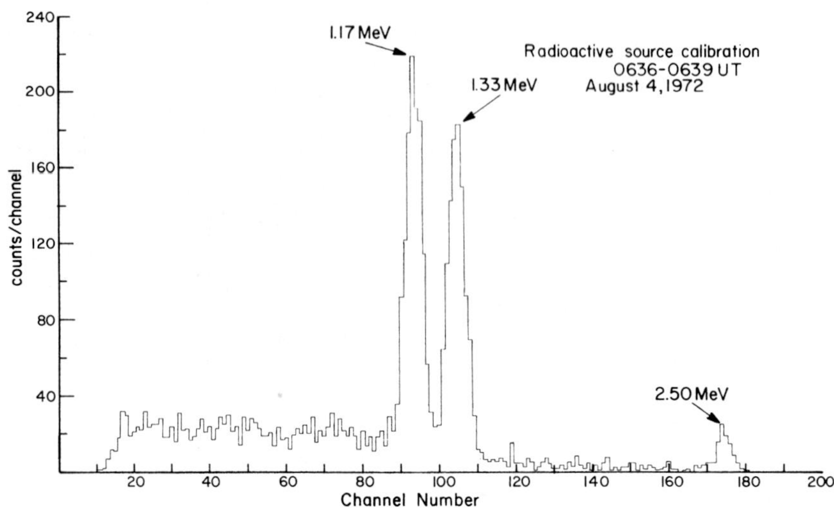

Fig. V-2. Typical inflight calibration spectrum, taken ~ 5 min after the 1972, August 4 3B flare.

of a well-defined peak to be determined to about 2% and verifies that the energy resolution is maintained at ~8% (referred to 662 keV). The energy resolution improves with energy approximately as $E^{-1/2}$. Figure V-2 shows a typical inflight calibration spectrum. The three calibration lines mentioned above are clearly shown. The γ-ray monitor has an auxiliary X-ray detector (not shown in Figure V-1) consisting of a NaI(Tl) crystal of diameter $1\frac{1}{4}''$ and thickness $\frac{1}{4}''$, covered by a 10 mil thick Be foil. The X-ray energy spectrum from 7.5 to 120 keV is measured by a four-channel analyzer. A complete X-ray spectrum is taken every 30 s.

During the normal mode of operation (see Figure V-1), γ-ray pulse height spectra are accumulated over 90 wheel rotations (taking approximately 3 min) in both solar and background quadrants. The γ-ray spectrum is analyzed with a 377 channel quadratic pulse height analyzer and stored in a buffer until readout. The solar and background directions do not change during the nighttime part of the orbit so the background quadrant looks at the sky during the night. A command allows the viewing quadrants to be shifted ninety degrees (90°) counterclockwise as shown in Figure V-1.

On 1972, August 4 γ-rays were observed during the beginning phase of the 0621 UT flare in close time association with a radio burst, an X-ray burst, and the Hα flash. In this case a ground level cosmic ray event was observed, delayed by ~8 h from the flare. On 1972, August 7 γ-rays were observed after the maximum phase of the 1500 UT flare just after the satellite emerged into daylight.

a. *The August 4 Event*

The flare activity on August 4 started with a precursor flare in the X-ray band (0.5 to 3 Å) at 0507 UT (Dere *et al.*, 1973). The precursor activity continued for about an hour until 0610 UT when the main flare started in the X-ray band, 7.5 to 15 keV, as recorded by the UNH X-ray detector. The main optical flare started in Hα at ~0617 UT (Solar Geophysical Data UAG-21, 1973). Before the OSO-7 satellite was eclipsed by the Earth at 0633.8 UT, excess γ-ray line and continuum emission was recorded by the UNH γ-ray detector. Strong radio emission also accompanied this event (Castelli *et al.*, 1973; Croom and Harris, 1973). There is very good correlation between the γ-ray continuum observed on OSO-7 and the impulsive radio emission (Suri *et al.*, 1975). The onset as well as the time of maximum are the same within a minute.

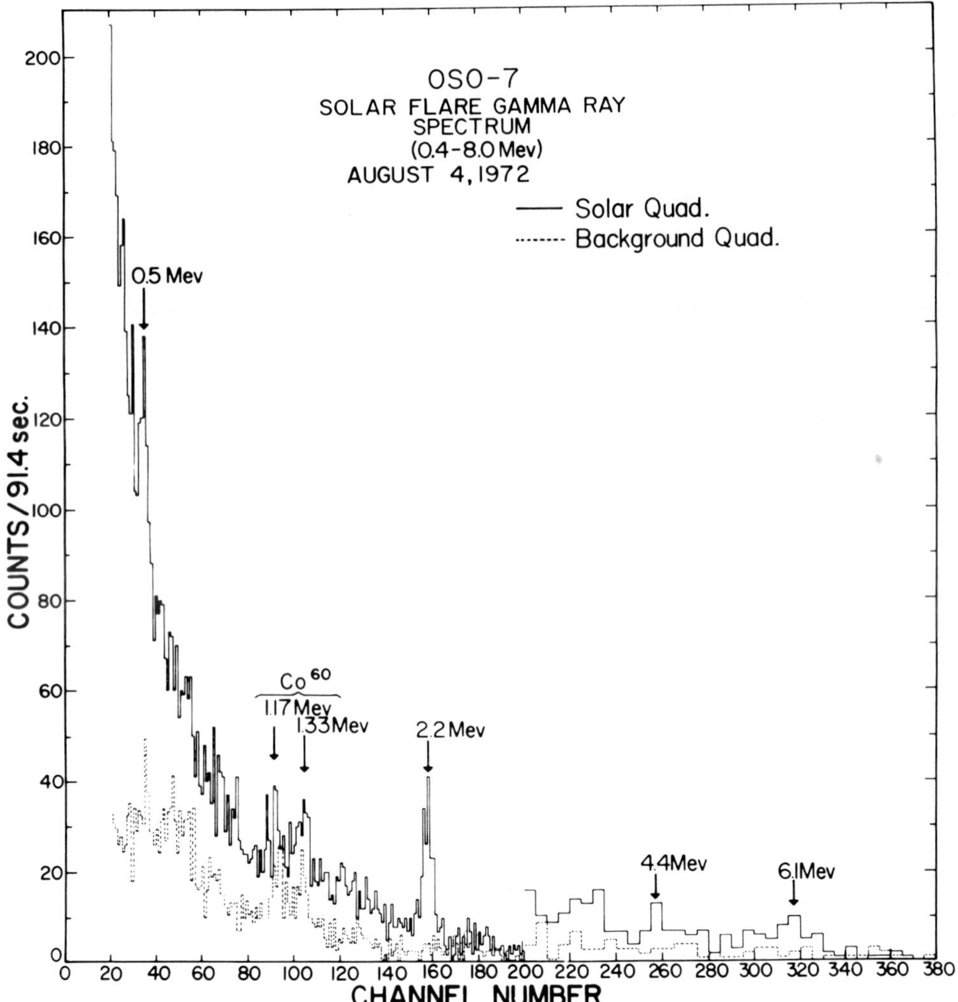

Fig. V-3. Solar quadrant γ-ray spectrum (solid histogram) obtained during the rising phase of the 1972, August 4 3B flare (0623:49 to 0633:02 UT). The simultaneously acquired background spectrum is also shown (dotted histogram). Ordinate: Solar quadrant total counts per channel in 91.4 s livetime. Background counts normalized to 91.4 s. Abscissa: Channel number. (From E.L. Chupp *et al.*: 1975, in S.R. Kane (ed.), *Solar Gamma-, X- and EUV Radiation*, IAU COSPAR Symposium No. 68, p. 341, D. Reidel Publishing Company. Used by permission.)

TABLE V-2

The γ-ray line flux above background is shown for the August 4 and August 7 solar flares. The measured energy position and error are based on inflight calibration data. (From E.L. Chupp *et al.*: 1975, in S.R. Kane (ed.), *Solar Gamma-, X- and EUV Radiation*, IAU COSPAR Symposium No. 68, p. 341, D. Reidel Publishing Company. Used by permission.)

Time of flare observations	γ-Ray flux at 1 AU (photons cm^{-2} s^{-1})			
3B (Hα) 1972, August 4 (0623:49-0633:02)UT	510.7 ± 6.4 keV	2.24 ± 0.02 MeV	4.4 MeV	6.1 MeV
Hα max – 0630 UT	(6.3 ± 2.0) × 10^{-2}	(2.80 ± 0.22) × 10^{-1}	(3 ± 1) × 10^{-2}	(3 ± 1) × 10^{-2}
3B (Hα) 1972, August 7 (1538:20-1547:33) UT	508.1 ± 5.8 keV	2.22 ± 0.02 MeV	4.4 MeV	6.1 MeV
Hα max – 1530 UT	(3.0 ± 1.5) × 10^{-2}	(6.9 ± 1.1) × 10^{-2}	< 2 × 10^{-2}	< 2 × 10^{-2}

Figure V-3 shows the time integrated solar and background γ-ray counting rate spectrum accumulated during the time interval ~0624 to 0633 UT. The ordinate shows the total number of counts accumulated in each channel during the total live time of 91.4 s for the solar quadrant. The total number of counts in each channel up to channel 200 is shown, and the sum of the counts in five consecutive channels thereafter. The background spectrum has been normalized to the live time shown in the figure. The flare spectrum shows a clear enhancement of the counting rate in both the 0.5 and 2.2 MeV spectral regions. The energy positions of these lines have been established from the calibration spectra and are at energies 510.7 ± 6.4 keV and 2.24 ± 0.02 MeV. The 2.2 MeV line is about 15σ above the continuum. The 0.5 MeV line is somewhat less significant, but there is no question about its presence at about the 4σ level. For the 0.5 MeV line, a contribution in the background quadrant has been subtracted.

The line features at 4.4 and 6.1 MeV are less significant (~3σ) and do not stand by themselves. Their presence is indicated in Figure V-3 because these are the most intense deexcitation lines from ^{12}C (4.4 MeV) and ^{16}O (6.1 MeV) and are expected to be produced in solar flares. Table V-2 gives a summary of the γ-ray flux values obtained from the excess counting rates above the γ-ray continuum accumulated in three full spectral scans. The excess counting rates in the peaks at 0.5 and 2.2 MeV were obtained by *first subtracting the background* quadrant counting rates from the solar quadrant counting rates and then fitting a function of the form

$$N(n) = A_1 + A_2 n + A_3 n^2 + B \exp[(n - n_0)^2/2\sigma^2] \qquad (V.1)$$

to the spectral data using 20–30 channels (n) around the γ-ray peak position. This function represents a Gaussian peak superimposed on a quadratic continuum. The normalizing peak number, B, the line width, σ, and the parameters, A_1, A_2, A_3, were varied to find the best fit as determined by a minimum in chi-squared. Table V-2 also gives the flux values at the Earth for the spectral features at 4.4 MeV (^{12}C) and 6.1 MeV (^{16}O).

b. *Time Profiles of the Positron Annihilation and Neutron Capture Lines – August 4 Event*

Figure V-4 shows the intensity-time profiles of the 0.5 and 2.2 MeV lines observed during the impulsive phase of the August 4 event. The time resolution of the instrument (3 min) and poor statistics (particularly for the 0.5 MeV line) do not allow us to draw any final conclusions about the history of the production of these lines. However, we can say that

(1) The production of these lines takes place in coincidence with the impulsive hard X-ray and γ-ray continuum.

(2) The 0.5 and 2.2 MeV line radiation rise to their maximum values in 3 to 6 min. The observed time history of the 2.2 MeV line shows that a reasonable value of the

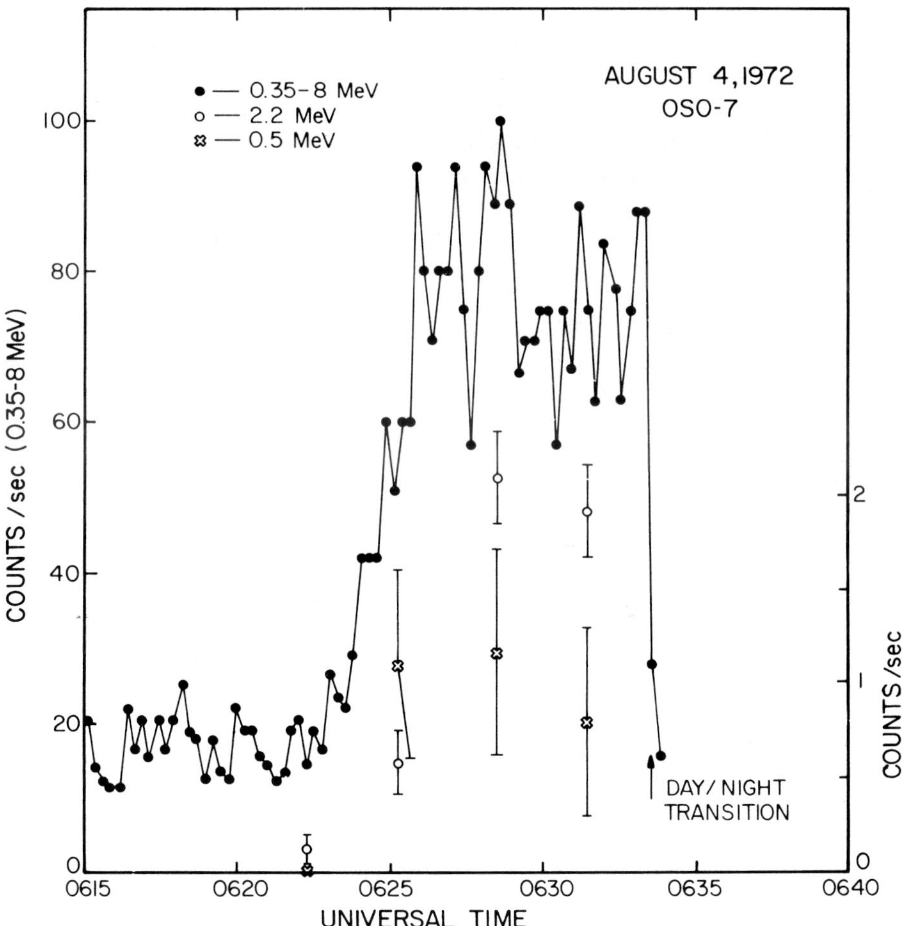

Fig. V-4. The time history of the 0.5 and 2.2 MeV lines observed during the impulsive phase of the 1972, August 4 event. (From E.L. Chupp *et al.*: 1975, in S.R. Kane (ed.), *Solar Gamma-, X- and EUV Radiation,* IAU COSPAR Symposium No. 68, p. 341, D. Reidel Publishing Company. Used by permission.)

capture time of the neutrons in the H of the photosphere is 100 ± 50 s (Reppin *et al.*, 1973) (see also Section III-3.1.2).

c. *Preflare Upper Limits – August 4 Event*

According to the flare model proposed by Elliot (1964, 1969, 1973) the flare energy is stored as energetic protons in the corona and then released to cause the flare. These energetic protons acquire their energy through a slow acceleration process, which could be operating above the flare site for hours or days. If this is the case, then a weak emission of γ-rays could be taking place prior to the onset of a flare.

The data were searched for γ-ray line radiation prior to the start of the August 4 event. No evidence was found for the emission of 0.5, 2.2, 4.4, and 6.1 MeV lines during the period 1437 to 2110 UT on August 3 and 0540 to 0618 UT on August 4, 1972. Data were rejected during the time the spacecraft repeatedly went through the South Atlantic Anomaly which introduces a serious γ-ray background (see Section VI-6.1.3). The preflare γ-ray line flux limit was $\sim 10^{-2}$ photons cm^{-2} s^{-1}. In fact, however, the present observations do not rule out the possibility that particles were accelerated before the flare and that γ-rays were produced in the manner described by Elliot but below the limit of detectability (Forrest *et al.*, 1975a).

d. *Shape of the* 0.5 MeV *Line – August 4 Event*

The possibility of observing thermal Doppler broadening in γ-ray lines produced during solar flares has been discussed by Kuzhevskii (1969) and Cheng (1972). The observations of these lines during the August 4 event allow us to put a limit on the temperature of the flare region in which these lines are produced. The line width due to thermal broadening is given by (Aller, 1963; Stecker, 1971)

$$\left(\frac{\Delta E_\gamma}{E_\gamma}\right)_{\text{thermal}} = 2\sqrt{[2kT(\ln 2)]/(mc^2)} \tag{V.2}$$

where m is the mass of system emitting the γ-ray. In the case of the August 4 flare, only the lines at 0.51 MeV and 2.23 MeV were intense enough to have a measurable line width. The broad features seen in Figure V-3 at 4.4 and 6.1 MeV were so weak that the counts over the whole resolution width of the instrument had to be used to obtain a flux estimate. As can be seen from Equation (V.2), the lowest (or most sensitive measure of) temperature that can be derived from a study of the line width is given by the lower mass emitters, and in this case that is the electron-positron system. The total observed counts in this line is ~ 100 over the impulsive rise of the flare, and these counts were distributed over ~ 10 channels of the pulse height analyzer in a Gaussian manner.

The 0.5 MeV peak flux obtained during the August 4 flare was obtained by subtracting the background quadrant data from the solar quadrant data, and then subtracting a fit to the γ-ray continuum below the peak. The remaining peak was best fitted by a Gaussian curve with a FWHM of 7.4%. This is nearly the measured resolution of the instrument at this energy, which actually was 8.8%. Since any thermal broadening of the line must be added to the instrument resolution $(\Delta E_\gamma/E_\gamma)_{\text{inst}}$, the total observed resolution is given by

$$\left(\frac{\Delta E_\gamma}{E_\gamma}\right)^2_{\text{obs}} = \left(\frac{\Delta E_\gamma}{E_\gamma}\right)^2_{\text{thermal}} + \left(\frac{\Delta E_\gamma}{E_\gamma}\right)^2_{\text{inst}} \tag{V.3}$$

If now we assume that there is no measurable thermal broadening, then we can estimate an upper limit on the temperature by adding to the instrumental resolution the measured error in the observed line width, to give the maximum expected observed line width for no thermal broadening. Since the 1σ error in the line width is $\sim 1.4\%$, that is, $\sim 20\%$ of the measured resolution, we can write

$$\left(\frac{\Delta E_\gamma}{E_\gamma}\right)_{\substack{\text{max}\\ \text{obs}}} = 0.088 + \begin{cases} 0.028 \\ 0.042 \end{cases} = \begin{cases} 0.116;\ 2\sigma \\ 0.130;\ 3\sigma\,. \end{cases} \tag{V.4}$$

Then from Equation (V.3) the upper limit contribution from any thermal broadening is given by

$$\left(\frac{\Delta E_\gamma}{E_\gamma}\right)^2_{\text{thermal}} \leqslant \left(\frac{\Delta E_\gamma}{E_\gamma}\right)^2_{\substack{\text{max}\\ \text{obs}}} - \left(\frac{\Delta E_\gamma}{E_\gamma}\right)^2_{\text{inst}} \tag{V.5}$$

$$\therefore \quad \left(\frac{\Delta E_\gamma}{E_\gamma}\right)_{\text{thermal}} \leqslant \begin{cases} 0.076;\ 95\%\ \text{confidence} \\ 0.096;\ 99\%\ \text{confidence}\,. \end{cases} \tag{V.6}$$

Using the value for $(\Delta E_\gamma/E_\gamma)_{\text{thermal}} = 0.096$ in Equation (V.2) gives an upper limit temperature, in the region where the positron-electron annihilation takes place, of $\lesssim 10^7$ K. An improvement in the temperature limit for the solar regions where γ-ray lines originate will require better statistics and detectors with better energy resolution.

e. *γ-Ray Continuum – August 4 Event*

In addition to the γ-ray line emission, γ-ray continuum emission was also observed extending up to 7 MeV as seen in Figure V-5 (Suri *et al.*, 1975). The differential photon spectrum derived from Figure V-3, along with that observed on TD-1A at lower energies (30 to 203 keV) from Van Beek (1973), is shown in the figure. The spectrum above ~ 300 keV was obtained by first subtracting the background quadrant counting rate spectrum from the solar quadrant counting rate spectrum. The counting rate contributions from known γ-ray lines at 0.5, 1.6, 2.2, 4.4, and 6.1 MeV were then subtracted, including the Compton continuum at lower energies associated with each photopeak.

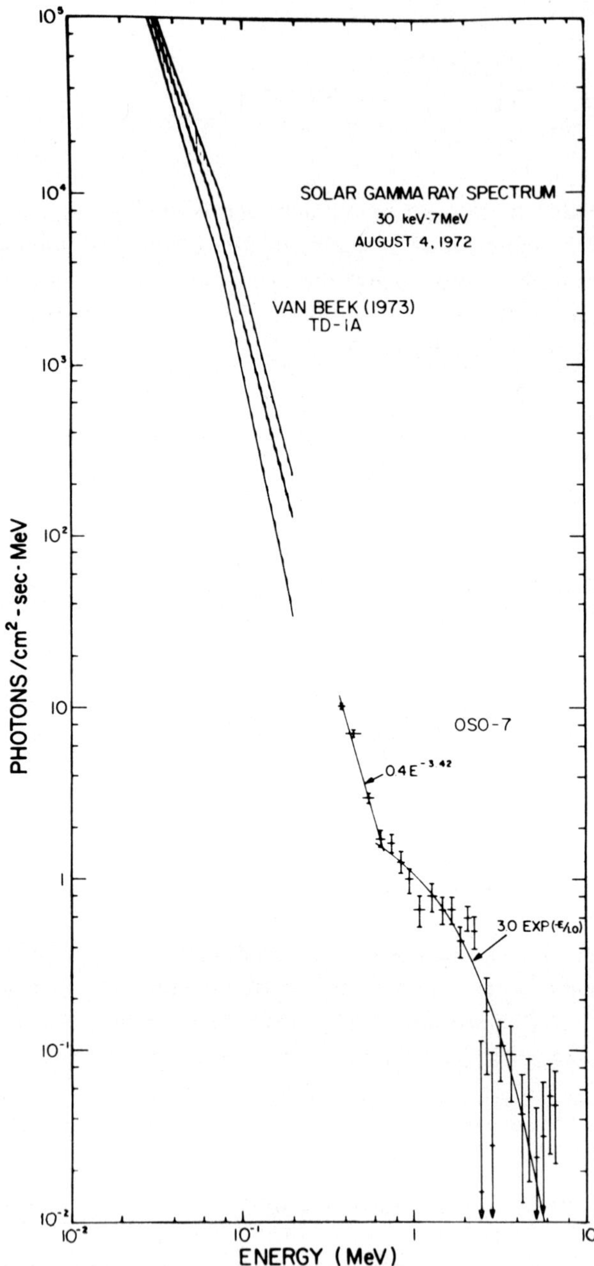

Fig. V-5. The differential photon spectrum observed on 1972, August 4, after line features and associated Compton continuum have been subtracted. (From A.N. Suri *et al.*: 1975, *Solar Phys.* **43**, 415, D. Reidel Publishing Company. Used by permission.)

The photon spectrum incident on the detector was then found by transforming this counting rate spectrum, with the known line contributions removed, through the detector response by the 'strip-off' method (Burrus, 1960). The first step consisted of obtaining a reasonably accurate measure of the Compton continuum and of the first and second escape peaks for various energies of interest for the OSO-7 detector. Response function data collected by Higbie *et al.* (1973) at several energies were used for this purpose. The Compton continuum correction was less than 20% of the observed counting rate at energies less than 0.7 MeV. However, the correction was about 30, 50, 44, 33, and 11% for energy bins 1 to 2, 2 to 4, 4 to 5, 5 to 6, and 6 to 7 MeV, respectively.

There are two effects which can give a spurious flattening of the observed γ-ray continuum spectrum between 1 to 2 MeV. The first one is the Compton scattering of the 2.2 MeV in the photosphere giving a degraded γ-ray spectrum which escapes from the Sun. Wang and Ramaty (1974) and Kanbach *et al.* (1975) have carried out Monte Carlo calculations on the transport of 2.2 MeV photons out of the photosphere. The results indicate that $\leqslant 20\%$ of the observed flux between 1 to 2 MeV could be due to Compton scattering of 2.2 MeV photons in the photosphere (personal communication from the above authors).

The second effect is the Compton scattering of 2.2 MeV photons in the Earth's atmosphere. Because of the broad angular response ($\sim 100°$) of the instrument, it is likely that the detector registered scattered atmospheric radiation during a part of the event just prior to the day/night transition when the Earth's atmosphere was in partial view of the detector. A detailed treatment of the problem is complicated; therefore, only a rough estimate of its effect on the observed continuum spectrum was made.

Considering the case of grazing incidence, Suri *et al.* (1975) estimate that the flux of scattered photons in the energy range 0.7 to 2 MeV falling on the detector is at most 6% of the observed γ-ray continuum flux between 0.7 to 2 MeV. It thus appears that the change in the spectral shape at $\sim$700 keV is real and not a local atmospheric effect.

In the energy range 360 to 700 keV, the differential photon spectrum in Figure V-5 was fit to a power law

$$\mathrm{d}J/\mathrm{d}E = 0.4E^{-3.42 \pm 0.3}\,(\text{photons cm}^{-2}\text{ s}^{-1}\text{ MeV}^{-1}) \qquad (V.7)$$

by the least squares method. For γ-ray energies in the range 0.7 to 7 MeV, however, a single power law is a poor fit. In this energy range the data were fit by the weighted least squares method to an exponential of the form

$$\mathrm{d}J/\mathrm{d}E = 3.0\,\exp(-E/E_0)\,(\text{photons cm}^{-2}\text{ s}^{-1}\text{ MeV}^{-1}) \qquad (V.8)$$

where $E_0 = (1.0 \pm 0.1)$ MeV.

A power law spectrum of index 3.4 in the energy range 360 to 700 keV is consistent with the observations made from TD-1A at lower energies as shown in Figure V-5. Over the energy range 29 to 203 keV, the TD-1A data (Van Beek, 1973) were fitted to a combination of two power law spectral distributions. Van Beek (1973) selected eight time intervals during the period 0623.5 – 0630.5 UT and determined power law parameters of the photon spectrum below and above the break. The spectral index varied from 2.7 to 3.5 below the break to 3.5 to 5.1 above the break. This is shown by the hatched area in Figure V-5. The solid line in the hatched area represents the spectral shape averaged over eight time intervals as given by Van Beek (1973).

The time averaged differential photon spectrum for the 1972, August 4 event over the energy range 30 keV to 7 MeV shows two basic features:

(a) A change in the slope at 80 to 100 keV (Van Beek, 1973) when the power law spectral index changes from 3 to 3.9.

(b) A change in the spectral shape at ~700 keV.

There remains the possibility that the latter feature is due to unresolved γ-ray lines. What is important here is that γ-ray line emissions from solar flares must be detected above a strong continuous spectrum which is due primarily to electron-proton bremsstrahlung. An interpretation of this continuum spectrum in relation to a nonthermal electron spectra causing the X-ray, γ-ray, and radio bursts has been given by Suri *et al.* (1975).

f. *August 7 Event*

The August 7 event was the second large flare of the August solar activity that gave evidence for the emission of γ-ray lines in solar flares. The flare began in Hα at 1455 UT when the OSO-7 spacecraft was behind the Earth. Approximately 40 min after the onset of the flare, the spacecraft emerged into sunlight and enhanced counting rates in the spectral region around 0.5 and 2.2 MeV were observed. The lines at 0.5 and 2.2 MeV were the only lines clearly evident in the solar quadrant compared to the background quadrant. Table V-2 gives a summary of the average γ-ray line fluxes for the August 7 event.

g. *Conclusions Regarding Solar γ-Rays*

The observations of γ-ray lines in solar flares just reviewed give general support to the calculations described in Section III-3.1. The relative intensities of the lines shown in Table V-2 are, in general, in accord with the calculations. However, as is clear from our discussion above, the observations are limited, in that the lines were seen only in the strongest flares. In addition, the counting statistics do not permit time histories to be determined for the prompt deexcitation lines from ^{12}C and ^{16}O. Nevertheless some general conclusions can be made:

(a) γ-ray-producing nuclear reactions begin in the first 200 s of a strong solar flare with the hard X-rays and before the optical maximum.

(b) Nuclear reactions occur for $\geqslant$600 s.

(c) The density of the region where electron-positron annihilation takes place is $\gtrsim 10^{12}$ (electrons cm^{-3}).

(d) The temperature of the region of positron-electron annihilation (and the positron production region, if it is the same) is $< 10^7$ K.

(e) The low energy primary cosmic γ-ray spectrum at the Sun necessary for γ-ray production is consistent with the prompt low energy cosmic ray spectrum seen at Earth and in space.

(f) The total cosmic ray particle energy deposited in a thick target dump is $< 10^{28}$ erg which is less than the optical energy in a large flare.

The relationship of the observed γ-ray fluxes to space probe observations of ^{3}He nuclei is discussed by Forrest *et al.* (1975a). Ramaty *et al.* (1975) also review in detail the expected fluxes of γ-rays for different parent solar cosmic ray spectral shapes and target conditions. These calculations, however, implicitly assume that all nuclear phenomenon in the solar atmosphere are due to non-thermal nuclear reactions and that the thermonuclear processes in flares are of no significance. Experimentally, our knowledge is sufficiently limited that some role for the thermonuclear reactions, such as ^{2}H (d, n) ^{3}He, cannot be ruled out (see also Sections II-2.4.3c and III-3.1).

5.2. Cosmic Observations (Point and Localized Sources)

We shall review here the important experimental efforts that have been made to detect γ-rays in the energy range 50 keV to 100 MeV from point and localized sources either within the galaxy or beyond. In the context used here, a localized source may be extended in one or two directions, such as a line source distributed along the galactic plane or a diffuse galactic nebula (e.g., M1 – the Crab Nebula).

5.2.1. SUPERNOVAE AND SUPERNOVA REMNANTS (SNR)

a. *Supernova Remnants*

The search for γ-ray lines from a SNR represents one of the most important applications of γ-ray line astronomy, because the successful measurement of a γ-ray line flux could provide direct verification of theories of nucleosynthesis, as discussed in Section III-3.2.1. Only a few experiments have been carried out to investigate this important problem, which we will describe briefly.

A search for nuclear lines below 1.5 MeV from the Crab Nebula has been made by Haymes *et al.* (1968) using a large area NaI spectrometer in an active collimator. Only upper limit fluxes were obtained for some of the predicted *r*-process lines discussed in Section III-3.2.1. The first attempt at using a high resolution solid-state γ-ray spectrometer to study the Crab Nebula was carried out by Jacobson (1968). The basic γ-ray detector used was a 4.6 cm^3 cooled Ge(Li) spectrometer in an active CsI(Tl) anticoincidence shield, also acting as a collimator with an opening solid angle of 0.077 sr (9° half-angle). This geometry was effective for γ-rays with energy less than ~300 keV. The instrument was carried by a balloon to an atmospheric depth of ~3.2 g cm^{-2} from Palestine, Texas ($P_c = 4.6$ GV) on 1967, July 28. The Crab Nebula passed through the opening aperture of the telescope, allowing a measurement of the Crab emission spectrum

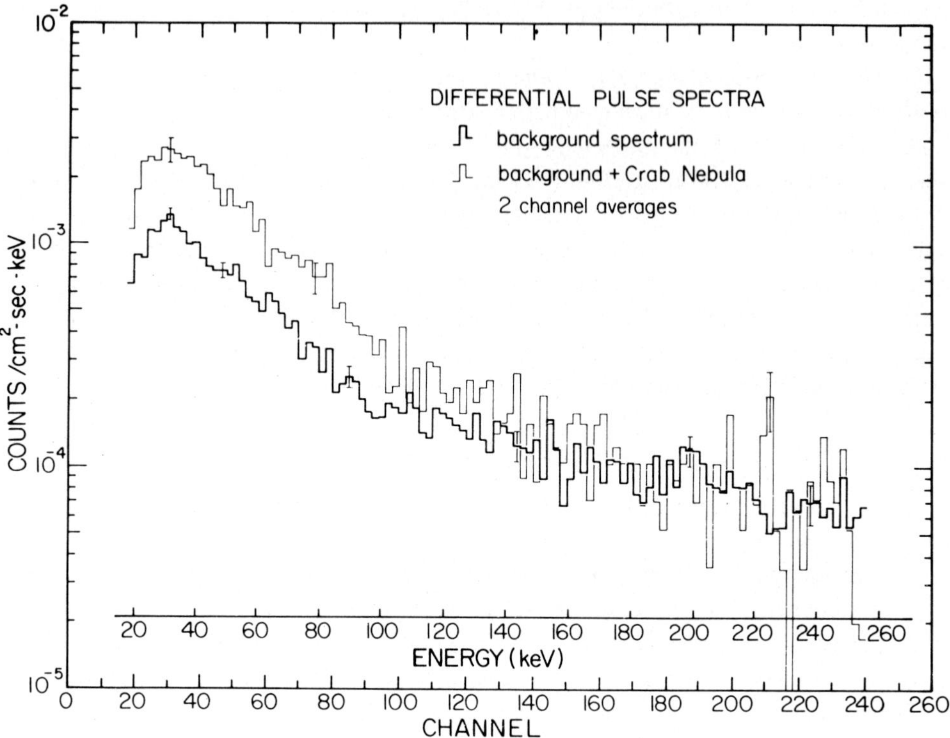

Fig. V-6. The high resolution γ-ray counting rate spectrum obtained with a Ge(Li) telescope when viewing the Crab Nebula. (From A.S. Jacobson, Ph.D. Thesis, University of California, San Diego, 1968.)

up to 300 keV. The counting rate spectra obtained are shown in Figure V-6 when viewing the Crab Nebula and also when the Crab Nebula was not in the field of view. The spectra correspond to two-channel averages. A hard X-ray continuum from the Crab Nebula is evident in the figure, and the contribution to the counting rate shown is consistent with the spectrum measured by Haymes *et al.* (1968) up to 560 keV. When the background spectrum is subtracted and the net Crab differential pulse spectrum is averaged over four channels, Jacobson (1968) finds that the only spectral feature that fits the criterion for a γ-ray line is one which is centred at channel 210 in Figure V-6. If this is a γ-ray line, then it has an energy of 224 ± 3 keV and an apparent intensity at the top of the atmosphere of $(6.6 \pm 2.0) \times 10^{-3}$ photons cm^{-2} s^{-1}. The line energy does not correspond to any of the predicted *r*-process lines. Jacobson (1968) has pointed out that the interpretation for this line should be considered carefully, since there was evidence for local neutron background activation in the Ge(Li) spectrometer in his experiment, and there was the possibility of time-dependent background features. The neutron background problem is discussed further in Section VI-6.1.5.

TABLE V-3

Upper limit fluxes for the Crab γ-ray lines compared with the *r*-process production and with modifications. (From A.S. Jacobson, Ph.D. Thesis, University of California, San Diego, 1968.)

r-process element	Energy (keV)	Predicted flux [Clayton and Craddock (1965)] $(cm^2\,s)^{-1}$	Predicted flux (this work) $(cm^2\,s)^{-1}$	Previous upper limits [Haymes *et al.* (1968)]	Upper limits (this work) $(cm^2\,s)^{-1}$
^{241}Am	59.6	5.7×10^{-5}	5.1×10^{-5}	3.9×10^{-3}	1.4×10^{-3}
X-rays	104	–	4.0×10^{-5}	—	1.0×10^{-3}
^{251}Cf	180	1.9×10^{-5}	7.6×10^{-5}	1.5×10^{-3}	1.1×10^{-3}

In Table V-3, the measured upper limits for some of the expected *r*-process lines given by Jacobson (1968) are compared with the predicted fluxes of Clayton and Craddock (1965) as shown in column 3. Jacobson (1968) has also revised some of these predicted fluxes using updated branching ratios for γ-ray-producing transitions. These are shown in column 4 of the table. The higher upper limits measured by Haymes *et al.* (1968) in the experiment mentioned above are given in column 5.

Very recently, a larger Ge(Li) spectrometer was constructed by Jacobson *et al.* (1975), and a further search has been conducted for γ-ray lines from the Crab Nebula. This instrument is similar to one designed for flight aboard the HEAO-C Earth-orbiting satellite, which is described briefly in Section VI-6.3.1. The balloon instrument uses four cooled Ge(Li) spectrometers each of 40 cm^3 in a 6.35 cm thick CsI(Na) collimator-shield. The main instrument element thus has a volume that is over 30 times that of the small version just described here. Jacobson *et al.* (1975) report that the sensitivity of this new balloon γ-ray spectrometer will be capable of detecting γ-ray lines from point sources with fluxes as low as $10^{-3} \rightarrow 10^{-4}$ photons $cm^{-2}\ s^{-1}$, Possibly by the time this monograph is published, this experiment will have given positive evidence for cosmic gamma lines (see Section VI-6.1.5b).

The OSO-7 experiment has also allowed a partial search to be made (Chupp *et al.*, 1974b) for γ-rays from a SNR, associated with the pulsar 0833-45, known as the Gum Nebula. According to the calculations of Ramaty and Boldt (1971) (as discussed in Section III-3.2.1) γ-ray lines are expected at several energies from this remnant. Therefore, data from the OSO-7 γ-ray monitor (Section V-5.1.1) were searched for radiation from this source.

The scan plane of the γ-ray monitor passed near the Gum Nebula for several weeks while γ-ray spectra were being accumulated in the normal solar and background quadrants. During the time period analyzed, the alternate quadrant (see Figure V-1) could receive γ-rays from over 50% of the Gum Nebula region as given by Kerr (1971). A plan of operation was adopted that switched data-taking from normal quadrants 1 (solar) and 3 (background) for a few orbits a day to alternate quadrants 2 and 4. Thus several spectra were available which should indicate any detectable fluxes from the Gum Nebula. In this case, a search was made for nuclear lines from ^{28}Si (1.78 MeV), ^{12}C (4.43 MeV), and ^{16}O (6.13 MeV) by comparing spectra taken from the Gum direction with equivalent non-Gum spectra. Under optimum viewing conditions, the detector receives radiation from ~80% of the Gum Nebula, if the full nebula is a source of γ-rays. No evidence for line emission exists at a flux level of $5 \times 10^{-2} \text{cm}^{-2} \text{ s}^{-1}$ for each of the lines at 1.78, 4.43, or 6.13 MeV. These flux limits allow limits to be placed on the relative amounts of Si($M_{Si}/M_{\odot}$) and C($M_C/M_{\odot}$) and O($M_O/M_{\odot}$) present in the nebula. The current results are $M_{Si}/M_{\odot} < 3$, $M_C/M_{\odot} < 1$, $M_O/M_{\odot} < 3$, if the cosmic ray flux currently existing in the nebula is that predicted by Ramaty and Boldt (1971). By summing spectra, it should be possible to reduce the mass limits by at least an order of magnitude; however, at the present time this analysis has not been completed.

b. *Supernovae*

The only published case when a γ-ray spectrometer with good energy resolution was in operation above the Earth's atmosphere when a nearby SN occurred, was in 1972 during the OSO-7 satellite mission (Chupp *et al.*, 1974b). A Type I SN, 1972e, was discovered on 1972, May 13, in NGC 5253, with a magnitude of 8.5. (Kowal, 1972). However, the time of its maximum magnitude may have been May 6, or earlier (IAU circular 2421). The galaxy NGC 5253 is an irregular dwarf which also hosted a Type I SN in 1895 (Sargent *et al.*, 1974) designated as 1895b. The event SN 1972e was the fourth extra-galactic SN in apparent brightness ever recorded according to the Palomar tabulation given by Sargent *et al.* (1974). The distance to NGC 5253 is ~3.5 Mpc (Sersic *et al.*, 1972), and therefore γ-ray lines could be expected from this event at an intensity of about 1/16 that predicted by the work of Clayton and collaborators summarized in Section III-3.2.1.

During this period, and before X-ray instruments on OSO-7 and Uhuru viewed the galaxy, upper limits were placed on the prompt X-ray emission from the SN above 7 keV by Ulmer *et al.* (1974) and in the 3 to 10 keV energy band by Canizares *et al.* (1974). From April through May 1972, the direction of view of the OSO-7 γ-ray spectrometer during satellite night was near NGC 5253. A search was made for specific unshifted γ-ray line emissions before and after the optical maximum at 0.51 MeV, 0.847 MeV and 2.2 MeV. In all cases, no evidence exists for enhanced emission with a limiting line flux at $\sim 10^{-2} \text{ cm}^{-2} \text{ s}^{-1}$ for a transit burst. In the case of the ^{56}Co γ-ray line at 0.847 MeV predicted by Clayton *et al.* (1969), spectra were summed for 1300 s or ten full data accumulations on May 3, 1972, which should be near the time of the SN light curve maximum. Since no evidence existed for a line at this energy, the limiting flux is $\leqslant 6 \times 10^{-3}$ (0.847 MeV photons $\text{cm}^{-2} \text{ s}^{-1}$). Finally, in view of the calculations mentioned in Section III-3.2.1, which suggest significant Compton broadening of the spectral lines predicted by Clayton *et al.* (1969), a search has been made for an enhanced continuum over the range 350 keV to 1 MeV. Specific searches were made on May 1, 3, and 23, all with null results, giving a limiting flux of 3×10^{-2} photons $\text{cm}^{-2} \text{ s}^{-1}$ in the 650 keV energy band specified above (Chupp *et al.*, 1974b).

The brief analysis described here has not given any positive indication of γ-ray lines or continuum from a recent, relatively nearby, SN. However, since the live time available for observing a transient burst (< 2000 s) was quite small, any such event might have been missed, even though the instrument is capable of detecting a burst of 1 MeV photons with an energy flux of $\sim 10^{-8}$ erg cm^{-2} s^{-1}. A SN of the apparent magnitude (8.5) of SN 1972e is relatively rare; therefore, in order to significantly increase the chances of detecting X-ray or γ-ray SN events, the sensitivity of instruments should be increased by at least two orders of magnitude and a balloon or satellite system developed with a 100% observational live time for any part of the sky. Such a system would greatly increase the prospects of detecting a transient burst or continuous line or continuum spectral emissions. All the possibilities of detecting γ-ray lines from these objects are not yet exhausted, however, since summing of selected spectra can reduce line flux limits to $\sim 10^{-3}$ photons cm^{-2} s^{-1}.

5.2.2. GALACTIC DISK AND CENTER

The possibility of a component of high energy γ-rays emitted predominately from the plane of the galaxy was first reported by Clark *et al.* (1968) using data from a directional detector on the OSO-3 satellite launched on March 8, 1967. The detector was a coincidence-type scintillation-counter telescope utilizing pair production in a multi-layer scintillation (CsI) detector. A directional Cerenkov counter recorded the pair electrons, and a lower NaI(Tl) sandwich detector recorded the remaining electron kinetic energy. This detector was responsive primarily to γ-rays above 100 MeV and had an effective axial sensitivity (efficiency × area product) of ~ 1.8 cm^2 at 100 MeV, rising to ~ 8 cm^2 at 400 MeV. The angular resolution of the detector was $\sim 24°$ FWHM. A detailed discussion of the operation and the calibration of this instrument is given by Kraushaar *et al.* (1972), who have given the final results for cosmic γ-rays from the OSO-3 experiment following a recalibration of the instrument.

In the 16mo of operation of this experiment *621* events were recorded due to cosmic γ-rays with energies above 50 MeV. These cosmic rays were attributed by Kraushaar *et al.* (1972) to two basically different sources – a galactic component and an apparent extra-galactic component (see Section V-5.3). The most important result of the experiment is the observation that the γ-ray directions are highly anisotropic with an $\sim 10\sigma$ excess contribution along the galactic equator and an additional pronounced excess in the region of the Galactic Center. Since the width of the galactic source was demonstrably less than the angular resolution (24° FWHM) of the γ-ray detector, the integral intensity of the galactic disk component was expressed as an 'equivalent line intensity' by using an experimentally determined geometric factor

$$G_{\text{line}}(E_0, b) = 4.1 \exp[-(b/15)^2]\, F(>E_0) \quad (\text{cm}^2 \text{ rad}) \tag{V.9}$$

where b is the angle between the line source and the instrument axis, and $F(>E_0)$ is a factor which depends on the shape of the incident energy spectrum and the lower limit, E_0, of the integral energy spectrum measured. For a π_0 γ-ray spectrum, $F(>E_0) = 0.86$, but is not strongly dependent on spectral shape for $E_\gamma > 100$ MeV. By assuming that the observed counting rate of all γ-ray events from high galactic latitudes ($b^{\text{II}} > 35°$) was due to an isotropic γ-ray flux, the net counting rate in the galactic plane was found.

Then, using the equivalent line geometric factor, the following galactic 'line' intensities were found for the indicated galactic coordinates:

General Galactic Emissions –
$(3.4 \pm 1.0) \times 10^{-5}$ photons cm^{-2} s^{-1} $rad^{-1} > 100$ MeV
$30° < l^{II} \leqslant 330°$; $-15° < b^{II} < 15°$

Galactic Center Emission –
$(1.3 \pm 0.3) \times 10^{-4}$ photons cm^{-2} s^{-1} $rad^{-1} > 100$ MeV
$-30° < l^{II} \leqslant 30°$; $-15° < b^{II} < 15°$.

These numbers are tabulated in Table V-4 and represent the latest available results of the OSO-3 experiment. It should be noted that the earlier OSO-3 results reported by Clark *et al.* (1968) were higher by about a factor of 3 and are corrected downward because of a recalibration of the instrument. Another satellite experiment on OGO-5 by Hutchinson *et al.* (1970) had given a high (but nevertheless positive) value of the galactic line flux at a glactic longitude well away from the Galactic Center region (see Table V-4).

Fichtel *et al.* (1968, 1969) have reported galactic γ-ray results from balloon flights of a digitized spark chamber flown from Palestine, Texas in late 1966. The instrument, described in detail by Ehrmann *et al.* (1967), consisted of two wire-grid spark chambers separated by a plastic scintillator detector operated in coincidence with the spark chambers. Below this system was a directional Cerenkov detector. This detector, the central scintillator, and an anticoincidence signal from a guard counter triggered the spark chamber for downward moving γ-rays. Each spark chamber contained 15 pair-producing plates consisting of 0.02 radiation lengths of gold plated on a thin aluminum base, an arrangement which minimized Coulomb scattering of the pair electrons. For γ-rays of energy ~100 MeV, the uncertainty in the arrival direction of a γ-ray was determined to ~3° and the maximum value of the area-time detection efficiency factor was ~9 cm^2 s at the same energy.

Fitchel *et al.* (1968, 1969) have studied the region of the celestial sphere near the Galactic Center during a balloon flight on 1966, December 10. A possible net positive galactic 'line' intensity was reported equal to $\sim(2.3 \pm 1.2) \times 10^{-4}$ cm^{-2} s^{-1} rad^{-1} for a galactic latitude range $-3° < b^{II} < 3°$ and for galactic longitude $-10° < l^{II} < 25°$, which is comparable to regions studied by OSO-3 (see Table V-4). The lower value of this flux as compared to the first reported result from OSO-3 of $(4.1 \pm 0.7) \times 10^{-4}$ cm^{-2} s^{-1} rad^{-1} by Clark *et al.* (1968) was not considered a disagreement, but the existence of the discrepancy led to a recalibration of the OSO-3 instrument and the 'line' intensity given in Table V-4 for OSO-3.

In early 1969, a multiplate spark chamber balloon experiment was carried out by Frye *et al.* (1969) from Parkes, Australia. This instrument, designed for photons >50 MeV, consisted of a single spark chamber with 31 stainless steel plates, followed by a plastic coincidence counter for the pair electrons and then a directional Cerenkov coincidence counter. The angular resolution of this instrument for γ-rays above 50 MeV was $\sim \pm 2.3°$. A search was made for a γ-ray line source near the Galactic Center with negative results. If the line source had a width of $<4.6°$, then the upper limit intensity is 3×10^{-5} photons cm^{-2} s^{-1} rad^{-1} for a source spanning a galactic longitude range of 60° at the Galactic Center. For a line source of width 30° centered on the galactic plane, the upper limit intensity is 7×10^{-5} photons cm^{-2} s^{-1} rad^{-1} for the same galactic longitude interval. The only way that Frye *et al.* (1969) could reconcile their results with the OSO-3 results was for the line source to have a half-width of 15° and for its intensity to be considerably less than the positive result given by Clark *et al.* (1968) and Kraushaar *et al.* (1972) (cf. Table V-4). This result is in decided conflict with OSO-3 results. As will be seen below, newer observations support the line source character of the Galactic Center emission. Frye *et al.* (1969) have, however, found positive evidence for point sources near the Galactic Center.

Kniffen and Fichtel (1970) have reported results on galactic γ-rays from a balloon flight of a large digitized spark chamber from Mildura, Australia. Since the instrument is similar to that used by Fichtel *et al.* (1968, 1969), we will not describe it here, except to note that it had an area larger by about a factor of ten but otherwise similar properties. The instrument was calibrated at the NBS synchrotron. The results from this experiment confirm the strong excess of γ-rays from the galactic plane in the direction of the Galactic Center, giving a 4.3σ excess above the background. The resulting average line intensity is $(2.0 \pm 0.7) \times 10^{-4}$ cm^{-2} s^{-1} rad^{-1} for $E_\gamma > 100$ MeV, a result which is larger than, but in substantial agreement with, the final OSO-3 result. This number is, therefore, also in disagreement with the result of Frye *et al.* (1969) discussed above, for essentially the same region around the Galactic Center. The other important outcome of this experiment was the failure to observe a statistically significant excess of γ-rays in the 50 to 100 MeV energy interval from the Galactic Center. Fichtel *et al.* (1972) have pointed out that this result is significant in determining the possible source mechanism for the Galactic Center flux. These authors give the 95% confidence level upper limit of the ratio

$$R = [J(50 \text{ to } 100\,\text{MeV})/J(>100\,\text{MeV})] < 0.50.$$

This ratio bears on the source mechanism for the Galactic Center γ-rays, since Fichtel *et al.* (1972) point out that Stecker's (1971) calculation indicates that the theoretical value for this ratio for a pure π^0 spectrum is 0.12. If the γ-ray spectrum were due to electron bremsstrahlung, then the ratio would be $R = 2.03$ for a galactic electron differential spectrum with a power law index of $\alpha = 2.6$. The same electron spectrum would give a value of the ratio $R = 0.74$ for either the Compton or synchrotron mechanisms (see Chapter II). Thus, Fichtel *et al.* (1972) conclude that the Galactic Center line γ-ray spectrum is due predominately to π^0 meson decay.

Frye *et al.* (1971) carried out another balloon flight in late 1969 with the instrument (Frye *et al.*, 1969) described above, this time from Longreach, Australia, where the effective geomagnetic cutoff is higher by about a factor of 2 (see Table V-4). Again, this group has found evidence for high energy γ-ray point sources near the Galactic

TABLE V-4

Summary of measurements of the high energy γ-ray flux from the galactic plane

Experiment	Launch site	Dates	Experimental group	Orbit Average P_c (GV)	Approximate Balloon altitude ($g\ cm^{-2}$)
1) OSO-3		3-8-67 to 7-68	Kraushaar *et al.* (1972) Clark *et al.* (1968)	$\sim 33°$	
2) OGO-5		3-4-68	Hutchinson *et al.* (1970)	Eccentric 24 $R_\oplus$	
3) Balloon	Palestine	12-10-66	Fichtel *et al.* (1969)		3.0
4) Balloon	Parkes	2-5-69 2-26-69	Frye *et al.* (1969)	4.5	2.86 mb 2.98 mb
5) Balloon	Mildura	10-23-69	Kniffen and Fichtel (1970) Fichtel *et al.* (1972)		2.9
6) Balloon	Longreach	11-26-69	Frye *et al.* (1971)	8.8	2.97 mb
7) Balloon	Lusaka	1-25-71 2-9-71	Bennett *et al.* (1972)	12.5	4.1 mb 11 mb
8) Balloon	Mildura	3-25-70	Dahlbacka *et al.* (1973)		2.8 mb
9) Balloon	Palestine	9-20-71	Browning *et al.* (1972)		127 k ft
10) Balloon	Panama	21-1-71	Share *et al.* (1974a)		3
11) SAS-2		11-15-72- 6-8-73	Kniffen *et al.* (1973); Fichtel *et al.* (1975)		
12) Balloon	Longreach	11-18-73	Sood *et al.* (1974)		3.5 – 6.0 mb

Experiment	Energy range	Angular resolution FWHM	Energy resolution FWHM	Galactic longitude	Galactic latitude	Flux $cm^{-2} sec^{-1} rad^{-1}$
1) OSO-3	> 100 MeV	24°		$330° < l^{II} < 30°$	$-15° < b^{II} < 15°$	$(1.3 \pm 0.3) \times 10^{-4}$
2) OGO-5	> 40 MeV	~ 30° (18°)		$45° < l^{II} < 75°$	$-30° < b^{II} < 20°$ (?)	$(9 \pm 5) \times 10^{-4}$
3) Balloon	> 100 MeV	~ 3°		$-10 < l^{II} < 25°$	$-3° < b^{II} < 3$	$(2.3 \pm 1.2) \times 10^{-4}$
4) Balloon	> 50 MeV	± 2.3°		$\lvert l^{II} \rvert < 30°$	$b^{II} < 4.6°$ $b^{II} < 15°$	$< 3 \times 10^{-5}$ $< 7 \times 10^{-5}$
5) Balloon	> 100 MeV			$-39° \leqslant l^{II} \leqslant 24°$	$-6° \leqslant b^{II} \leqslant +6°$	$(2.0 \pm 0.7) \times 10^{-4}$
6) Balloon	> 100 MeV			$-36° < l^{II} < 36°$	$-2.3° < b^{II} < 2.3°$	$< 2.7 \times 10^{-5}$
7) Balloon	> 200 MeV	± 5° (200 MeV)	180%	$-20° < l^{II} < 50°$	$-9° < b^{II} < +0°$	$(2.8 \pm 1.4) \times 10^{-4}$
8) Balloon	> 100 MeV			$-16° < l^{II} < +6$	$\Delta b^{II} = \pm 6°$	$< 2 \times 10^{-4}$
9) Balloon	> 100 MeV			$5° < l^{II} < 30°$		$< 8 \times 10^{-5}$
10) Balloon	> 15 MeV	1° – 2°	< 25%	$330° < l^{II} < 30°$	$\Delta b^{II} \sim 3°$	$2.4\text{-}4.6 \times 10^{-4}$
11) SAS-2	> 100 MeV	2.5° (100 MeV)		$330° < l^{II} < 30°$	$\Delta b^{II} = \pm 10°$	$(0.96 \pm 0.14) \times 10^{-4}$
12) Balloon	> 500 MeV	2° (500 MeV)	30% (500 MeV)	$330° < l^{II} < 10°$	$-10° < b^{II} < 10°$	$(4.1 \pm 1.5) \times 10^{-5}$

Center, but no evidence for a diffuse emission from the Galactic Center region that could be considered a line source corresponding to the OSO-3 result. They have analyzed their data under three assumptions to establish upper limits to a Galactic Center line source. In the first case they have assumed a line of width $-2.3° < b^{\mathrm{II}} < +2.3°$ over a latitude range $\Delta l^{\mathrm{II}} = \pm 36°$, which corresponds to the width of the HI region from radio observations. Their 95% confidence level upper limit to the line source in this region is 1.3×10^{-5} photons cm^{-2} s^{-1} rad^{-1} with a positive contribution from a point source G_γ 341 + 1 removed, or 2.7×10^{-5} photons cm^{-2} s^{-1} rad^{-1} if the contribution from the point source is included. Second, in analyzing the data for the Galactic Center region observed by Kniffen and Fichtel (1970) and Fichtel *et al.* (1972), Frye *et al.* (1971) found an upper limit line flux of 5.3×10^{-5} cm^{-2} s^{-1} rad^{-1}, and with the point source contribution removed, the upper limit for the diffuse line flux is 3×10^{-5} cm^{-2} s^{-1} rad^{-1}. Finally, Frye *et al.* (1971) have analyzed their data for the specific Galactic Center region surveyed by OSO-3 (see Table V-4). In this case the upper limit to the line source intensity is 5.4×10^{-5} cm^{-2} s^{-1} rad^{-1}, which is about a factor of 2 below the revised OSO-3 intensity. In Table V-4 we give the lowest upper limit to a diffuse line source deduced by Frye *et al.* (1971). The main conclusion from this work is that the OSO-3 observations can be considered as the combined effect of several point sources, and if a line source exists, it must be at a level $\lesssim 10^{-5}$ cm^{-2} s^{-1} rad^{-1}.

Bennett *et al.* (1972) have flown a spark-chamber/Cerenkov-counter detector to study galactic γ-rays at energies above 200 MeV. The detector area was 125 cm^2 with an efficiency of 19% at 200 MeV and an angular resolution of 5° at the same energy. The results of this experiment support the evidence for a galactic line source above 200 MeV which is consistent with an extrapolation of the lower energy results of OSO-3 and Fichtel *et al.* (1972) (see Table V-4). Only upper limits were found for point γ-ray sources.

An improved technique for studying high energy γ-rays (> 100 MeV) has been devised by Dahlbacka *et al.* (1973). A combination nuclear-emulsion /spark-chamber/ Cerenkov-counter system was used for determining the celestial coordinates of an incident photon to ±45 arc min. The emulsion acted as the pair-production medium, and the electron tracks in the spark chamber were traced back to the emulsion. Thus, accurate location of the pair event could be determined, as well as the opening angle of the pair before much multiple scattering had occurred. The effective geometric factor of the instrument was 84 cm^2 sr. The detector was flown from Mildura, Australia in early 1970 for a brief (~2.4 h) period. This short exposure unfortunately limited the sensitivity of the instrument, which was designed to detect point sources of intensity $\sim 2 \times 10^{-6}$ cm^{-2} s^{-1}. This experiment gave upper limit results on the Galactic Center line emission which are about at the level reported from OSO-3. The limit and corresponding galactic coordinates are shown in Table V-4.

Browning *et al.* (1972) have reported on Galactic Center γ-ray observations (> 100 MeV) using an optical spark chamber triggered by a scintillation-counter / Cerenkov-counter telescope. The results of a flight from Palestine, Texas on 1971, September 20 did not reveal any evidence for a diffuse line emission from a Galactic Center region corresponding to a portion of that scanned by OSO-3. The pertinent results are shown in

Table V-4. Positive evidence for point sources was obtained, however, and are discussed briefly in Section III-3.2.1.

The nuclear-emulsion spark-gap technique has also been used by Share *et al.* (1974a) to study the galactic γ-rays down to ~10 MeV (see Dahlbacka *et al.*, 1973; May and Waddington, 1969). The basic instrument properties are high angular resolution (1° → 2°) and an energy resolution of 25% above 10 MeV. The efficiency of the instrument was a maximum of ~5% at ~50 MeV and the area of the emulsion stack was 650 cm^2. A balloon flight of this instrument from Parana, Argentina in late 1971 gave strong evidence for a narrow band of diffuse emission from the Galactic Center of width ~3° in galactic latitude, covering a 60° longitude interval centered at the Galactic Center. The result for an estimated integral line flux above 15 MeV depends on the assumed spectral shape for the incident γ-rays. If a π^0 γ-ray spectrum shape is assumed, the line flux is 4.6×10^{-4} cm^{-2} s^{-1} rad^{-1} (> 15 MeV); and if an integral power law spectrum of shape E_γ^{-1} is assumed, then the line flux is 2.4×10^{-4} cm^{-2} s^{-1} rad^{-1} (>15 MeV). Both results are consistent with the positive results above 100 MeV reported by Karushaar *et al.* (1972) and Fichtel *et al.* (1972). It is interesting to note that this experiment failed to confirm the existence of the reported γ-ray point sources of Frye *et al.* (1971) and Browning *et al.* (1972) near the Galactic Center.

Except for the OSO-3 and OGO-5 experiments mentioned earlier, the experiments studying the galactic diffuse γ-ray flux were all carried out with balloons, and the results are in conflict as far as verifying a line source. The most advanced satellite γ-ray telescope (Kniffen *et al.*, 1973) was flown on the second Small Astronomy Satellite (SAS-2) from 1972, November 15 to 1973, June 8. The basic instrument was similar to the double spark chambers flown by Kniffen and Fichtel (1970) and Fichtel *et al.* (1972). In this case, each spark chamber consisted of 16 magnetic core spark modules interleaved with thin (0.03 radiation length) W pair-conversion plates. Two such chambers were separated by four central plastic scintillators to provide triggering of the spark chambers in anti-coincidence with a scintillator dome surrounding the chambers. Below the bottom chamber was a directional Cerenkov counter to exclude γ-rays coming from below. The sensitivity (or effective area) of this instrument at 100 MeV was ~60 cm^2 for photons incident along the detector axis. This can be compared with the ~2 cm^2 sensitivity of the OSO-3 instrument. The angular resolution at 100 MeV is also much improved over the OSO-3, being ~2.5° at 100 MeV, deteriorating to 5° at ~40 MeV, and improving slightly (1.5°) up to 1000 MeV (Fichtel *et al.*, 1975).

The results of this experiment cover the general diffuse radiation, Vela bursts, localized sources, and the galactic disk radiation of interest here. The essential result in regard to the galactic disk radiation is that enhanced emission is observed along the entire galactic plane, as was the case with the OSO-3 instrument, but in this case the higher angular resolution shows the radiation to be confined to a disk of half-width ~6° in latitude in the direction of the Galactic Center (Kniffen *et al.*, 1973; Fichtel, 1974). Figure V-7 shows the relative intensity of γ-rays (> 100 MeV) as a function of latitude for a 60° longitude band centered at galactic latitude $l^{\mathrm{II}} = 0$. The data from the OSO-3 experiment are shown for comparison (Kraushaar *et al.*, 1972). Fichtel (1974) has in-

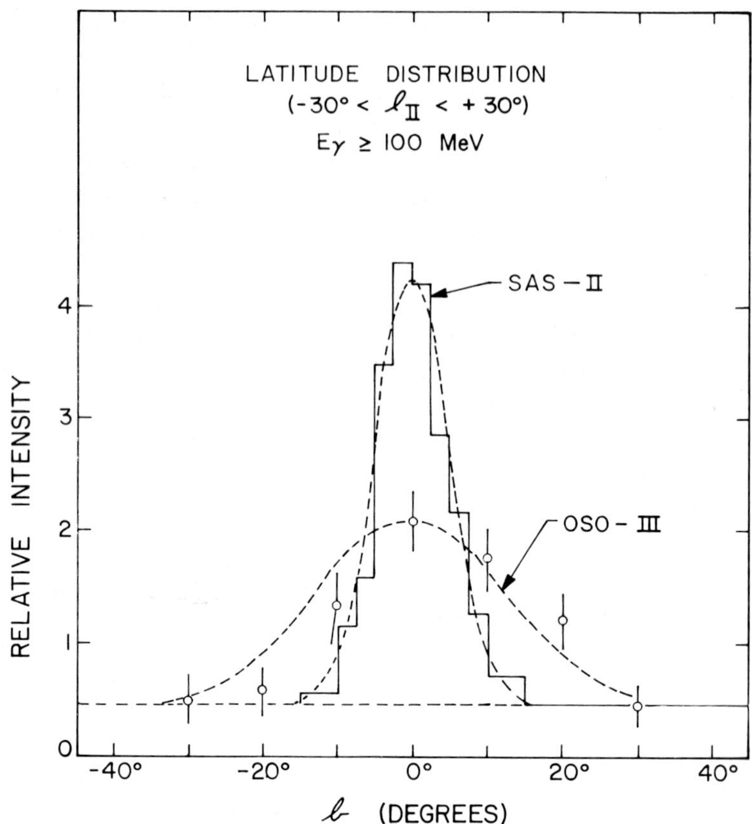

Fig. V-7. Distribution of high energy ($E_\gamma > 100$ MeV) γ-rays summed from $l^{II} = 330°$ to $l^{II} = 30°$ as a function of b^{II}. The OSO-3 data are those of Kraushaar *et al.*, (1972). The dashed curve through the SAS-2 data is a Gaussian distribution with $\sigma = 4.5°$. This distribution still includes a substantial experimental angular uncertainty, so the real distribution of γ-rays is narrower. (From D.A. Kniffen *et al., Astrophys. J. (Letters)* **186,** L105. Copyright 1973, The American Astronomical Society, by permission of the University of Chicago Press.)

dicated that the uncertainties in the knowledge of the pointing direction of the SAS-2 instrument are such that a pure line source (infinitesimal width) would be broadened to have a (1σ) half-width (HWHM) of 3.5 ± 0.5°. Therefore, the real width in latitude of the galactic plane γ-rays may be considerably smaller than 6°. The SAS-2 longitude distribution for γ-rays (> 100 MeV) along the galactic plane is shown in Figure V-8 (Fichtel, 1974), in which the γ-ray flux is integrated over a latitude range $\Delta b^{II} = \pm 10°$. The diffuse background as seen by SAS-2 is shown by the dashed line. Also shown is the OSO-3 data of Kraushaar *et al.* (1972) for a longitude interval not covered by SAS-2. In this case the data is integrated over a galactic latitude range $b^{II} = \pm 15°$ and is therefore somewhat higher than might be expected. It should be noted that the γ-ray flux is given in relative units where one unit is $\sim 10^{-4}$ photons cm^{-2} s^{-1} rad^{-1}. It is of considerable interest to note that the intensity of the line source near the Galactic Center is rela-

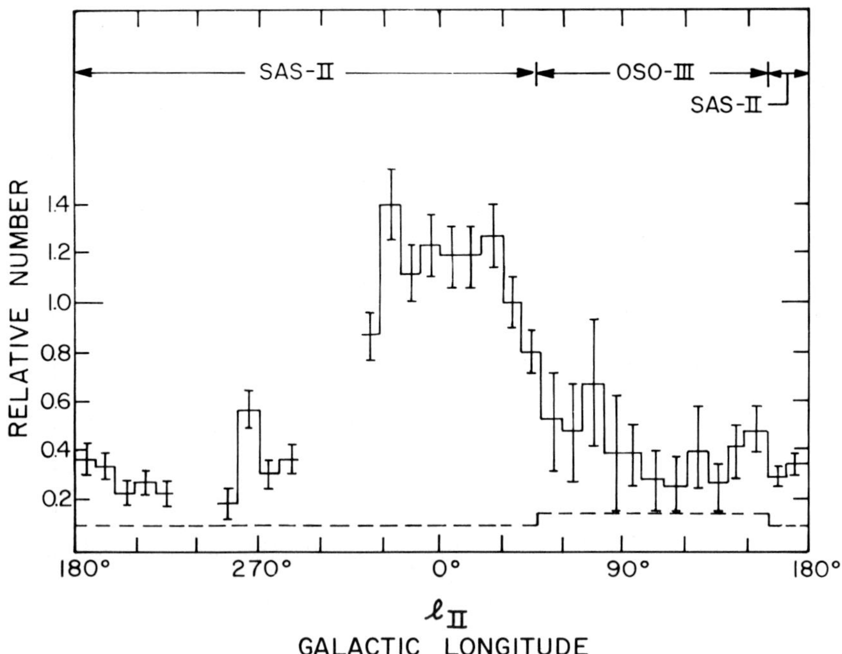

Fig. V-8. Distribution of high energy (> 100 MeV) γ-rays along the galactic plane. The data marked OSO-3 is that of Kraushaar *et al.* (1972), and that marked SAS-2, of Kniffen *et al.* (1973) and Thompson *et al.* (1974). The diffuse background level is shown by a dashed line. It is higher in the case of OSO-3 than SAS-2 because the OSO-3 data is summed from $b^{II} = -15°$ to $b^{II} = +15°$ and the SAS-2 data from $b^{II} = -10$ to $b^{II} = +10$. The ordinate scale is approximately in units of 10^{-4} photons cm^{-2} rad^{-1} s^{-1}. (From C.E. Fichtel: 1974, *GSFC X-662-74-57.*)

tively flat over a range ($320° < l^{II} < 40°$). Fichtel (1974) has pointed out that this lack of a peak at the 'center' negates any theory which attempts to explain the general enhancement by a strong source at the center. Also, these observational data strongly contradict the balloon observations of Galactic Center point sources discussed above (e.g., Frye *et al.*, 1971; Browning *et al.*, 1972). The value of the Galactic Center line flux above 100 MeV from SAS-2 is given in Table V-4 and is consistent with the latest OSO-3 value if one takes into account the difference in the latitude band over which the data were summed.

Sufficient data have now been obtained so that the γ-ray spectral shape from the galactic plane can be estimated. In Figure V-9 the integral γ-ray diffuse line flux measurements are plotted vs. energy. This curve, taken from Share *et al.* (1974a), summarizes several of the results shown in Table V-4 with the addition of the integral spectrum measured during a balloon flight by Fichtel *et al.* (1972). The integral flux value of Share *et al.* (1974a) depends on spectral shape. The upper point at 15 MeV shown by a solid circle is for an assumed 100% π^0 spectrum and gives the upper extrapolated dotted line shown. If the integral photon spectrum is $\propto E_\gamma^{-1}$, then the Share *et al.* (1974) measurement at 15 MeV gives the flux shown by the lower open circle which normalizes the

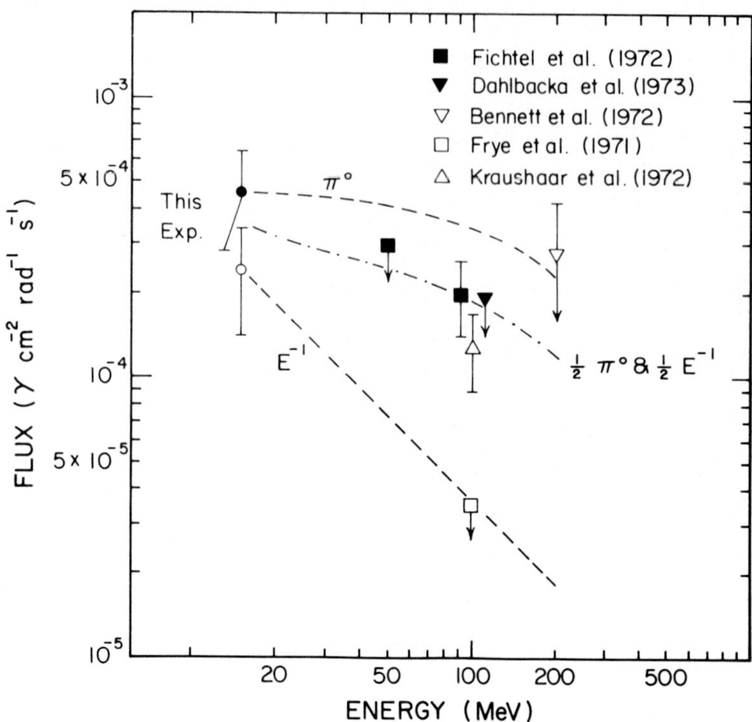

Fig. V-9. Measurements of the spectral flux of γ-rays from the galactic plane near the center of the galaxy. The results of Share *et al.* (1974a) for assumed π^0 decay (filled circle) and power law (open circle) spectra are extrapolated to higher energies. Extrapolation of a combined spectrum is given by the dot-dash curve. (From G. H. Share *et al.*, *Astrophys J.* **187**, 45. Copyright 1974, The American Astronomical Society, by permission of the University of Chicago Press.)

spectral shape. If the spectral shape is due to an equal contribution from a π^0 γ-ray spectrum and a power law spectrum, then the Share *et al.* (1974a) measurement is shown by the intermediate dot-dash line. The measured SAS-2 spectrum falls slightly below this curve (Kniffen *et al.*, 1973).

A recent balloon experiment has given positive evidence for the galactic line source above 500 MeV. Sood *et al.* (1974) have flown a multi-gap spark chamber/directional-Cerenkov detector from Longreach, Australia on 1973, November 18 to a pressure altitude of $3.5 \rightarrow 6$ mb. The instrument had an angular resolution of $\pm 2°$ at 500 keV, a sensitive area of 1130 cm^2, and an efficiency of $\sim 30\%$, giving a sensitivity of 339 cm^2. The resulting integral γ-ray line flux ($>$500 MeV), averaged over the Galactic Center region of width 60° in longitude and 20° in latitude, is $(4.1 \pm 1.5) \times 10^{-5}$ cm^{-2} s^{-1} rad^{-1} (see Table V-4).

From the experimental point of view one might question the existence of a Galactic Center line source, since although two satellite experiments and three balloon experiments confirm its existence, two balloon experiments attribute the excess Galactic Center

γ-flux to point sources which are not generally confirmed. The SAS-2 observations clearly appear to establish the existence of a narrow source ($< 6^\circ$ HW) lying in the galactic plane. Improved angular resolution may, however, show structure from several sources (see, for example, Samini *et al.*, 1974).

In summary, then, the general characteristics of the galactic line γ-radiation are:

Galactic Center Region –
$I_\gamma(> 100\,\text{MeV}) \simeq (1 \to 2) \times 10^{-4}$ photon $\text{cm}^{-2}\ \text{s}^{-1}\ \text{rad}^{-1}$

General Galactic Disk –
$I_\gamma(> 100\,\text{MeV}) \simeq (1 \to 2) \times 10^{-5}$ photons $\text{cm}^{-2}\ \text{s}^{-1}\ \text{rad}^{-1}$.

The energy spectrum for the radiation from the galactic disk appears to be consistent with a two-component model composed of a π^0 decay spectrum and a differential power law of the form $\propto E_\gamma^{-2}$. Several other aspects of the interpretation of this radiation are given in Section III-3.2.4, and Chapter VII gives reference to the latest developments.

5.2.3. GALACTIC CENTER γ-RAY LINES

Besides the observations of γ-ray lines associated with solar flares (see Section V-5.1) and the possibility that the OSO-3 and SAS-2 galactic plane γ-rays are due partly to π^0 decay, we have the additional possibility of nuclear lines from the Galactic Center. Johnson *et al.* (1972) and Johnson and Haymes (1973) have reported the observation of a line with an energy just under 500 keV from the direction of the Galactic Center in 1970 and 1971. A recent report however places this Galactic Center line at 530 ± 11 keV (Haymes *et al.*, 1975) and also gives evidence for a γ-ray line at 4.6 ± 0.1 MeV, also from the Galactic Center. These reported positive observations of γ-ray lines are, of course, of crucial significance to our subject. We, therefore, will describe in some detail how the early measurements were made since the same basic approach was used in the more recent work of Haymes *et al.* (1975).

In November 1970 a large area ($75\,\text{cm}^2$) actively collimated scintillation spectrometer was used to scan the Galactic Center region and the bright X-ray source GX 5-1 (Schnopper *et al.*, 1970) for γ-rays in the energy range 23 to 930 keV (Johnson *et al.*, 1972). The balloon experiment carried out from Paraná, Argentina ($P_c = 11.5$ GV) utilized a large NaI(Tl) γ-ray spectrometer (4 inches diameter by 2 inches thickness) in a large NaI(Tl) antiçoincidence shield collimator, which formed a crude telescope with a 12° half-angle. The instrument had an energy resolution of 15% at 511 keV. By use of an oriented gondola, spectra (over the above energy interval) were taken while viewing the source direction GX 5-1 during the time interval GMT 1135 to 1850, on November 25, 1970. During this time interval and for alternate 10 min periods, the detector was rotated 180° away from the source direction in azimuth. This procedure gave alternate 10 min background measurements with the detector axis at the same zenith angle the source had during the prior 10 min interval. Since the source rises in the east and sets in the west, the zenith angle of the source first decreases until meridian transit and then increases again, while the background direction rises in the west and sets in the east. This procedure insures that source and background measurements pairs are made at nearly the same zenith angle. The acceptance angle of the telescope is sufficiently large (24° FWHM) that the source is not completely removed from the field of view during the background measurement. This is especially true near the time of meridian transit since GX 5-1 had a zenith angle then of 7°.

The source spectrum (in terms of counting rate/channel) is found for a source observation interval by subtracting the average counting rate/channel during the two adjacent 10 min background measurements, giving the counting rate spectrum from the source, assuming the background is independent of azimuth. For each source measurement, correction is made for atmospheric absorption, detector absorption and efficiency and

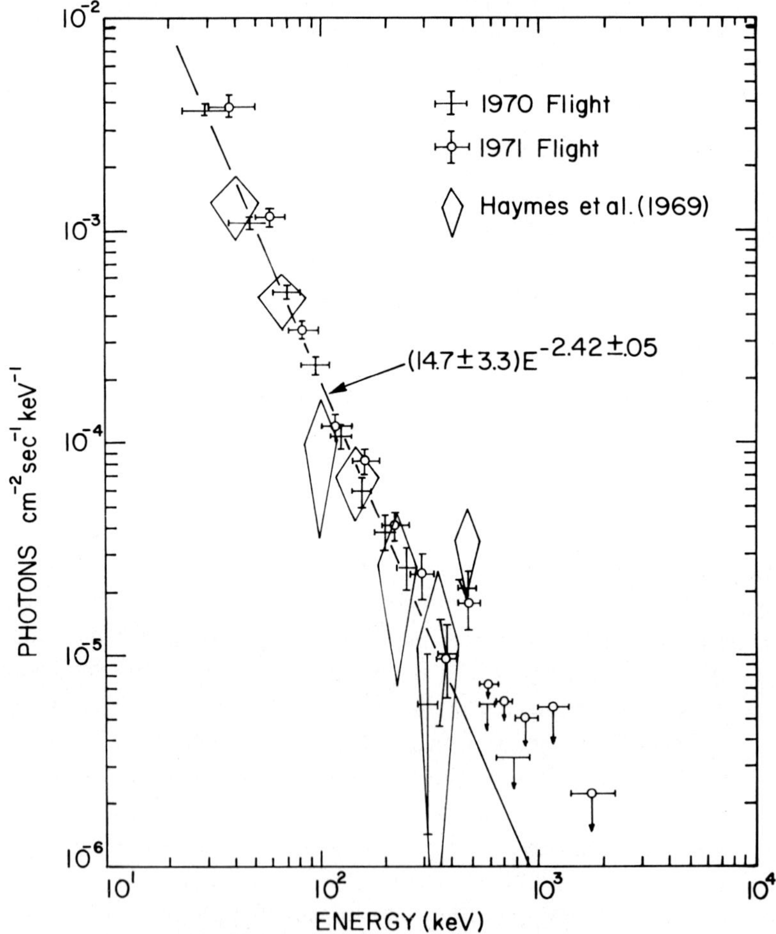

Fig. V-10. The differential photon flux from the Galactic Center for the 1969, 1970 and 1971 flights; the solid line is the best fit spectrum for the combined 1970 – 1971 data. (From W.N. Johnson and R.C. Haymes, *Astrophys. J.* **184,** 103. Copyright 1973, The American Astronomical Society by permission of the University of Chicago Press.)

then all source measurements are averaged, giving the Galactic Center data points shown in Figure V-10 for the 1970 flight. The points at different energies are for groups of channels about 1.5 times the FWHM resolution of the detector. The two important results indicated by the 1970 data points are the apparent continuum flux expressed by

$$N(E) = (10.5 \pm 2.2)\ E^{-(2.37 \pm 0.05)}\ (\text{photons cm}^{-2}\ \text{s}^{-1}\ \text{keV}^{-1}) \qquad (\text{V}.10)$$

(Johnson *et al.*, 1972) and the data point indicated (1970) at about 500 keV. At energies below 160 keV others have shown (Buselli *et al.*, 1968; Lewin *et al.*, 1971) that individual sources in the direction of the Galactic Center give measurable spectra so that shown in Figure V-10 is the sum of all sources in Sagittarius.

Of most significance to the present monograph is the data point above the continuum at about 500 keV. This feature was studied in more detail in the raw counting rate spectrum and indicated an excess counting rate over the continuum for about 20 analyzer channels. The best fit to the spectrum was made using a Gaussian line centered at 473 keV. Based on these data Johnson *et al.* (1972) gave a line flux at this energy of $(1.8 \pm 0.5) \times 10^{-3}$ photons cm^{-2} s^{-1}.

The existence of this spectral feature was confirmed by an additional balloon flight experiment on November 20, 1971 from Paraná, Argentina, by Johnson and Haymes (1973). This later flight was timed to coincide with a lunar occultation of the X-ray source GX 3 + 1 to help identify the source of the feature seen in the 1970 flight. The 1971 experiment had two nearly identical scintillation telescopes operating, while the 1970 experiment, also with two telescopes, had only one operating. The angular resolution of the telescopes was ~24° (FWHM) for each, while the energy resolutions at the standard energy (662 keV) were 16% FWHM and 11% FWHM. Occultation of GX 3 + 1 lasted ~1h 14 min from 1353 UT until 1507 UT on 1971, November 20. The measurements were made at an atmospheric depth of 3.4 to 3.2 g cm^{-2}. Source and background measurements were made in a manner similar to that already described for the first flight, except for the occultation period, when three source observations were taken totaling 40 min, and a total of 20 min of background observations were taken. The data points for 1971 are shown in Figure V-10 along with the net Galactic Center spectrum for the combined 1970 and 1971 flights. The spectral shape is given by

$$N(E) = (14.7 \pm 3.3) E^{-(2.42 \pm 0.05)}\ (\text{photons cm}^{-2}\ \text{s}^{-1}\ \text{keV}^{-1})$$

A detailed plot of the Galactic Center and GX 3 + 1 counting rate spectra for both detectors with a least square fitting of the data gave a spectral feature at 485 ± 36 keV for the 1971 data. This fitted line width is stated to be somewhat broader than the detector's energy resolution (Johnson and Haymes, 1973).

By comparing the net spectra obtained when the source GX 3 + 1 was not occulted, and when it was occulted, it was found that this source did not provide the major portion of the continuum above 40 keV (see Figure V-10). The spectral region surrounding 485 keV did, however, exhibit a small flux depression (~2.3σ) during the occultation;

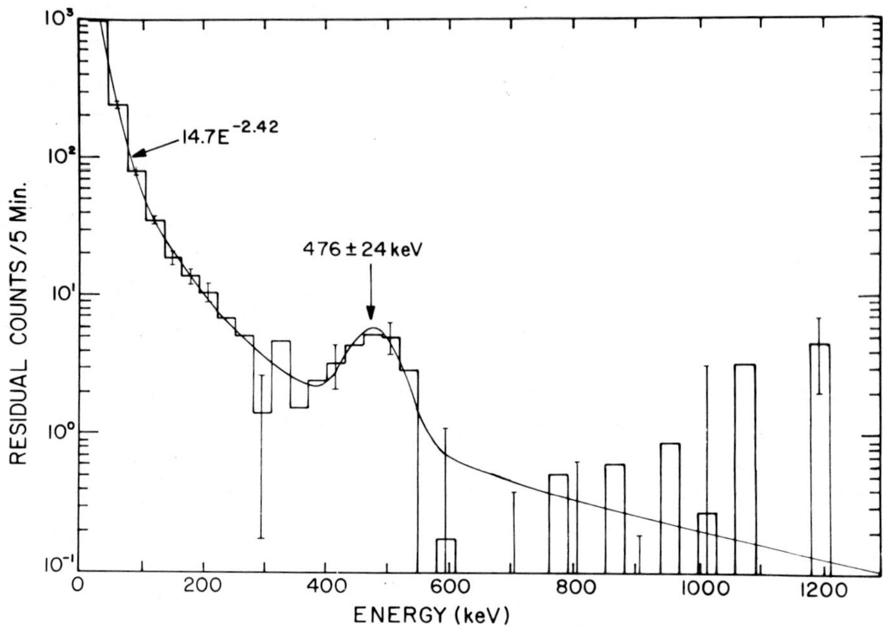

Fig. V-11. A histogram representation of the measured flux from the Galactic Center region, as determined by a weighted average of the 1970 and 1971 observations. The pulse height channels have been combined into consecutive 30 keV energy intervals. Also shown is the best fit power law continuum with the best Gaussian photopeak superimposed at an energy of 476 keV. The displayed error bars represent $\pm 1\sigma$. (From W.N. Johnson and R.C. Haymes, *Astrophys. J.* **184,** 103. Copyright 1973, The American Astronomical Society, by permission of the University of Chicago Press.)

Johnson and Haymes (1973) point out that the occultation statistics are so poor that the effect could be due to a slight flux of penetrating continuum form GX 3 + 1.

In Figure V-11 we show the result obtained by the Rice group for the net counting rate spectrum from the Galactic Center for the average 1970 and 1971 data. The best fit to this spectrum gives a photopeak at 476 keV. It is of interest to review the intrinsic intensity of the continuum and line source as given by Johnson and Haymes (1973). For the continuum between 30 keV to ~2 MeV the total energy flux received from the direction of the Galactic Center is 1.1×10^{-8} erg cm^{-2} s^{-1} which is about half of that received from the Crab Nebula. If indeed the Galactic Center, at 10 kpc, is the source of this continuum, then the γ-ray luminosity is 1.3×10^{38} erg s^{-1} (30 keV to 2 MeV). The actual source of this γ-ray continuum cannot be determined from the Rice measurement (Johnson and Haymes, 1973), since the instruments do not have sufficient angular resolution. Efforts to compare γ-ray spectra with higher resolution results at lower energies do not shed much light on the source (see a discussion by Johnson and Haymes (1973) on the continuum γ-rays). The observed intensity of the spectral feature interpreted as a single line at 476 keV superimposed on the continuum is $(1.8 \pm 0.5) \times 10^{-3}$ photons cm^{-2} s^{-1} which is at a 3.6σ significance level, although the feature is

5.3σ above a power-law continuum fit to all the 1970 and 1971 data. The reason for the apparent reduced significance in the flux value quoted above is the inclusion of the error associated with the fitted power law spectrum (Johnson and Haymes, 1973). If the source of this assumed γ-ray line is the Galactic Center, then the luminosity in the 476 keV line is $\sim 1.6 \times 10^{37}$ erg s^{-1}. It is also considered possible by these workers that the apparently larger width for the line ($\sim$ 18% FWHM) than the instrument width of $\leqslant$ 16%, could allow the existence of more than one line. Finally, the occultation data on GX 3 + 1, mentioned above, are consistent with this X-ray source as the origin of the line, but it is not conclusive proof.

One important point should be considered before this flux is considered a true celestial γ-ray line, and that is the requirement for an independent observation of the line using a different technique. The 1970 and 1971 observations of this line by the Rice group were made with essentially the same methods of measuring backgrounds and analyzing the data. Therefore, an unknown systematic effect cannot be ignored even though the Rice group has been very careful to investigate all imaginable such effects. In spite of the importance of these reported observations to γ-ray astronomy, no other balloon experiment results have yet (at mid-1975) been reported to verify this observation except for the additional results from Rice reported by Haymes *et al.* (1975), which are discussed below.

Some inconclusive information on this line is available now from the OSO-7 mission (see Section V-5.1), and we shall discuss these observations briefly. In the case of the OSO-7 mission, the wide angle γ-ray spectrometer scanned the Galactic Center for several weeks in 1972. By comparing these spectra with spectra obtained when the Galactic Center region was not in an optimum position for viewing, it was possible to search for γ-ray line emissions that could be associated with the Galactic Center over an energy range 300 keV to 9 MeV (Suri *et al.*, 1974). In Figure VI-17 (Section VI-6.1.3), we show the γ-ray counting rate spectrum observed with the OSO-7 spectrometer when viewing the Galactic Center for an integrated instrument live time of $\sim$4h. Also shown is a comparison spectrum obtained when the spectrometer was viewing another region of the sky, but otherwise under identical conditions. This comparison spectrum is shifted down by a factor of 2 for clarity in the figure; however, by careful scrutiny, it can be noted that the two spectra are identical to within the statistical fluctuation error. Looking at the portion of the spectra near the expected location of the reported 476 keV Galactic Center line just mentioned, there is no evidence for a line; however, there are spectral features nearby which tend to mask a line. By a careful study of the statistical fluctuations in this region of the spectrum, we can conclude that the intensity of a γ-ray line at 476 keV with the resolution width of the instrument would be $< 2 \times 10^{-3}$ photons cm^{-2} s^{-1}, which is just the value of the reported Galactic Center line flux. This experiment therefore cannot confirm or refute the balloon data discussed above and a satellite or balloon instrument of higher flux sensitivity will be required. It was possible, however, to investigate one of the interpretations of the reported 478 keV line as offered by Fishman and Clayton (1972) (see Section III-3.2.4). This model interprets the line as due to the decay of an excited state of ^{7}Li at 478 keV. The ^{7}Li is produced by cosmic ray spallation of heavier nuclei and can be excited by interactions with lower energy cosmic rays in an equilibrium situation. This so-called nuclear de-excitation model predicts γ-ray line fluxes at several other energies as well (see Section III-3.2.4). The OSO-7 data shown in Figure VI-16 show no evidence for a line feature at 473 to 485 keV, whether compared with the background spectrum or not. The absence of the predicted observable lines (Fishman and Clayton, 1972) at 4.43 MeV ($^{12}C^*$) and 1.63 MeV ($^{20}Ne^*$) gives limiting fluxes for these lines of 1.7×10^{-3} photons cm^{-2} s^{-1} and 2.0×10^{-3} photons cm^{-2} s^{-1}, respectively. Kozlovsky and Ramaty (1974) have also recently considered the production of the 478 ^{7}Li line by α-α reactions (see Section III-3.1).

The physical interpretation of the '476 keV feature' is still unresolved, but, besides the ^{7}Li hypothesis just mentioned, several other mechanisms have been suggested, which includes a proposed gravitational red shift of 0.51 MeV photons leaving neutron stars. This was discussed in Section III-3.2.2 with regard to neutron stars (see also Guthrie and Tademaru, 1973). One of the most promising explanations (if the line is real) is that proposed by Leventhal (1973a). Since positron annihilation in the interstellar medium takes place predominantly through the bound state of positronium (see Section II-2.4.4), then the annihilation spectrum consists of the line at 0.51 MeV and a three photon continuum extending to zero photon energy (see Figure II-27). If the γ-ray spectrometer measuring such a spectrum has a resolution of $\sim 15\%$ (as was the case in the Rice experiment), then Leventhal (1973a) has shown that the folding together of the positronium annihilation spectrum and the instrument resolution will shift the peak close to the 480 keV position indicated by the experiment. This explanation, however, requires an excessive value for the 0.51 MeV line intensity from the Galactic Center (see Section III-3.2.4).

The evidence for a Galactic Center line appears now to be drastically altered by new observations reported by Haymes *et al.* (1975). On April 2, 1975 a γ-ray telescope similar to that used by Johnson *et al.* (1972) and Johnson and Haymes (1973) was used to study the Galactic Center region from a balloon launched at Rio Cuarto, Argentina. The new instrument had only slightly improved properties over the previous one with a sensitive area of 182 cm^2, an energy resolution of 12% at 511 keV and a 15° (FWHM) acceptance cone. The method of taking Galactic Center and background spectra was essentially the same as that used by Johnson *et al.* (1972). The resulting difference spectrum for a 219 min observation of the Galactic Center again gave evidence for a γ-ray line near 0.5 MeV but the best fit Gaussian was centered at 530 ± 11 keV with a flux of $(8.0 \pm 2.3) \times 10^{-4}$ photons cm^{-2} s^{-1}. This may be compared with the earlier value reported by Johnson and Haymes (1973) at 476 ± 24 keV with a flux of $(1.8 \pm 0.5) \times 10^{-3}$ photons cm^{-2} s^{-1}. Haymes *et al.* (1975) argue that these differences are real and offer three possible hypotheses to explain them:

(a) Since the solid angle in the latest experiment was 0.4 of that used previously, the lower flux may be evidence that the Galactic Center source is larger than the 15° beam width. This could give support to the suggestion that detectable nuclear γ-rays are produced by low energy cosmic rays interacting with interstellar matter.

(b) The apparent time variation of the flux values and the line energy may be a result of nova activity in the Galactic Center region as discussed by Clayton and Hoyle (1974).

(c) The average energy of the '0.5 MeV line' may depend on the galactic coordinates being observed. Because there was a slight difference in the viewing directions for the different observations, there may be evidence for such a correlation.

Finally this latest experiment gave evidence for a significant line feature located at 4.6 ± 0.1 MeV with a flux of $(9.5 \pm 2.7) \times 10^{-4}$ photons cm^{-2} s^{-1}. This is attributed to excitation of the first excited state of ^{12}C at 4.43 MeV. Within the next year there will most certainly be several experiments carried out by other groups to investigate further the galactic γ-ray spectrum for evidence of nuclear lines.

5.3. Diffuse γ-Ray Flux Observations (100 keV to 100 MeV)

The first indication of a possible extraterrestrial γ-ray flux in the MeV energy region was provided by the remnant background counting rate observed in a small NaI spectrometer on the Ranger 3 and 5 spacecraft. These observations reported by Arnold *et al.* (1962) and Metzger *et al.* (1964) were described briefly in the Introduction. At higher γ-ray energies (>100 MeV), the OSO-3 γ-ray detector (Kraushaar *et al.*, 1972) also gave evidence for an excess background counting rate when the directional detector looked away from the Earth (see Section 5.2.2). In addition, a solar X-ray telescope on OSO-3 gave evidence (Schwartz *et al.*, 1970; Schwartz, 1970) for a remnant background counting rate above ~7 keV. There has been a natural tendency to explain the background counting rates in these experiments as due to an isotropic cosmic γ-ray source. The reason for this is that, in all of the experiments mentioned above, care was taken to investigate backgrounds due to local charged particle effects, so any counting rate remaining after corrections was attributed to an external photon flux. In the region below ~100 keV, Schwartz and Gursky (1973) and Schwartz (1974) have given strong evidence for the true cosmic nature of the background radiation. Schwartz (1974) points out that two spectral shapes are needed to fit the data in the energy region 2 to 82 keV. In terms of a differential photon spectrum below 25 keV, the best power law fit is $F(E_\gamma) = 8.5E^{-1.40}$ (photons cm^{-2} s^{-1} keV^{-1} sr^{-1}), and above 30 keV the best power law fit is $F(E_\gamma) = 167E^{-2.38}$ (photons cm^{-2} s^{-1} keV^{-1} sr^{-1}). These spectra cross at ~20 keV. The low energy photon flux is isotropic in intensity over the sky to a few percent as measured by the Uhuru X-ray telescope ($\lesssim$ 10 keV) with a field view of $\sim 5° \times 5°$. It is usually concluded that the isotropy indicates that the source is of extragalactic origin.

In the energy region from 100 keV to 50 MeV, there is currently considerable uncertainty in the spectral shape of the apparent diffuse flux, and its very existence is in question. An extrapolation of the power law spectrum observed above 30 keV (given above) to the MeV energy range falls far below the spectrum measured at the higher energies (> 1 MeV). The situation today is not much different than in 1972, when Pal

(1972) reviewed the measurements existing at that time in the light of their possible cosmic nature. At that time he concluded that there was not conclusive proof for the existence of a cosmic γ-ray flux above 1 MeV. In spite of this negative note, it is important to review briefly the current experimental situation because of the astrophysical significance of such a flux.

5.3.1. OBSERVATIONS (100 keV TO 10 MeV)

Several further attempts at measuring a diffuse γ-ray flux have been made using small omnidirectional alkali halide γ-ray spectrometers which were similar to those used in the early Ranger 3 and 5 observations. Satellite measurements were carried out by Vette *et al.* (1970) on ERS-18, Golenetskii *et al.* (1971) on Cosmos 135 and 163, Konstantinov *et al.* (1971) on Cosmos 135 and Mazets *et al.* (1975) on Cosmos 461. Measurements have also been made by Trombka *et al.* (1973) during the Apollo 15 lunar mission. Measurements using balloon-borne instruments at photon energies below 10 MeV have also been carried out by several groups, e.g., Vedrenne *et al.* (1971), Schönfelder and Lichti (1974) and, most recently, by Fukada *et al.* (1975).

The Apollo 15 measurements were obtained during the trans-Earth phase of the mission using a 7 cm diameter by 7 cm length NaI(Tl) spectrometer surrounded, except at the photomultiplier end, by a thin (1 cm) plastic scintillator anticoincidence shield. The raw energy loss spectra obtained with this spectrometer in various modes of operation are discussed in Section VI-6.1.4 along with further details of the experiment. The energy loss spectrum obtained when the detector was extended on a boom from the Appollo spacecraft was used to obtain an equivalent photon spectrum using a matrix inversion technique and a measured response function (Adler and Trombka, 1970). Several additional corrections were applied to this spectrum including a correction for background lines seen by the detector, a correction due to spallation of the NaI(Tl) by cosmic rays and a correction for background from the spacecraft. In Table VI-5 is shown the contribution of these corrections to the energy loss spectrum. The photon spectrum that is obtained after all these corrections are made is shown as the solid curve in Figure V-12, which extends from 0.3 to 27 MeV. A final correction due to attenuation of any external flux is also included so this curve represents an upper limit to the diffuse γ-ray flux or else an actual flux, if all background corrections have been made (see Sections VI-6.1.2 and VI-6.1.4).

Figure V-12 also shows the extrapolation of the lower energy diffuse X-ray radiation to the MeV energy range according to a functional form given by Pal (1973), which is $\propto E^{-2.1}$. This spectral shape is slightly flatter than the $E^{-2.38}$ dependence given above by Schwartz (1974); however, it is clear that in the 1 to 10 MeV energy range the Apollo flux is as much as a factor of 5 above the highest reasonable extrapolation of the energetic diffuse X-ray spectrum. Below 1 MeV the measurements on Ranger 3 and 5 (Metzger *et al.*, 1964) and on ERS-18 (Vette *et al.*, 1970) are in close agreement with those made on Apollo 15 shown in Figure V-12. Above 1 MeV the earlier results from both Ranger 3 and 5 and ERS-18 were found to be in error (see Trombka *et al.*, 1973).

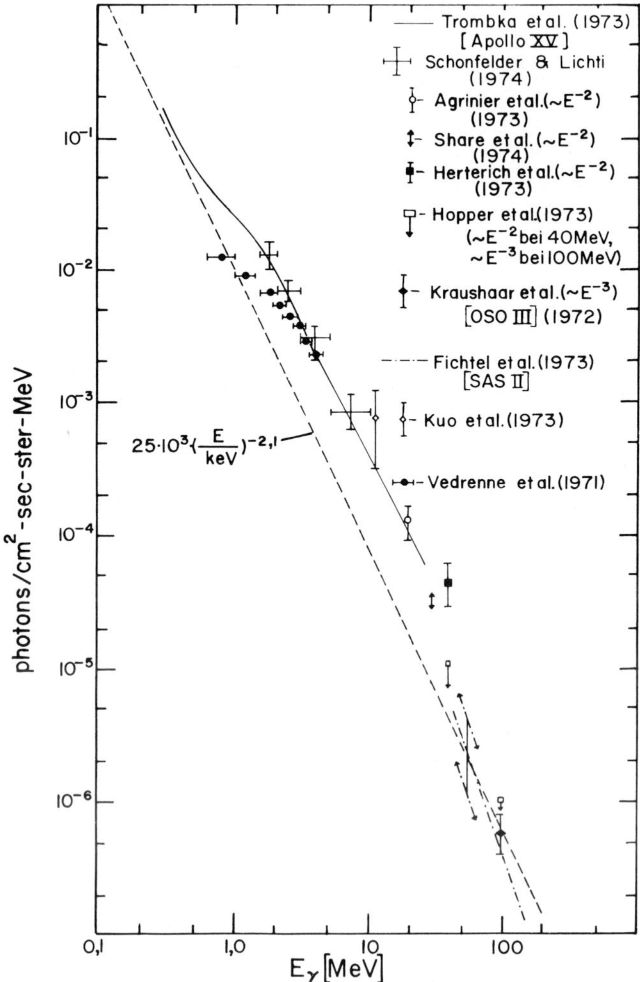

Fig. V-12. A collection of the more recent reports of a diffuse flux of cosmic γ-rays. The results (+) of the double Compton telescope experiment of Schönfelder and Lichti (1974) are compared with other experiments. Dashed line: extrapolation from the X-ray range (according to Pal, 1973 and Dennis *et al.*, 1973). The other experimental results are from Agrinier *et al.* (1973); Fichtel *et al.* (1973); Herterich *et al.* (1973); Hopper *et al.* (1973); Kraushaar *et al.* (1972); Kuo *et al.* (1973); Share *et al.* (1974b); Trombka *et al.*, (1973) and Vedrenne *et al.* (1971). (From V. Schönfelder and G. Lichti, *Astrophys. J. (Letters)* **191**, L1. Copyright 1974, The American Astronomical Society by permission of the University of Chicago Press.)

At energies below 10 MeV, balloon measurements of diffuse cosmic γ-rays are also of particular interest. However, because of the presence of the atmospheric γ-rays, it is difficult to measure the weak cosmic γ-ray flux in such an environment. Vedrenne *et al.* (1971) have attempted a measurement by using an omnidirectional stilbene γ-ray neutron spectrometer with pulse shape discrimination and have studied the latitude dependence of the detector counting rate near the top of the atmosphere. At the lowest latitude (~10° N) they found that the ratio of γ-ray flux to neutron flux increased sharply above a pressure altitude of 20 mb, whereas this ratio is nearly constant at the higher latitudes where measurements were made, i.e., 46° N and 62° N. Vedrenne *et al.* (1971) therefore concluded that an extraterrestrial γ-ray flux was evident at the lowest latitudes. The results of their experiment are also shown in Figure V-12 which indicates agreement with the Apollo 15 results at ~4.5 MeV but a significant disagreement below this energy.

The cosmic diffuse γ-ray flux has recently been investigated by Schönfelder and Lichti (1974) using a balloon-borne Compton telescope flown from Palestine, Texas on February 27, 1973 to an atmospheric depth of 2.45 g cm^{-2}. The instrument, which utilizes double Compton scattering and time-of-flight in a coincidence arrangement to separate out neutron effects, is described briefly in Section VI-6.3.2. If it is assumed that the atmospheric contribution near the top of the atmosphere depends linearly on atmospheric depth and an extraterrestrial γ-ray flux is absorbed exponentially, then the total counting rate in a directional telescope would vary as

$$R = at + be^{-t/\tau} \qquad \text{(V.11)}$$

where a and b are constants which, in general, depend on photon energy, t is the atmospheric depth in g cm^{-2}, and τ is the absorption mean free path for γ-rays in air. The growth curve of the Compton telescope vs. atmospheric depth is shown in Figure V-13

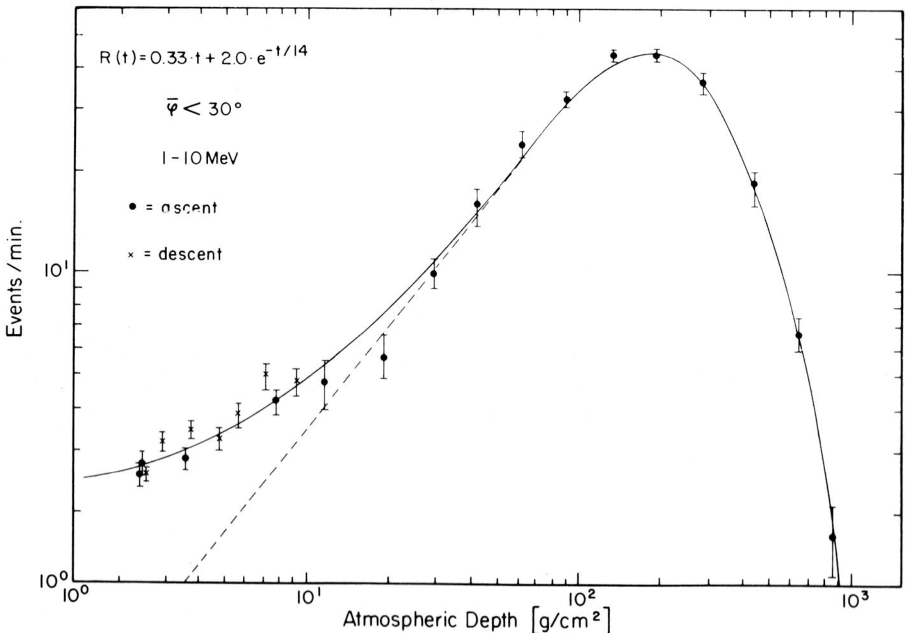

Fig. V-13. The atmospheric growth rate curve measured with a Compton telescope from which a cosmic γ-ray flux is deduced (1 to 10 MeV). (From V. Schönfelder and G. Lichti, *Astrophys. J. (Letters)* **191**, L1. Copyright 1974, The American Astronomical Society, by permission of the University of Chicago Press.)

for incident photons in the energy interval 1 to 10 MeV. For this energy interval, the telescope accepted only γ-rays which made an angle of less than 30° with the vertical. The solid curve in Figure V-13 gives the best fit to the data points shown. Near the top of the atmosphere the best fit parameters for the functional form given in Equation (V.11) are

$$a = (0.33 \pm 0.03) \text{ min}^{-1} \text{ g}^{-1} \text{ cm}^2,$$
$$b = (2 \pm 0.2) \text{ min}^{-1}, \text{ and}$$
$$\tau = 14(+6, -3) \text{ g cm}^{-2}.$$

By plotting such growth curves for four energy intervals between 1.5 and 10 MeV, the extraterrestrial counting rate at zero atmospheric depth ($t = 0$) was found. Taking into account the effective viewing solid angle of the telescope and its measured response function, the energy spectrum of the cosmic flux was deduced. The results for the four energy bands are shown in Figure V-12 and each falls close to the Apollo 15 spectral shape. These results of Schönfelder and Lichti (1974) were corrected downward slightly for the effect of multiple Compton scattering of primary γ-rays in the atmosphere above the telescope. This process enhances the number of γ-rays which enter the 60° telescope aperture and amounts to a correction $\lesssim 8\%$ for photons with energies above 1.5 MeV. An additional correction which must also be applied to the Compton telescope spectrum is a background caused by atmospheric neutrons which can simulate γ-ray events. The basic interactions contributing to this background are discussed in Section VI-6.1.5. In this case, Schönfelder and Lichti (1974) have stated that the fluxes in Figure V-12 should be lowered by 14% below 3 MeV and 50% between 3 and 10 MeV.

5.3.2. OBSERVATIONS (10 MeV TO 100 MeV)

At the higher photon energies the first evidence for a remnant cosmic γ-ray flux was obtained on the OSO-3 experiment, which was described in Section V-5.2.2. Kraushaar *et al.* (1972) have given a value for the integral flux of γ-rays above 100 MeV as

$$I_B = (3.0 \pm 0.9) \times 10^{-5} \text{ cm}^{-2} \text{ s}^{-1} \text{ sr}^{-1}. \tag{V.12}$$

This value is based essentially on the observed OSO-3 detector counting rate at high galactic latitudes ($b^{II} > 35°$), using the sensitivity of their telescope for an isotropic flux. This value obtained for the background flux is somewhat sensitive to assumptions concerning the division of the observed OSO-3 counting rate (see Kraushaar *et al.*, 1972) into galactic disk and isotropic sources. The OSO-3 flux is shown in Figure V-12 expressed as a differential flux in energy at 100 MeV assuming the spectral shape is $\propto E^{-3}$.

The SAS-2 experiment, which was also described in Section V-5.2.2, has confirmed the existence of a background γ-ray flux above 30 MeV from the sky which is generally consistent with the OSO-3 result. This observation by Fichtel *et al.* (1973) is also shown in Figure V-12 as a dot-dash line. If the differential photon energy spectrum is represented as a power law of the form $\propto E^{-\alpha}$, then the exponent $\alpha = 2.7$ (+0.4, −0.3), which is somewhat steeper than the low energy spectrum (~30 keV) discussed above, where $\alpha \simeq 2.4$. The integral flux above 100 MeV, based on the SAS-2 differential spectrum, is

$$I_B(>100\,\text{MeV}) = 2.8(+0.9, -0.7) \times 10^{-5}\ \text{photons cm}^{-2}\ \text{sr}^{-2}\ \text{s}^{-1} \quad (V.13)$$

which is consistent with the OSO-3 result just discussed. It is interesting to note that the spectral shape for the diffuse background at higher energies seems to be significantly steeper than the Galactic Center γ-ray spectrum in Section V-5.2.2. This may indicate different source mechanisms for the two forms of radiation. Fichtel *et al.* (1973) point out that the energy spectrum and isotropy of this radiation needs to be determined before theoretical models can be tested.

An important measurement of the 'diffuse' γ-ray flux at 10 MeV has been reported by Kuo *et al.* (1973) using a balloon-borne spark chamber flown from Palestine, Texas to an atmospheric depth of 3 g cm^{-2}. We will not describe the instrument here, but will only note that the spark chamber uses much thinner plates than the usual spark chambers operating at higher energies. Kuo *et al.* (1973) point out that in their chamber 10 MeV pair electrons appear similar to a 100 MeV pair in conventional chambers. Also, the energy response of the chamber is limited to a range from 4 MeV to 20 MeV. During a balloon flight with a 5156 s float period on 1968, August 19, the chamber recorded 149 electron pairs with an opening angle of $<60°$, which they interpret to be due to γ-rays of external origin. If the γ-rays were all of atmospheric origin and followed a $\sec\theta$ zenith angle distribution, then Kuo *et al.* (1973) deduce a vertical atmospheric flux of

$$I_\gamma = (1.1 \pm 0.1) \times 10^{-3}\ \text{photons cm}^{-2}\ \text{s}^{-1}\ \text{sr}^{-1}\ \text{MeV}^{-1}$$

at 10 MeV and 3 g cm^{-2}. This flux value is compared with the average value of the measured atmospheric γ-ray flux which Kuo *et al.* (1973) have found by extrapolating several other measurements to the conditions of their experiment. From their paper, the extrapolated value of the atmospheric flux at 3.1 g cm^{-2} and at 10 MeV is 5×10^{-4} photons cm^{-2} s^{-1} sr^{-1} MeV^{-1}. Since this is much lower than the above vertical flux value, they conclude that the difference is due to a cosmic diffuse flux at 10 MeV of magnitude

$$I_c = (8.0 \pm 4.5) \times 10^{-4}\ \text{photons cm}^{-2}\ \text{s}^{-1}\ \text{sr}^{-1}\ \text{MeV}^{-1} \quad (V.14)$$

where the error quoted is based on the range of the extrapolated atmospheric flux value (see Kuo *et al.*, 1973). This result is also shown on Figure V-12 slightly above the curve from Apollo 15 given by Trombka *et al.* (1973).

A few other measurements above 10 MeV have been performed by balloon experiments and are also shown in Figure V-12. Share *et al.* (1974b), using an emulsion spark chamber, find an *upper limit* value for the differential flux at 30 MeV equal to 3.5×10^{-5} photons cm^{-2} s^{-1} sr^{-1} MeV^{-1}. This was found by extrapolating to the top of the atmosphere an upper limit integral flux over 10 MeV to 30 MeV using a differential spectral shape $\propto E^{-2}$. The differential flux at 10 MeV, in this case, also falls well below the Kuo *et al.* (1973) value. Herterich *et al.* (1973) give a value at 40 MeV of 4.4 ×

10^{-5} photons cm^{-2} s^{-1} sr^{-1} MeV^{-1} using a balloon-borne spark chamber. Finally, Hopper *et al.* (1973) give upper limit values at 40 and 100 MeV of 1.6×10^{-5} and 1.0×10^{-6} photons cm^{-2} s^{-1} sr^{-1} MeV^{-1}, respectively, using a balloon-borne spark chamber.

5.4. Transient γ-Ray Bursts*

Apparent cosmic bursts of γ-rays (>200 keV) with time durations ranging from 0.1 to 30 s were first reported by Klebesadel *et al.* (1973) and have presented the newest challenge to γ-ray astronomy. Even though the origin of these bursts is unknown at the present time, there is a good possibility that they could be produced in SN events where nuclear lines may be produced. We will therefore review here the present observational knowledge of these events.

The Nuclear Test Ban Treaty in 1963 initiated the development of the Vela satellite series, operated by the Los Alamos Scientific Laboratory, whose primary purpose was to monitor for evidence of clandestine nuclear tests in space. One of the possible emissions from such tests is a short period burst of γ-rays with energies above 100 keV. Since 1967 widely-spaced Vela satellites have detected γ-ray bursts which were neither of man-made origin nor from the Sun or Earth. The first reported observations were made by γ-ray detectors on the Vela satellites denoted as 5A, 5B, 6A, and 6B which were nearly equally spaced in angle in a circular orbit about the Earth with a geocentric radius of about 1.2×10^5 km (Klebesadel *et al.*, 1973). Each spacecraft carried six CsI scintillation counters (each of volume 10 cm^3) which responded to discrete energy loss events in the energy range 200 keV to 1 MeV for Vela 5, and 300 keV to 1.5 MeV for Vela 6 spacecraft. The γ-ray detection efficiency for these devices is reported to be between 17% and 50%. Of particular significance is the fact that the instruments by themselves had no anticoincidence system which allows discrimination against charged particle induced events; however, the passive shielding around the scintillators is such as to prevent *direct* penetration to the CsI detector of electrons with energies below 750 keV or protons with energies below 20 MeV. Of course the charged particles can produce, in any material, secondary γ-rays which can penetrate to the detectors. To specifically identify events as due to γ-rays, it was therefore necessary to resort to the near simultaneous recording of bursts in the separated Vela spacecraft.

Since this system was designed only to detect bursts, the output pulses from the six detectors were summed and fed as a pulse train into the logic circuitry. When a statistically significant increase in the counting rate occurs, a trigger initiates the recording of discrete counts in a series of quasi-logarithmically increasing time channels (Klebesadel *et al.*, 1973). Since this event is recorded along with an absolute time, photon-like events and their origin can be deduced from the degree of simultaneity of similar bursts seen in the widely separated spacecraft.

* The γ-ray bursts discussed in this section, often referred to as Vela bursts, are those detected in coincidence by Vela satellite instrumentation or other satellite instrumentation in coincidence with a Vela detector.

It is clear that the required trigger must be generated when a sufficiently large fluctuation occurs in the counting rate over the average background. The background rates $B\,(\mathrm{s}^{-1})$ are $150\,\mathrm{s}^{-1}$ and $20\,\mathrm{s}^{-1}$ in Vela 5 and 6, respectively. Thus a trigger could be provided when the number of counts in a time interval Δt exceeds $N = n\sqrt{B\Delta t}$, or $N_5 > 12.2\, n\sqrt{\Delta t}$ and $N_6 > 4.5\, n\sqrt{\Delta t}$, where the subscripts refer to spacecraft number, and n is the number of standard deviations of the background fluctuation required to provide the trigger. Here n and Δt are unspecified, but clearly a compromise must be made, since if n is too small there will be numerous false triggers, and if n is too large only the largest γ-ray burst would provide a trigger. At this point, it should be mentioned that a burst detection scheme as described here biases selection of burst events

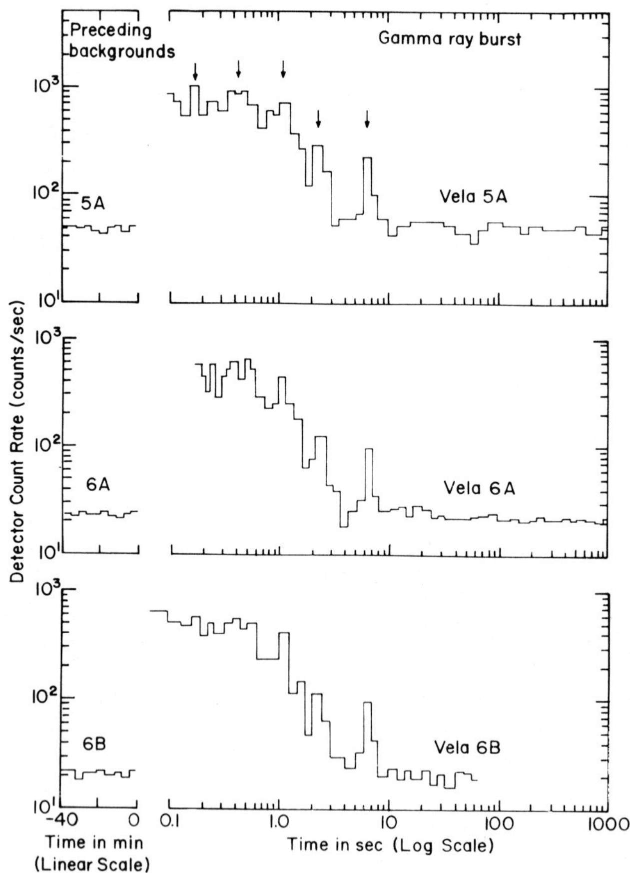

Fig. V.14. The counting rate as a function of time for the γ-ray burst of 1970, August 22 as recorded at three Vela spacecraft. Arrows indicate some of the common structure. Background counting rates immediately preceding the bursts are also shown. The Vala 5A rates have been reduced by 100 counts per second (a major fraction of the background) to emphasize structure. (From R. W. Klebesadel *et al.*, *Astrophys. J. (Letters)* **182**, L85. Copyright 1973, The American Astronomical Society, by permission of the University of Chicago Press.)

to that *class* which have rise times of the order of Δt, the time band in the trigger circuit. For Vela type measurements $\Delta t \ll 1$ s.

The initially published Vela results (Klebesadel *et al.*, 1973) gave information on bursts which were recorded by at least two spacecraft with a deviation from simultaneity of 4 s or less. (Note: The *maximum* time delay between spacecraft events for bursts generated by photons is 0.8 s.) Sixteen events met this criteria and of these two were recorded by 4 Vela spacecraft. Figure V-14 shows an example of a burst observed by three Vela spacecraft on 1970, August 22, taken from Klebesadel *et al.* (1973). The ordinates give the three detector counting rates at the times indicated on the abcissa. Before the start of the event, the time scale is linear to show 10 measurements of the background counting rate made at 4 min intervals. After the burst begins, the time base is logarithmic in time. In this case, the counts obtained in the first 4 s of the burst have an integrated flux density in the energy range 200 keV to 1 MeV of $\sim 8 \times 10^{-5}$ erg cm^{-2}, from Vela 5 and $\sim 6 \times 10^{-5}$ erg cm^{-2} in the energy range 300 keV to 1.5 MeV, from Vela 6 (Klebesadel *et al.*, 1973). As can be seen from Figure V-14, there is clearly a common time structure seen in the three independent records at several times but especially at 6 s after the start of the burst. This structure had been used to adjust the three records in time relative to the start of the event. In the case of the peak centered at 6.5 s shown in Figure V-14, the time-integrated flux densities are 10^{-5} erg cm^{-2} and 4×10^{-6} erg cm^{-2} from Vela 5 and Vela 6, respectively, indicating a softer spectrum than in the first 4 s of the burst. It is of particular interest that no known nova or supernova occurrences could be related in time or direction to any of the bursts based on their preliminary timing data. Again the Earth and Sun are ruled out as sources of these apparent γ-ray bursts.

Strong *et al.* (1974) have published a preliminary catalogue of all γ-ray bursts observed by the Vela satellites between 1967, July 2, and 1973, June 10. The method of identifying events for this catalogue was essentially the same as that described above and gives a list of 23 non-solar events. Table V-5 is a reproduction of this list, where the events are identified by calendar date and Universal Time of occurrence in seconds given in columns two and three. This is the time of arrival of the signal at the Earth in most cases, but unless the source direction is known it is not possible to give this accurately. The remainder of the table gives information on possible location of the sources, a point which needs clarification. In the lower part of Table V-5 some additional events which have been identified are listed in a slightly different format. Most of these additional events were found by correlation of Vela data with that from other types of spacecraft carrying different instrumentation as discussed below.

Since the ultimate question about these γ-ray bursts is really: "Where do they come from?", we will digress briefly to discuss the method of determining source locations using time delays between pairs of satellites. In Figure V-15 is shown the basic geometry, where we consider a burst source at an arbitrary location on the celestial sphere and use as a reference the orientation of the line connecting two satellites or spaceprobes, P_1 and P_2, whose absolute positions are known in space in reference to the Earth. The source

TABLE V-5

A list of cosmic γ-ray bursts observed through 1973. (From I. B. Strong *et al.*: *Astrophys. J. (Letters)* **188**, L1. Copyright 1974 The American Astronomical Society, by permission of the University of Chicago Press), (also, private communication.)

Event Number	Date			UT (sec)	Position coordinates				Circle of position			Estimated flux (erg cm^{-2})
					Equatorial (α, δ)		Galactic (l^{II}, b^{II})		Equatorial	Galactic		
	Yr.	Mo.	Day		Pos. 1	Pos. 2	Pos. 1	Pos. 2	α, δ	l^{II}, b^{II}	Radius (deg)	
67-1	67	7	2	51568	No directional information possible							
69-1	69	7	3	26233	. . .	. . .	. . .	. . .	191, + 20	294, + 83	29	2 × 10^{-5}
69-2	69	10	7	26791	. . .	. . .	. . .	. . .	236, + 33	52, + 52	19	2 × 10^{-4}
69-3	69	10	17	11927	. . .	. . .	. . .	. . .	27, − 27	215, − 77	86	2 × 10^{-5}
69-4	69	10	17	78113	. . .	. . .	. . .	. . .	156, + 1	244, + 46	83	4 × 10^{-5}
70-1	70	6	14	18416	Solar event							
70-5	70	7	10	19066	. . .	. . .	. . .	. . .	. . .	. . .	. . .	. . .
70-2	70	8	22	60571	144, + 61	209, − 29	153, + 44	320, + 31	. . .	. . .	. . .	1 × 10^{-4}
70-3	70	12	1	72059	. . .	. . .	. . .	. . .	91, − 29	235, − 22	76	4 × 10^{-5}
70-4	70	12	30	25337	120, + 10	149, − 30	212, + 20	264, + 19	. . .	. . .	. . .	3 × 10^{-4}
71-1	71	1	2	69056	216, + 60	(225, + 3)	103, + 54	(1, + 50)	. . .	. . .	. . .	1 × 10^{-4}
71-6	71	2	27	62855	. . .	. . .	. . .	. . .	. . .	. . .	. . .	. . .
71-2	71	3	15	40827	. . .	. . .	. . .	. . .	4, − 21	72, − 80	82	1 × 10^{-5}
71-3	71	3	18	55685	69, + 12	115, − 71	185, − 22	283, − 22	. . .	. . .	. . .	1 × 10^{-4}
71-4	71	4	21	11919	. . .	. . .	. . .	. . .	247, + 34	55, + 43	83	3 × 10^{-6}
71-5	71	6	30	63059	. . .	. . .	. . .	. . .	27, − 30	228, − 77	52	5 × 10^{-4}
72-1	72	1	17	63556	104, + 9	136, − 29	206, + 6	255, + 12	. . .	. . .	. . .	7 × 10^{-5}
72-2	72	3	12	57195	277, + 1	298, + 35	31, + 4	71, + 4	. . .	. . .	. . .	5 × 10^{-5}
72-3	72	3	28	49588	. . .	. . .	. . .	. . .	283, + 22	53, + 9	86	1 × 10^{-4}
72-6	72	4	27	39512	. . .	. . .	. . .	. . .	. . .	. . .	. . .	. . .
72-4	72	5	14	13591	176, + 78	. . .	127, + 39	. . .	. . .	. . .	. . .	2 × 10^{-4}
72-5	72	11	1	68206	1, + 19	309, − 56	121, − 44	342, − 37	. . .	. . .	. . .	7 × 10^{-6}
73-1	73	5	7	29072	342, + 56	(254, − 33)	107, − 3	(351, + 6)	. . .	. . .	. . .	6 × 10^{-5}
73-2	73	6	10	75582	. . .	. . .	. . .	. . .	25, − 36	255, − 76	61	1 × 10^{-4}

Remarks to Table V-5 (above)

Event 67-1. 2 peaks, ~ 2 s apart.

Event 69-1. 2 distinct peaks totalling ~ 0.5 s.

Event 69-2. Structured pulse ~ 3 s. Similar one beginning at ~ 5 s.

Event 69-3. Single 0.1-s spike. Weak continuing flux.

Event 69-4. Double peak ~ 0.5 s wide. Another at ~ 2.5 s.

Event 70-2. Broad, structured 3-s pulse. Another at ~ 6 s.

Event 70-3. Structured 1-s burst.

Event 70-4. Strong structured 1.5-s burst. Weaker one begins ~ 2 s.

Event 71-1. Long event. No clear structure. Decays over ~ 10 s. or more.

Event 71-2. Complex 3-s pulse. Another large one at ~ 5 s.

Event 71-3. 4-s pulse. Weaker one begins ~ 5 s.

Event 71-4. Highly structured 1-s pulse.

Event 71-5. Sawtooth, 2-s pulse. Second pulse at ~ 5 s.

Event 72-1. Structured, 1-s burst. Broad pulse centred at ~ 6-7 s.

Event 72-2. Complex. ~ 6-s burst.

Event 72-3. Complex, ~ 6-s burst. Another beginning ~ 8 s.

Event 72-4. Complex set of peaks lasting ~ 4 s. Weaker ones at ~ 7 s and ~ 28 s.

Event 72-5. Short, 0.1-s spike.

Event 73.1. Complex, structured, ~ 3-s burst.

Additions (private communication, R.W. Klebesadel)

Date	UT	Observations
69-07-19	03:46:46	OSO-5; 1 Vela
72-12-18	20:27:39	2 Velas; Imp-7, 1972-076B
73-01-25	15:15:00	2 Velas
73-03-02	23:27:58	2 Velas; Imp-7; SAS-2
73-06-06	07:07:28	Imp-7; SAS-2
73-06-06	18:47:14	SAS-2; Imp-7; 2 Velas
73-07-21	08:55:18	Velas; Imp-7
73-07-25	17:07:57	Velas; Imp-7
73-12-17	08:09:08	Imp-7, Vela
73-12-23	00:39:47	Imp-7, Vela

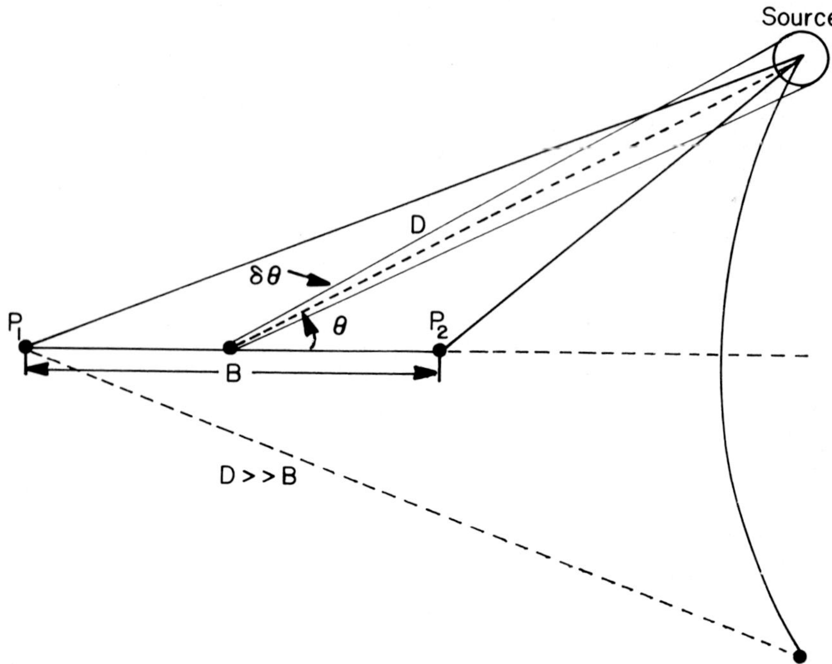

Fig. V-15. The burst location geometry is shown for two observing sites separated by the base line B.

is at a distance D from the base line of length B. In the case that $D \gg B$ then the arrival time difference for light signals at P_2 and P_1 is given by

$$t_{\mathrm{D}} = (B/c) \cos \theta . \tag{V.15}$$

Thus the measurement of the time delay between the signals directly locates the source as lying on a cone of half-angle θ, if we know t_{D} with no error. The uncertainty, Δt_{D}, in the time delay gives an uncertainty $\Delta\theta$ in the cone's half-angle equal to

$$\Delta\theta = \Delta t_{\mathrm{D}}/[(B/c) \sin \theta] . \tag{V.16}$$

Therefore, the source lies on an annular ring on the celestial sphere of angular width $\Delta\theta$ and radius θ. In Figure V-16 we give a plot of $\Delta\theta$ vs. the base line B, which ranges from a fraction of the Earth's radius to ~7 AU. This covers a range of base lines for conceivable experiments conducted on balloons and deep spaceprobes and for several values of θ and Δt_{D}. Observation of the same burst by a third satellite would give, in general, two intersecting rings on the celestial sphere which would further limit the source location to the regions of intersection.

The next development in the γ-ray burst saga was the identification of six of the Vela bursts in the Imp-6 data records by Cline *et al.* (1973). The Imp detector consisted of a 2.25 inch diameter by 1.5 inch thick CsI (Tl) crystal surrounded with a thin 4π plastic scintillator viewed by a single PM tube and operated as a Phoswich, which gives charged particle rejection. Pulse height analysis was carried out by a 14 channel

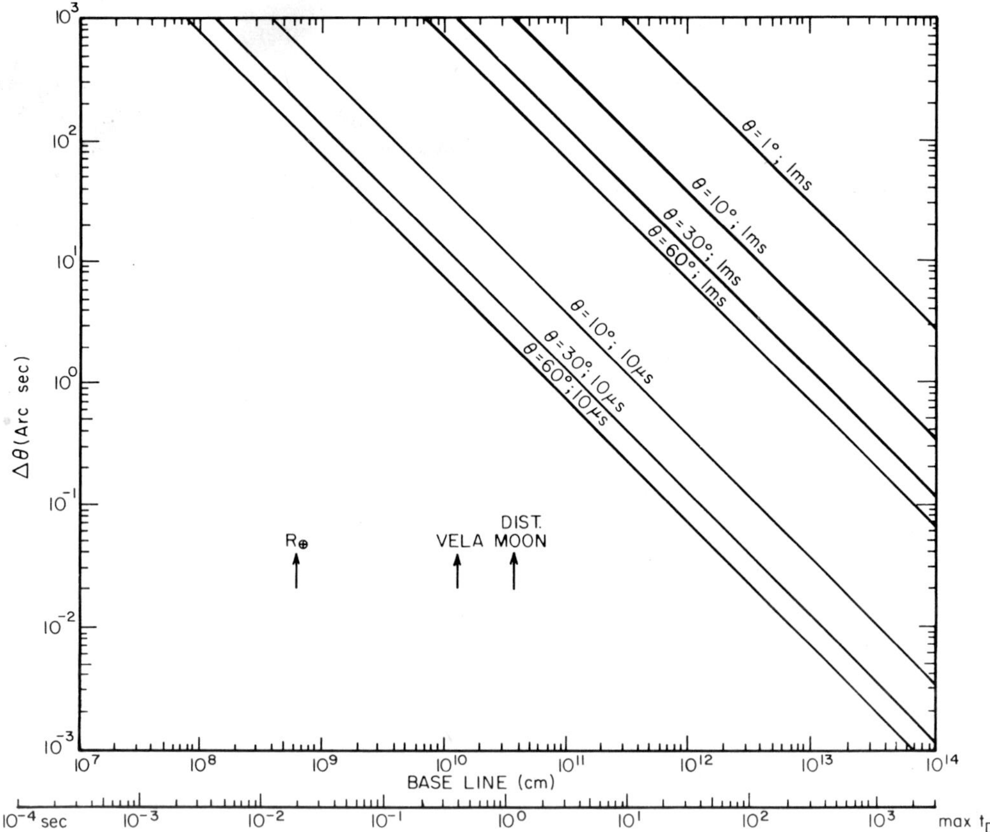

Fig. V-16. The angular uncertainty, $\Delta\theta$, of the burst location circle in the celestial sphere vs. the base line, B, between two observing sites. See Equation (V.16).

analyzer with simultaneous storage in all channels. In addition, several other counting rates were recorded including the total intensity, charged particle intensity and γ-ray intensity. Spectral accumulation times were 5.1 s and were taken at 6.3 s periods during the daytime portion of the orbit. The detector therefore has a 50% time coverage, but since the satellite is spinning and covers all the sky, the bursts may be modulated with the spin period of 12.6 s.

As an example, during the Vela event on 1971, June 30, the Imp-6 detector gave a clear indication that the events are truly due to photons since the average charged particle shield rate increased by only ~10% while the γ-ray crystal rate increased by nearly 100% over several seconds. Also, during one 5 s interval, the photon counts in the energy range 140 to 475 keV increased from a background level of 400 counts to about 5000 counts, a very dramatic event indeed, and apparently one of the strongest recorded by Imp-6. In the case of six overlapping Vela-Imp events, the Vela data indicated two distinct pulses of a few seconds duration and separated by several seconds. The Imp detector was able to record pulse height spectra for several of these separate pulses. It was found that the energy spectra of the burst are fairly well fit by the form

$$dn/dE = I_0 \exp(-E/E_0) \text{ (photons cm}^{-2} \text{ keV}^{-1}\text{/burst.)} \quad (V.17)$$

The product $E_0 I_0$ gives the total number of photons cm^{-2} arriving at the Earth during the burst. If allowance is made for the possible directions of the sources, which is not well known at this time, and attenuation of photons by the Imp spacecraft then all spectra are consistent with $E_0 \cong 150$ keV. The Imp spectra also show (Cline *et al.*, 1973), for those cases where more than one pulse height spectrum per pulse was obtained, that there was no evidence for significant changes in the spectral shape (E_0). Also there was no evidence for line structure during the Imp events within the energy and time resolution of the instrument. The intensity spectrum for a photon number spectrum of the form given in Equation (V-17) above is: $E\,dn/dE = I_0 E \exp(-E/E_0)$ (erg cm^{-2} keV^{-1}/burst), and the total energy flux above a photon energy E (keV) is:

$$E_T = 1.6 \times 10^{-9}\, I_0 (E_0^2 + E_0 E)\, e^{-E/E_0} \text{ (erg cm}^{-2}\text{)}. \quad (V.18)$$

Values for I_0 (cm^{-2} keV^{-1}) and E_0(keV) are given in Cline *et al.* (1973) for the Vela-Imp bursts. Typical values for the energy content of these bursts above 100 keV are in the range $3 \times 10^{-6} \rightarrow 5 \times 10^{-4}$ erg cm^{-2}. Table V-5 gives the energy content of the 23 Vela bursts as deduced from those observations.

One of the largest events also was seen by the two UCSD X-ray detectors on the OSO-7 satellite on 1972, May 14, (Wheaton *et al.*, 1973). In this case the burst was seen by three Vela satellites in 120 000 km orbits, Imp-6 in an eccentric 200 000 km apogee orbit (at a geocentric radius of 43 000 km for the burst) and by OSO-7 in an approximately 500 km altitude, 33° inclination orbit. Figure V-17 shows the energy spectra for both pulses of the 1972, May 14, event from the OSO-7 solar X-ray detector (11 to 346 keV) and from Imp-6 (> 108 keV), where the latter is normalized to the OSO-7 spectral data in the energy range 100 to 346 keV, since the two satellites did not accumulate data over the same time interval.

We can note two important conclusions from the spectra shown:

(1) There is no evidence for strong line structure within the resolution of the detectors, and;

(2) The second pulse appears to be harder than the first pulse.

It is not known if corrections have been made for the changing efficiency as a function of the burst direction with respect to the X-ray telescope axis. In the lower energy band of the X-ray telescope ($10.6 \rightarrow 14.4$ keV) the maximum flux was nearly 1 photon cm^{-2} s^{-1} keV^{-1} compared to a quite time 2σ upper limit of 2×10^{-3} photons cm^{-2} s^{-1} keV^{-1} from the event direction. The data on this event are the most complete and clearly establish the γ-ray character of this event and the fact that neither the Sun nor the Earth are direct sources. Wheaton *et al.* (1973) also point out for the 1972, May 14 event, that no X-ray sources, supernovae, or flare stars lie close to the OSO-7 source position. They mention, however, the intriguing possibility that QSO 3C 249.1 is 4° from the UCSD and 2° from the Vela center positions. Also many galaxies and stars lie within the positional uncertainty of this event. Since ~0.1 s time variations occur in the sudden pulses in this event, one can infer that the physical dimensions of the emitting

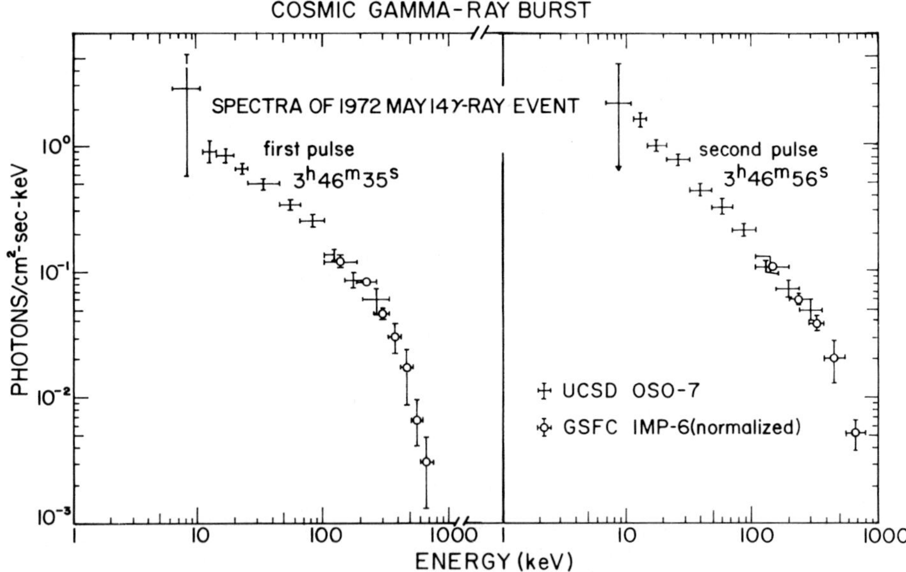

Fig. V-17. The spectra of the γ-ray event of 1972, May 14, during the two pulses or peak intensity points. The open circles are the GSFC Imp-6 data normalized to the data points from the UCSD OSO-7 solar X-ray telescope. (From W.A. Wheaton *et al.*, *Astrophys. J. (Letters)* **185**, L57. Copyright 1973, The American Astronomical Society, by permission of the University of Chicago Press.)

source could not be larger than $\sim 3 \times 10^9$ cm. Quasi-stellar object (QSO) time variations, on the other hand, are longer (~ 1 day), so the source dimensions must be much larger (e.g., 10^{-3} pc for a QSO).

Subsequent to the initial reports providing verification of the occurrence of the Vela γ-ray bursts, other satellite experiments have provided valuable information. Some of the more important later observations correlated with Vela events have been made by experiments on Uhuru, OGO-5, Cosmos 461, U.S. Satellite 076B, Apollo 16, SAS-2, Imp-7 and OSO-6. Uhuru, which was launched on 1970, December 12, has detected 5 of the reported Vela events (Koch *et al.*, 1974). The events detected in this case were not primary γ-ray events as in the case of the scintillation counter measurements described above from Vela, Imp-6, and OSO-7. The X-ray satellite Uhuru used 2 separate proportional counter detectors with viewing directions separated by 180° and with $5° \times 5°$ and $0.5° \times 0.5°$ collimator fields of view. Detection of γ-ray bursts by these proportional counters is primarily through the Compton electrons produced in their walls of thickness of a few g cm^{-2}. The electron energy loss detected is in the range $2 \rightarrow 20$ keV; however, flux determinations are difficult, since the effective thickness of spacecraft material for the γ-rays is greatly different for different source locations. In general, the following correlations were made with the Vela events:

(a) Uhuru events occurred simultaneously with those of Vela within the minimum time resolution of ~ 1 s.

(b) The time structure during the separate pulses were similar within statistics.

(c) The relative counts in the events observed by Vela and Uhuru detectors were approximately the same from event to event.

Finally, these observations assisted in identifying the directions of some of the sources. In particular, in the case of Vela event 71-1 on 1971, January 2, the only Vela position allowed is at galactic coordinates $l^{\mathrm{II}} = 103°$ and $b^{\mathrm{II}} = +54°$ (see Table V-5).

The University of Chicago cosmic electron experiment on OGO-5 also obtained clear evidence (L'Heureux, 1974) for the Vela event 71-5, which occurred on 1971, June 30. This experiment was designed to study electrons in the energy range 10 to 200 MeV, and in its normal mode of operation is insensitive to a Vela-type γ-ray burst. Since the rates from the individual detectors in the system are monitored separately, a means is available to detect such phenomena. The two subdetectors used in the study were: (1) a solid state detector Si(Li) of thickness 800 μm and area 10 cm^2 with a γ-ray threshold of 150 keV, which could detect a 1 s burst with a flux of 30 photons cm^{-2} s^{-1}, and (2) a 5 cm $\times$ 6 cm CsI(Tl) scintillation crystal having a threshold of $\sim$5 MeV, which could detect a 1 s burst with a flux greater than 50 photons cm^{-2} s^{-1}. In addition, it was believed by L'Heureux (1974) that the photomultiplier tube used to detect the light in a gas Cerenkov counter also responded to the γ-ray burst through the direct production of electrons in its photo-cathode or dynode structure. The PM threshold was set at $\sim$0.5 photoelectrons.

During the lifetime of this experiment from March 4, 1968, until July 1971, one Vela cosmic event was detected (71-5) while OGO-5 was at a distance greater than 60 000 km from the Earth in its highly eccentric orbit. Only the solid state detector with the 150 keV threshold and the Cerenkov photomultiplier responded to this burst. The signal to noise ratio was, however, three during the burst, which occurred within only one 1.153 s data sampling period. The OGO-5 flux for this event was 100 to 200 photons cm^{-2} s^{-1} above 150 keV with an upper limit of 50 photons cm^{-2} s^{-1} for the flux above 5 MeV, since the CsI detector did not respond. This flux value is also consistent with the differential energy spectrum reported by Cline *et al.* (1973) for this event. Failure of this experiment to detect any other Vela bursts was undoubtedly due to the smaller size (by more than a factor of 3) of the other events and the fact that photons arriving at large angles to the main detector axis would be absorbed in detector and spacecraft material. Unfortunately, no clear directional information can be obtained for this event from OGO-5.

A NaI(Tl) scintillation spectrometer on Cosmos 461 has also observed the Vela burst (72-1) on 1972, January 17 (Mazets *et al.*, 1973). The detector threshold was 50 keV, and its area was $\sim$57.5 cm. This detector responded to a three-pulse sequence starting at 63555.5 s UT, which lasted for 37 s. Mazets *et al.* (1973) report that the total energy flux in this event is $\sim 3 \times 10^{-5}$ erg cm^{-2} in the energy region 50 to 300 keV. Because of screening of the satellite by the Earth, the source could not lie in a circle on the celestial sphere of radius 68° centered at equatorial coordinates $l = 205°$, $\delta = +7°$.

An important characteristic of the Vela events is the short scale time structure observed in some events. Klebesadel (1974) has reported time structure as short as 16 ms, which is as short as the Vela instrumentation can measure. Imhof *et al.* (1974) have given a detailed time history of a Vela event observed on 1972, December 18, observed on board the polar orbiting satellite 1972-076B. The instrumentation on this satellite consisted of a 50 cm^3 cooled Ge(Li) γ-ray spectrometer in a plastic scintillator anti-coincidence shield, but the spectrometer was also collimated to $\pm 45°$ by a W well. The axis of the collimator was oriented at 105° to the spin axis of the satellite, which had a 5 s period. The high resolution γ-ray spectra were accumulated and read out every 32 ms and the counts in the anticoincidence shield were accumulated during alternate 32 ms intervals. The signals from a second anticoincidence shield on a second Ge(Li) spectrometer (which did not function) were accumulated during the other 32 ms intervals. Even though the Ge(Li) spectrometer had excellent energy resolution of 3.5 keV at all energies shortly after launch on 1972, October 2, the energy resolution had degraded seriously before the 1972, December 18, burst occurred so this important advantage was lost. However, the high time resolution properties of the instrument gave valuable new information on the event. In particular, the Vela instrument was triggered at 7365.75 s on 1972, December 18, but the 076B Ge(Li) and shield detectors recorded at least two microbursts before the Vela 6B trigger. These two microbursts had widths of about 60 ms, which indicates that the width of the γ-ray emitting region must have been less than $\sim 2 \times 10^9$ cm. In addition during the whole event, five microbursts appear to have occured in 1.8 s. Also the combination of the high time resolution and energy resolution properties of the instrument permitted the measurement of approximate energy spectra for time intervals as short as 0.1 s during the bursts. The general conclusion from the spectral measurements is that the microbursts recorded before the Vela 6B trigger were harder than those observed at later times. Another Vela event was also recorded by the shield detector on 076B on 1973, July 21, giving further evidence for fine time structure (<100 ms) during these events (Imhof *et al.*, 1975).

A photon burst was detected on 1972, April 27 with the γ-ray spectrometer on Apollo 16 while in *cis*-lunar space on trans-Earth trajectory (Metzger *et al.*, 1974). This event was confirmed by Vela 6A records but was not initially reported by Klesbesadel *et al.* (1973); however, it is listed in Table V-5. This is a example of a Vela γ-ray burst discovered, in a sense, after the fact. The γ-ray spectometer on Apollo 16 consisted of a 7.0×7.0 cm cylindrical crystal of NaI(Tl) surrounded by a 0.8 cm thick plastic scintillator for charged particle rejection. The pulse height spectrum in the NaI(l) crystal was fed to a 512 channel pulse height analyzer and the data were read out in real time with a single data frame time of 328 ms. In addition the Apollo 16 instrumentation included a seven-channel X-ray spectrometer of 60° field of view and a time resolution of 8 s.

The event was recorded by both Apollo 16 X-ray and γ-ray spectrometers but in Figure V-18 shows only the high time resolution counting rate profile from the γ-ray spectrometer, which covers the energy range from 0.067 to 5.1 MeV. The main portion of the event is considered by Metzger *et al.* (1974) to begin with the peak of the first main impulse occurring at 10:58:29:7 UT just before the strong peak shown in Figure V-18 at 10:58:30.2 UT. This peak is followed by two successive strong peaks occurring

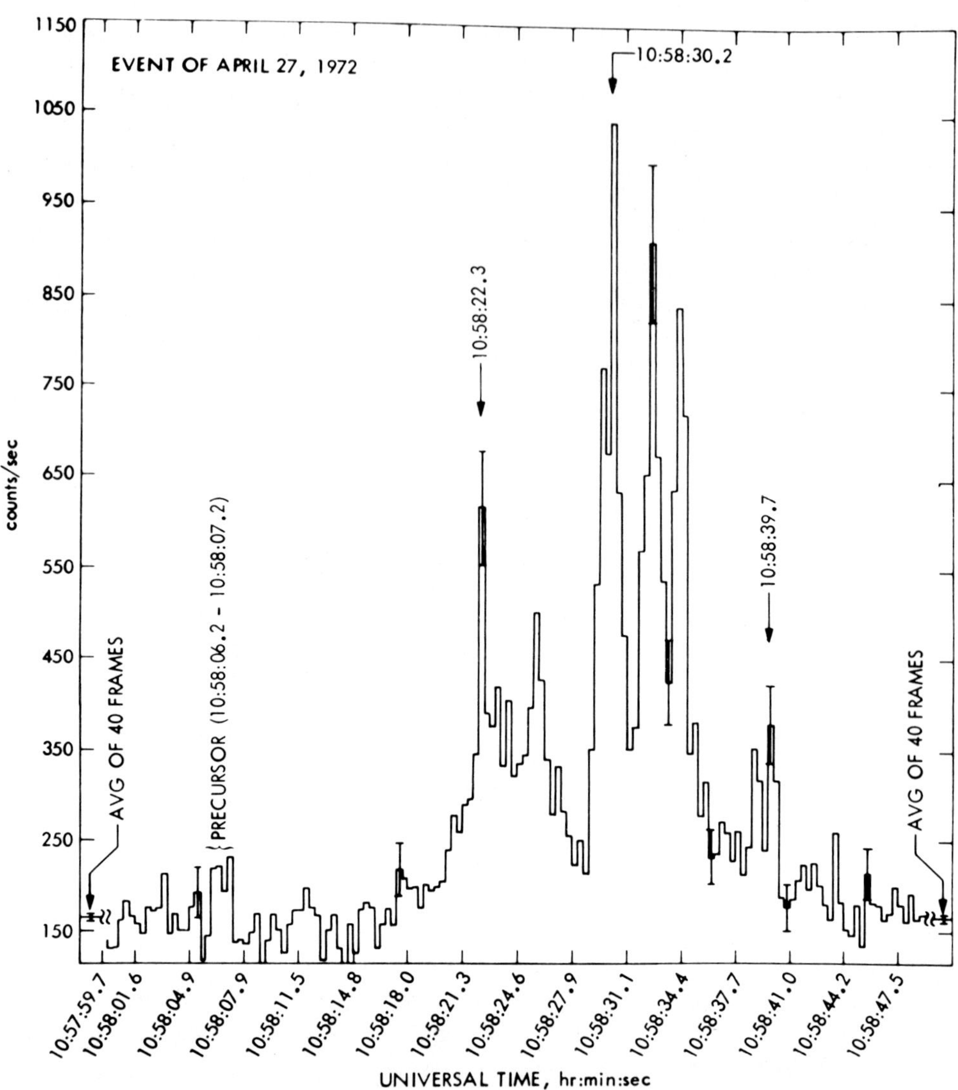

Fig. V-18. The time profile is shown for the γ-ray burst observed by Apollo 16 instrumentation on 1972, April 27. The event was also located in the Vela 6A records. (From A.E. Metzger *et al.*, *Astrophys. J. (Letters)* **194**, L19. Copyright 1974, The American Astronomical Society, by permission of the University of Chicago Press.)

within 4 s. Both subsequent and prior to these three main bursts are several smaller bursts and there is evidence for a precursor to the main event occurring from 10:58:06.2 to 10-58:07.2 UT. In addition by comparison of the Apollo 16 time profile shown with that from Vela 6A, Metzger *et al.* (1974) have shown that the latter's instrument triggered on the second of three main bursts, which started at about 10:58:32 UT. Velas 5A and 5B were not in the readout mode at the time of the event and Vela 6B

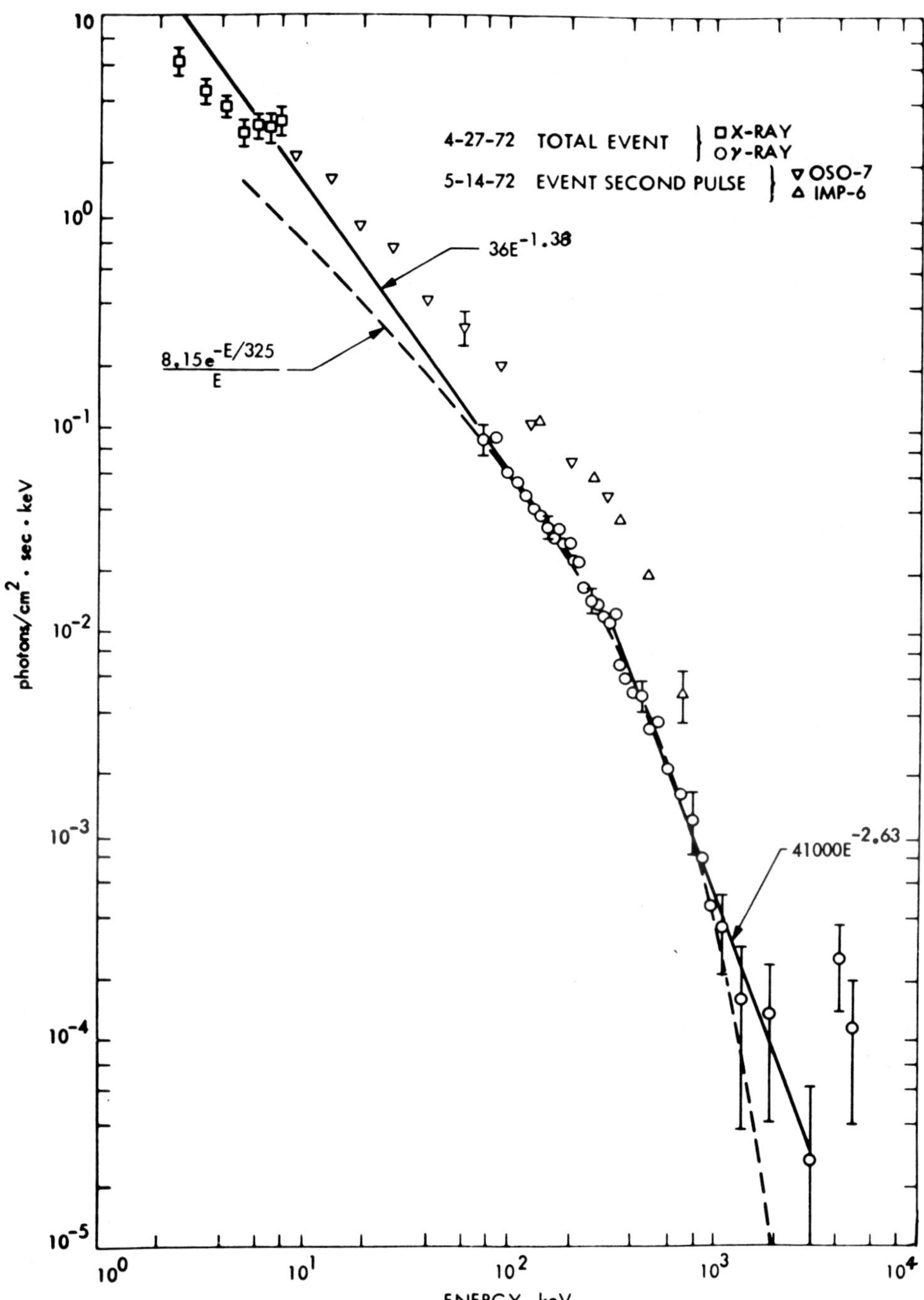

Fig. V-19. The incident photon spectrum is shown for the total event of 1972, April 27, for the energy range 2.0 keV to 5.1 MeV. The spectrum for the second pulse of the event on 1972, May 14 observed by Wheaton *et al.*, (1973) is also shown. (From A.E. Metzger *et al.*, *Astrophys. J. (Letters)* **194**, L19. Copyright 1974, The American Astronomical Society, by permission of the University of Chicago Press.)

apparently required a higher signal. The extensive time history, for this event from Apollo-16 and the demonstration that small precursors and other time structure can be masked by the Vela trigger requirements suggest the apparently complex nature of this phenomenon. The photon energy spectrum for the full event on April 27 is shown in Figure V-19 and is based upon data from the Apollo 16 X-ray and γ-ray spectrometers. In order to convert the measured energy loss spectrum into a photon spectrum, manual unfolding and matrix inversion techniques were used as well as knowledge of the γ-ray detector's response for photons incident at 20° to its normal. [The probable direction of the source of the energetic photons was determined from an analysis of the time response to the event as measured by the Apollo 16 X-ray spectrometer and is centered at $\alpha = 22^h.9$ and $\delta = -68°$ (Trombka *et al.*, 1974).] The general shape of the spectrum follows that given for the event of May 14, 1972 observed by Wheaton *et al.* (1973) and also shown in Figure V-19. Of particular interest for this event is the lack of evidence of any γ-ray spectral features that would indicate nuclear processes at work in the source region. The upper limit flux established for nuclear lines in the spectrum is 5×10^{-2} photons cm^{-2} s^{-1} at 0.51 MeV and 1×10^{-1} cm^{-2} s^{-1} at 2.2 MeV; however, these rather high values do not absolutely exclude nuclear processes. The total energy in the spectrum was 2×10^{-4} erg cm^{-2} above 2 keV and 1.5×10^{-4} erg cm^{-2} above 70 keV. It should also be noted that the Apollo 16 event shows substructure on the time scale of 50 ms similar to the observation of Imhof *et al.* (1974).

A general characteristic of the γ-ray bursts appears to be the constancy of the average shape of the differential energy spectrum for all events for which sufficient data was obtained. This evidence has been primarily strengthened by the observations made by Cline and Desai (1975) on Imp-7. The Imp-7 γ-ray detector and spectral analysis scheme were identical to that in Imp-6 as discussed above (Cline *et al.*, 1973); however, the Imp-7 spin rate and the data collection duty cycle was greatly improved. The main consequence of these differences is that the Imp-7 detector was more nearly omnidirectional than the one on Imp-6 and therefore the γ-ray energy loss spectra obtained were more directly representative of the actual photon spectrum incident on the spacecraft.

Based upon a study of nine of the eleven recorded gamma ray bursts during 1972 and 1973, Cline and Desai (1975) have demonstrated that the average photon number spectral shape is of the form:

$$\begin{aligned} dN/dE &\propto \exp(-E_\gamma/150); \quad 100 \rightarrow 400\,\text{keV} \\ dN/dE &\propto E_\gamma^{-2.5}; \quad 400 \rightarrow 1100\,\text{keV} \end{aligned} \tag{V.19}$$

where E_γ is the photon energy in keV. This approximate spectral shape is also consistent with that observed by detectors on OSO-6 (Palumbo *et al.*, 1974), on OSO-7 (Wheaton *et al.*, 1973) and on Apollo 16 (Metzger *et al.*, 1974). It has also been mentioned above that the event of 1972, December 18 appeared to show a change of spectral shape with time during the event (Imhof *et al.*, 1974); however, the Imp-7 observation of the average spectral shape for this event fits that given in Equation (V-19).

The Imp-7 observations have also indicated that the event rate is $8 \pm 2\,yr^{-1}$ rather than the earlier reported rate of $5 \pm 1\,yr^{-1}$. This increase is primarily due to the lower event size threshold of the Imp detector, which permitted the smaller events to be identified in the Imp data and then confirmed in the Vela records.

There are many other confirming observations of the Vela events that could be discussed (e.g., Fichtel *et al.*, 1974; Share *et al.*, 1974c; Share, 1975; Pizzichini *et al.*, 1975) but the apparently significant characteristics of this phenomenon have been obtained by the methods discussed above (see Section III-3.2.6). Probably the most important new piece of information necessary to help identify the nature of the basic mechanism initiating these events is the precise localization of the sources in space, even for one event. This will require a measurement with angular accuracy of much less than the few degree positions now known – down to an arc second is desired. When this is achieved, it is possible that observations can be made at other wavelengths (optical, radio, etc.) and the key to understanding the phenomena may be found. Observation of nuclear γ-ray lines would help to establish the radiation mechanism(s) operative in the sources. Finally, it is important to emphasize that all searches for coincident phenomena have thus far been unsuccessful (cf. Baird *et al.*, 1975, for a VHF radio search and Grindley *et al.*, 1974, for an optical search). Chapter VII and Notes Added in Proof should be consulted for references to the latest developments in this field.

CHAPTER VI

EXPERIMENTAL CONSIDERATIONS FOR NUCLEAR γ–RAY ASTRONOMY

In this chapter considerations pertinent to the design of experiments will be discussed. Instrument backgrounds present a basic limitation to the flux sensitivity for any detector scheme; however, the *a priori* determination of the background for a nuclear γ-ray line experiment is an extremely foreboding task. Therefore we present some of the important basic limitations affecting the background in an experiment, as well as some important results obtained in actual practice by various groups. Improvements can only be made with a knowledge of these basic limitations. A short discussion is also given of the flux sensitivity and energy and spatial resolution requirements for the field. Finally, the most significant detection methods that are in use now, or are being developed for forthcoming experiments, will be reviewed. A recent review by Peterson (1975) gives a discussion of some experimental techniques used in X-ray astronomy and complements some of the discussion here.

6.1. Background Factors

Consider an arbitrary γ-ray detector or element of such a detector placed in a balloon or satellite environment. In general the detector will see a net background counting rate, $\mathrm{d}B(E)/\mathrm{d}E$ (cts s^{-1} MeV^{-1}), where E is the energy loss in the γ-ray detector that would correspond to some channel number equivalent to a full energy loss by a γ-ray of energy E. The limiting γ-ray line flux that can be measured from a source in the direction θ, ϕ is then

$$F_{\mathrm{min}} \leqslant \frac{n}{\mathrm{S}(\mathrm{E}, \theta, \phi)} \sqrt{\frac{2 \cdot \mathrm{d}B(E)/\mathrm{d}E \cdot \Delta E}{T_{\mathrm{obs}}}} \qquad \text{(VI.1)}$$

where ΔE is the FWHM energy resolution of the instrument at energy E, T_{obs} is the observing time of the measurement and $S(E, \theta, \phi)$ is the sensitivity of the detector for a photon of energy E entering the detector from the source in the direction (θ, ϕ). Sensitivity as used here is the full energy efficiency at energy E times the effective area the detector presents in the direction (θ, ϕ). For inorganic scintillators or solid state detectors, full energy efficiency is the interaction probability for a photon of energy E times the photofraction or, equivalently, the probability that the full energy of the photon is lost in the crystal. The parameter n in Equation (VI.1) is effectively the number of standard deviations of the background fluctuations that corresponds to a detectable signal and is often taken as 2 or 3 by many workers, but no serious experiment should be designed for which n is less than 5 unless one has hope for a long shot. The expression (VI.1) is derived on the assumption that the background is measured also for the time T_{obs} (see also Section VI-6.2), and that the measurement of a source counting rate yields a null result.

Ultimately one is interested in achieving the lowest background conditions possible for experiments which will be carried out on balloons or in near-Earth satellites. In the nuclear line region for a balloon experiment near the top of the atmosphere and at mid-latitudes or near the equator, there are several sources of background that are important. Table VI-1 shows a classification of several sources which can contribute to the background $[\mathrm{d}B(E)/\mathrm{d}E]$ in high resolution γ-ray line detectors such as NaI(Tl) or Ge(Li). The external neutral radiations, neutrons and γ-rays, would give the lowest backgrounds possible for any experiment. Assuming these were the only radiation present, shielding could be devised to eliminate most of the background except for that coming through the aperture of the telescope. Thus, the internal radioactivity would be essentially the only additional contribution to the background. These conditions can be nearly attained at the bottom of the atmosphere in low-level counting laboratories. Unfortunately, in the balloon or satellite environment, the charged cosmic rays dramatically alter the situation since their energies are above nuclear reaction thresholds, and interactions in any material around the detector or in the detector itself can yield prompt emission of γ-rays or neutrons or, what is even worse, activation of the detector and surrounding materials can take place.

In Sections VI-6.1.3, VI-6.1.4, and VI-6.1.5, we describe some actual measurements of the quantity $[\mathrm{d}B(E)/\mathrm{d}E]$ with various detectors in different environmental situations.

TABLE VI-1

Classification of background sources for γ-ray astronomy experiments

External neutral radiation (Ambient and induced by charged particles)	
Diffuse cosmic γ-rays	(Section V-5.3)
Atmospheric γ-rays	(Section VI-6.1.1)
Atmospheric neutrons	(Section VI-6.1.1)
Locally produced γ-rays[a] (not in primary detector)	(Section VI-6.1.2)
Locally produced neutrons[a] (not in primary detector)	(Section VI-6.1.2)
Bremsstrahlung from primary and secondary electrons (including reentrant albedo)	
Internal radiation	
Intrinsic radioactivity of detector material	(Section VI-6.1.1)
Activation of detector materials[a]	(Section VI-6.1.2)

[a]These sources of radiation are a result of charged particle interaction in any local matter as a result of ambient primary and secondary cosmic rays, the Van Allen radiation, or solar cosmic rays.

A useful method of treating this problem has been discussed by Peterson *et al.* (1973b) and Ling (1974), and we shall describe their approach, but use slightly different notation. Consider a detector with an acceptance solid angle $\Delta\Omega$ and a full energy sensitivity of $S(E, \alpha)$ cm^2 for photons of energy E incident at angle α to the detector axis from a volume element at a distance r away, which emits $A_0(E, h')$ photons of energy E (unit energy)$^{-1}$ (g of air)$^{-1}$ s^{-1}. Assuming that the detector is in the atmosphere at a depth h (g cm^{-2}), viewing at a zenith angle θ so background and source photons are incident at angle $\alpha = 0$, then the counting rate (counts s^{-1}) of the detector due to all background sources in a band width ΔE at E (corresponding to the full energy peak in the instrument over the instrument resolution width) is

$$\frac{dB(E)}{dE}\Delta E = \int_{\Delta\Omega}\int_{\Delta E}\int_0^{h \sec\theta} S(E,0)\frac{A_0(E,h')\rho(h')}{4\pi}\exp\left[-r'/\lambda(E)\right] dr\, dE\, d\Omega$$
$$+ \int_{\Delta\Omega}\int_{\Delta E} S(E,0)F_D(E)\exp\left[-h\sec\theta/\lambda(E)\right]\, dE\, d\Omega$$
$$+ \text{(Background rate from activation, neutrons, electrons)} \qquad \text{(VI.2)}$$

where $\rho(h')$ is the atmospheric density at the depth of the source element, $r' = \int_0^r \rho(r)\, dr$ (g cm^{-2}), and $\lambda(E)$ is the absorption mean free path for photons of energy E measured in g cm^{-2}.

The first integral is the contribution due to the ambient atmospheric γ-flux in the form of a depth dependent source function $A_0(E, h')$ expressed as the number of γ-rays of energy E in ΔE emitted per second from a gram of air at a particular latitude. Peterson *et al.* (1971) have found that A_0 near the top of the atmosphere is independent of depth for photon energies up to $\sim$300 keV and is a monotonically decreasing function of energy, $A_0(E)$, which simplifies the integration. Above this energy the source function becomes more complex since the depth independence is not justified, and $A_0(E)$ must be replaced by $A_0(E, h')$. In principle, $A_0(E, h')$ can be directly measured from balloon flights at different atmospheric depths and geomagnetic cutoffs and the appropriate change made in the first integral. Ling (1974) has recently discussed the empirical determination of this source function from balloon flight data taken with a directional γ-ray telescope and gives results useful for intermediate latitudes. In general, it should be remembered that $A_0(E, h')$ will be dependent on the design of the particular detector used to determine it.

The second integral is the contribution from the diffuse cosmic γ-ray flux which penetrates through the atmosphere to the detector and may be estimated from the best current knowledge of this radiation (see Section V-5.3). A contribution from any point source in the field of view of the detector is not included in Equation (VI.2). The last term is the most difficult to evaluate from any *a priori* considerations, since it depends greatly on the detector design itself and thus cannot be generalized. This residual detector counting rate can, in principle, be measured by blocking the aperture of the telescope as described by Peterson *et al.* (1971).

Once relation (VI.2) has been determined for a given instrument configuration, then the flux sensitivity of the experiment for detecting a given γ-ray line is given by Equation (VI.1). This procedure for determining the background counting rate for a γ-ray telescope has been used successfully by the UCSD group (cf. Peterson *et al.*, 1971; Ling, 1974). The necessary calculations are laborious, but computer programs have been developed to study γ-ray telescopes of standard designs.

The approach described above is somewhat oversimplified, since it assumes that an atmospheric or cosmic photon of energy E_γ incident on the detector will deposit its energy (with some probability) also at E_γ in the detector. This, of course, is not always true, particularly above ~300 keV where Compton scattering dominates and above 1.02 MeV where pair production becomes important and escape peaks occur. This complication can be taken into account by using the measured response matrix of the detector $R_{ij} \equiv R(E_i, E_j)$, where E_i is the incident photon energy and E_j is the energy deposited in the detector. Then the resulting energy loss spectrum in the detector is $E_j = R_{ij}E_i$. Trombka (1970) has applied this approach to unfolding atmospheric γ-ray energy loss spectra obtained with a NaI(Tl) spectrometer in the energy range 100 keV to 2 MeV. It should be noted here that the true detector background must be known before any conclusions concerning an external flux, such as F_D, can be made.

6.1.1. AMBIENT NEUTRAL BACKGROUNDS

A direct approach in determining the background in a particular experiment requires a knowledge of the energy spectrum of the neutral radiation components incident on the telescope. We will therefore review briefly here the available data on the ambient radiation including the cosmic diffuse γ-rays, the atmospheric γ-ray and neutron spectrum, and the natural radioactivity of materials used to construct telescopes. Activation of instruments due to the ambient charged particle environment will be discussed in Section VI-6.1.2.

a. *The Diffuse γ-Ray Flux*

The current status of the diffuse γ-ray flux measurements was discussed in Section V-5.3. There it was pointed out that the true existence of a diffuse cosmic component appears to be safely established only below ~82 keV according to Schwartz (1974). The detailed shape of the energy spectrum and its spatial distribution have not yet been established, since measurements of this radiation have not been made with directional instruments of high energy resolution.

For the design of new experiments, we recommend a conservative approach. Figure V-12 gave a recent summary of most of the relevant measurements of this radiation from 100 keV to 100 MeV. The dotted line in the figure, which fits approximately the measurements in the X-ray region (not shown) and also at the highest energies (~100 MeV), is given by $F_D = 25\ E_\gamma^{-2.1} (\gamma\ \mathrm{cm}^{-2}\ \mathrm{s}^{-1}\ \mathrm{sr}^{-1}\ \mathrm{MeV}^{-1})$. It is in the energy region 1 to 10 MeV where the largest experimental uncertainty lies, and future measurements might lower the apparent diffuse flux in this region of the spectrum. From the experimental point of view, the measurements of Kuo *et al.* (1973) are of great interest. Whereas the measurements below 10 MeV were made with scintillation detectors, Kuo's measurement was made with a spark chamber and purports to confirm that the diffuse flux at this energy is much higher than the simple power law extrapolation from lower energies. In fact, as discussed in Section V-5.3, this 10 MeV diffuse value was found by subtracting an average value of the atmospheric flux from the total measured flux. This gave a value for the diffuse intensity at 10 MeV of $(8.0 \pm 4.5) \times 10^{-4}$ cm^{-2} s^{-1} sr^{-1} MeV^{-1}. As Kuo *et al.* (1973) point out, this large value has important experimental consequences, since at 4.5 GV cutoff and at a balloon depth of $\sim 1\ \text{g cm}^{-2}$ the atmospheric flux at 10 MeV is 2×10^{-4} cm^{-2} sr^{-1} MeV^{-1} or lower (see Figure VI-5), which is $\sim 20\%$ of the diffuse value. At ~ 12 GV cutoff, the atmospheric component would only be $\sim 10\%$ of the diffuse flux or less (see Figure VI-6). Thus, the cosmic diffuse flux and not the atmospheric secondary γ-rays could be the limiting background (in the ~ 10 MeV energy region) for any investigations of emission from 'point' sources such as the Crab Nebula.

b. *Atmospheric γ-Rays*

Detailed knowledge of the properties of atmospheric γ-rays near the top of the atmosphere is sparse. As was discussed in the last section, such γ-ray measurements may include a major contribution from the cosmic diffuse radiation and, therefore, separation of the background into atmospheric and diffuse components may be very difficult. In recent years, however, there has been a major effort to isolate the truly atmospheric γ-rays, since proper subtraction of this background is necessary for any balloon-borne γ-ray astronomy experiment. It should be noted that the atmospheric secondary γ-rays have been used to calibrate 'in flight' the OSO-3 high energy γ-ray experiment (Kraushaar *et al.*, 1972). For this latter purpose, Stecker (1973b) has also estimated the angular distribution of γ-rays (>100 MeV) escaping from the atmosphere.

Here we wish to review briefly the theoretical and experimental knowledge available which refers primarily to the vertical continuous spectral flux near the top of the atmosphere. Recent reviews have been given by Thompson (1974), Daniel and Stephens (1974), and Kinzer *et al.* (1974). The two-component spectral model for atmospheric γ-rays ~ 10 MeV was originally developed by Beuermann (1971). The decay of π^0 mesons produced in the atmosphere by the primary cosmic rays contributes one spectral component with a broad peak at ~ 70 MeV, while the bremsstrahlung from primary, secondary, and reentrant albedo electrons contribute the second component. These calculations were made for a cutoff rigidity of zero and 4.5 GV and have been extended by Daniel and Stephens (1974) to other rigidities. Puskin (1970) has also carried out Monte Carlo calculations for the energy range 300 keV to 10 MeV, and Ling (1974) has done semi-empirical calculations for photon energies below 10 MeV.

Thompson (1974) most recently carried out Monte Carlo calculations for the atmospheric γ-ray spectrum above 30 MeV for all atmospheric depths and all zenith angles. These calculations were compared with balloon-borne spark chamber measurements made from Palestine, Texas in 1971. The work is very valuable since, at least for the

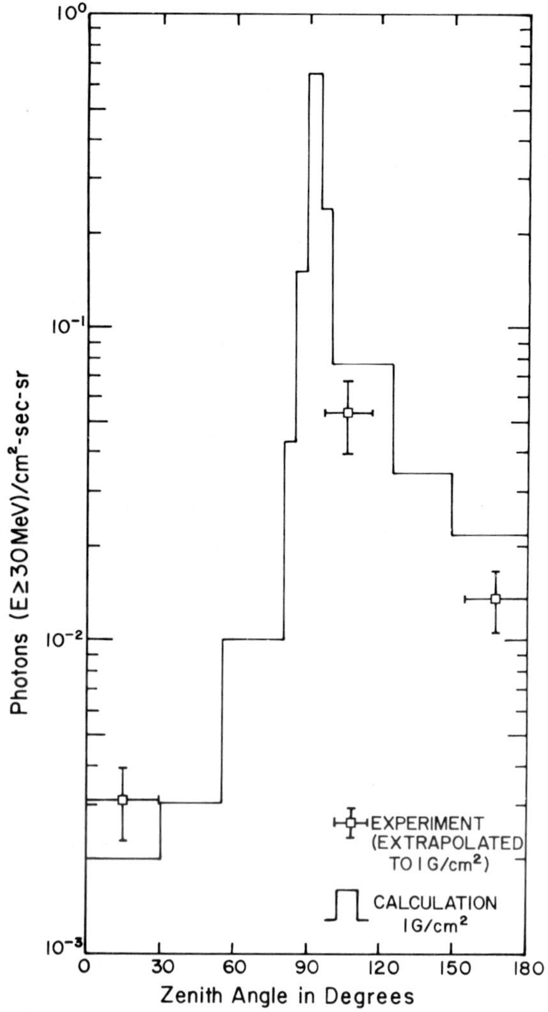

Fig. VI-1. The calculated directional intensity of γ-rays ($\geqslant 30$ MeV) at 1 g cm^{-2} vs. zenith angle compared with measurements at ~ 4.5 GV cutoff. (From D.J. Thompson, *J. Geophys. Res.* **79**, 1309, 1974, copyrighted by American Geophysical Union.)

higher photon energies, complete angular distribution information is available. In Figure VI-1 the calculated directional intensity of γ-rays (>30 MeV) at an atmospheric depth of 1 g cm^{-2} is compared with extrapolated measurements at $\sim$4.5 GV cutoff for three angles also made by Thompson (1974) using the γ-ray spark chamber described by Ehrmann *et al.* (1967) and Fichtel *et al.* (1969). The strong flux near the horizon is a consequence of both the increased atmospheric depth in that direction and the forward momentum of the π^0 parent of the pair of γ-rays. In Figure VI-2 a more complete summary of the experimentally observed angular distribution of γ-rays as a function of

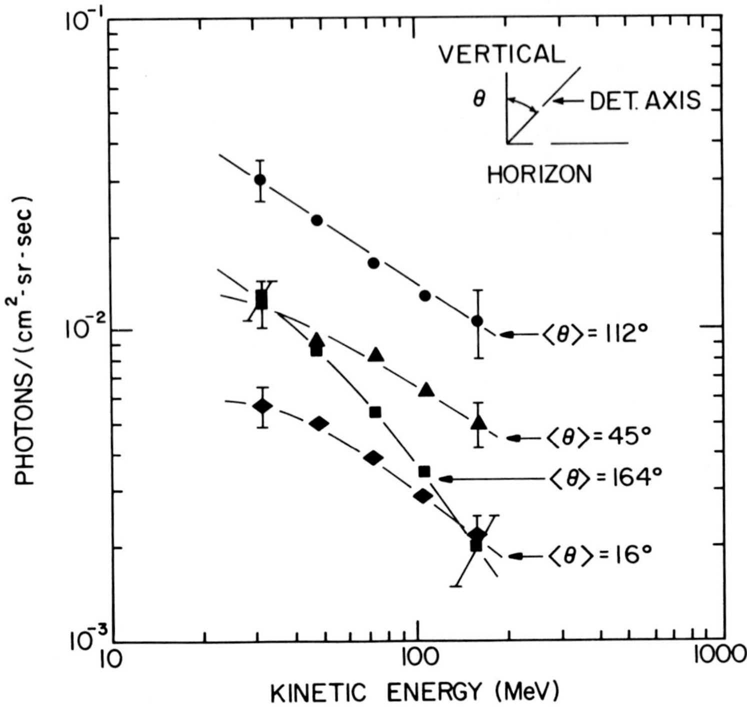

Fig. VI-2. The directional atmospheric flux above 30 MeV is shown vs. γ-ray energy at several zenith angles. (From C.E. Fichtel *et al., Astrophys. J.* **158**, 193. Copyright 1969, The American Astronomical Society. Used by permission of the University of Chicago Press.)

photon energy is shown from Fichtel *et al.* (1969) for an atmospheric depth of 3 g cm^{-2} at the Palestine latitude for γ-ray energies above 30 MeV. Thompson (1974) has pointed out that the strong peak in the γ-ray flux from the direction of the horizon near the top of the atmosphere shifts continuously to smaller zenith angles as the atmospheric depth increases.

In the photon energy range <10 MeV, a region of direct importance for nuclear γ-ray line work, information on the directional γ-ray flux is very limited. At these lower energies near the top of the atmosphere, the atmospheric flux measured by a *strongly* directional detector should be dependent on the zenith angle of view simply because the source volume contributing to the counting rate increases as thicker atmosphere is viewed. A *sec* θ zenith angle dependence is, however, too strong at small atmospheric depths because of the curvature of the atmosphere. This effect is not seen in most scintillation counter γ-ray telescopes operating below 10 MeV since the counting rate is dominated by local background sources and leakage of atmospheric γ-rays through the shield.

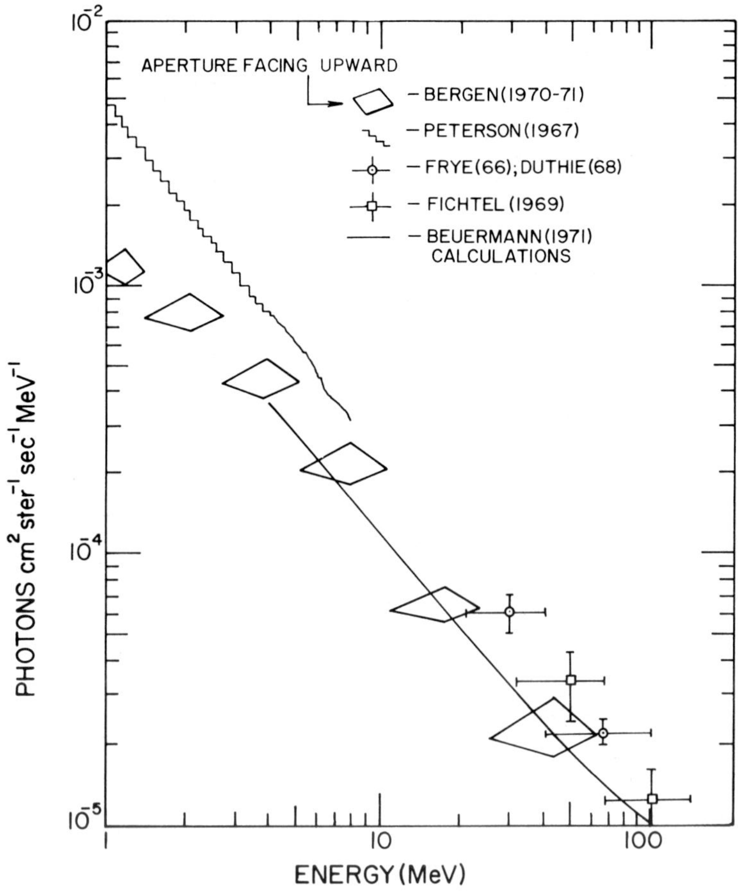

Fig. VI-3a. Differential energy spectrum of secondary photons at 2.6 GV geomagnetic cutoff and 2.4 g cm^{-2} residual atmosphere for $\theta = 0°$. (From J. Scheel and H. Röhrs: 1972, *Z. Physik* **256**, 226, by permission Springer-Verlag, Berlin.)

In this energy range useful observations have been made by Scheel and Röhrs (1972) using a standard collimated CsI(Tl) spectrometer. This instrument allowed directional measurements at zenith angles of 0°, 90°, and 180° at an atmospheric depth of 2.4g cm^{-2} and a cutoff of 2.6 GV. The directional differential flux from these measurements is given in Figures VI-3a and VI-3b. In Figure VI-3a, which gives the vertical (downward) flux of γ-rays, the equivalent isotropic flux inferred from Peterson *et al.* (1967) is at least a factor 5 larger than the directional measurement of Scheel and Röhrs (1972) even though the higher cutoff (4.5 GV) for the Peterson *et al.* (1967) measurement was compensated for. This, of course, is exactly what is expected near the top of the atmosphere for an average directional flux if the 90° flux is enhanced as discussed above. Also, from Figure VI-3b, the upward flux ($\theta = 180°$) near the top of the atmosphere is

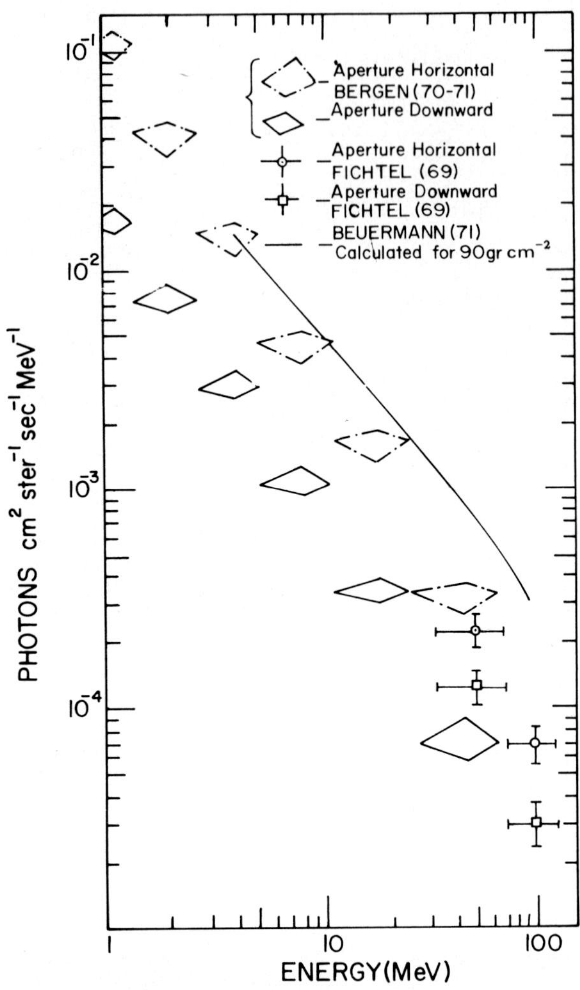

Fig. VI-3b. Differential energy spectrum of secondary photons at 2.6 GV geomagnetic cutoff and 2.4 g cm^{-2} residual atmosphere for $\theta = 90°$ and $\theta = 180°$ (From J. Scheel and H. Röhrs: 1972, *Z. Physik* **256**, 226, by permission Springer-Verlag, Berlin.)

higher than the assumed isotropic flux, and at $\theta = 90°$ the 1 MeV flux is a factor of 20 greater than the equivalent isotropic flux shown in Figure VI-3a. These results, even though not fully confirmed by other measurements, are plausible simply because of the larger atmospheric depth at 90° and the greater flux from below. These observations suggest that extreme care should be taken when conducting experiments near the top of the atmosphere when cosmic sources are observed at different zenith angles. No reliable information is available yet on the azimuthal distribution although some dependence may be expected because of the known east-west effect for the primary cosmic rays. A measurement of the vertical γ-ray flux (<10 MeV) at 4.5 GV cutoff has also become available from double Compton telescope experiments (Schönfelder and Lichti 1975; White *et al.*, 1973) (see Figure VI-5).

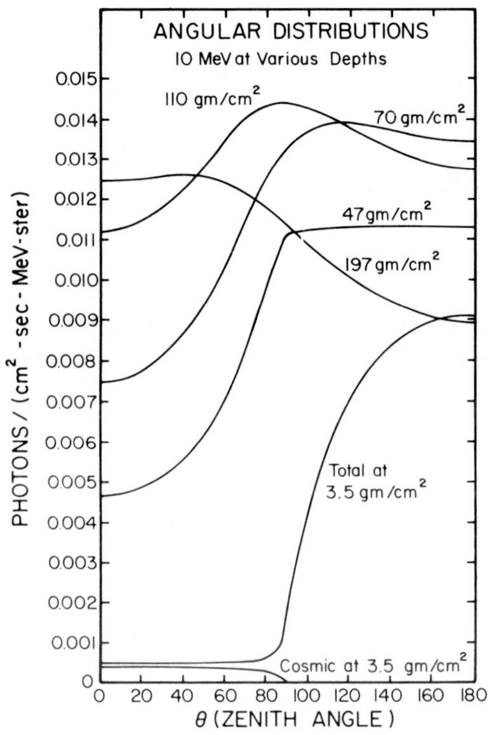

Fig. VI-4. The total directional photon intensity (including diffuse cosmic γ-rays) at 10 MeV is shown as a function of zenith angle for several values of atmospheric depth. These curves are a result of empirical model calculations and measured fluxes at $\lambda = 40°$. (From J.C. Ling, Ph.D Thesis, University of California, San Diego, 1974.)

Empirical estimates of the angular distribution of the low energy γ-rays (<10 MeV) at balloon altitudes have been made by Ling (1975). Figure VI-4 shows a plot of the angular distribution of 10 MeV photons at several different values of atmospheric depth. These curves are a result of model calculations of Ling (1974) using an empirical source function for atmospheric γ-rays determined from measurements of Peterson *et al.* (1973b) at geomagnetic latitude 40° N. Similar results are also available for 1 MeV photons in Ling (1974) as well as angular distributions at 3.5 g cm^{-2} for photon energies from 300 keV to 10 MeV. In general these results show the strong enhancement of the directional intensity at zenith angle ~90° near the top of the atmosphere in qualitative agreement with the measurements of Scheel and Röhrs (1972) at a lower geomagnetic cutoff. It should be noted, however, that these results have not been confirmed in detail by experiment; however, the absolute value of the atmospheric flux at 3.5g cm^{-2} in Figure VI-4 generally confirms the observation of Kuo *et al.* (1973), which gave a diffuse cosmic vertical intensity several times larger than the atmospheric γ-ray intensity at 10 MeV. It is also of interest to note in Figure VI-4 the behavior of the angular distribution at larger atmospheric depths giving a peak in the angular distribution along the horizon at 110 g cm^{-2}, whereas at 1 MeV the angular distribution is nearly isotropic at the same atmospheric depth. Ling (1974) attributes this behavior to the fact that at 10 MeV the photon interaction mean free path is greater than at 1 MeV.

Valuable recent results have been obtained in the energy range 10 MeV to 200 MeV by spark chambers which, because of electron track visualization, are able to give directional information but with limitations at the lowest energies. Kinzer *et al.* (1974) have utilized a spark chamber in conjunction with nuclear emulsions, a proportional counter, and scintillation counter to measure the depth dependence at cutoff rigidities of 4.5 GV and 11.5 GV. It was found that the secondary γ-ray depth dependence (for the vertical flux) is essentially linear as shown by the calculations of Beuermann (1971) and Thompson (1974) near the top of the atmosphere. These measurements give the following growth rates above the transition maximum for photons in the energy range 10 to 200 MeV:

$$\begin{aligned} &(2.54 \pm 0.56) \times 10^{-3}\ \gamma\ \text{cm}^{-2}\ \text{s}^{-1}\ \text{sr}^{-1}\ \text{g}^{-1}\text{cm}^{2};\ 4.5\,\text{GV} \\ &(1.17 \pm 0.32) \times 10^{-3}\ \gamma\ \text{cm}^{-2}\ \text{s}^{-1}\ \text{sr}^{-1}\ \text{g}^{-1}\text{cm}^{2};\ 11.5\,\text{GV}. \end{aligned} \tag{VI.3}$$

Figure VI-5 shows the essential result of these measurements expressed as (γ cm^{-2} s^{-1} sr^{-1} MeV^{-1} g^{-1} cm^2) for a geomagnetic cutoff of ~4.5 GV. An estimate of the vertical flux can be obtained simply by multiplying by the atmospheric depth of interest. The Beuermann calculations (1971) are also shown. It should be noted that the other results shown may not be directly compared with the results of Kinzer *et al.* (1974) if they contain a cosmic diffuse contribution or a depth independent background. Figure VI-6 illustrates the Kinzer *et al.* (1974) results, again with their summary of other measurements at cutoff $\geq$ 11.5 GV. One major conclusion from this work is that the two component (π^0 and bremsstrahlung) spectrum fits the data reasonably well but the relative importance of the π^0 component at lower energies (~25 MeV) depends on the cutoff and may be more significant at the higher cutoff.

c. *Intrinsic Activity*

The natural radioactivity of the materials making up the γ-ray spectrometer telescope can be an important background contribution. It does not appear that this source has yet been the limiting factor in any γ-ray line experiments; however, it could become important as experiments are designed to measure lower and lower γ-ray line fluxes. We will discuss briefly the natural activity in some of the more common materials used in constructing γ-ray spectrometers. A more complete discussion may be found in Kreger and Mather (1967). A detailed discussion of natural radioactivity may be found in Eisenbud (1973).

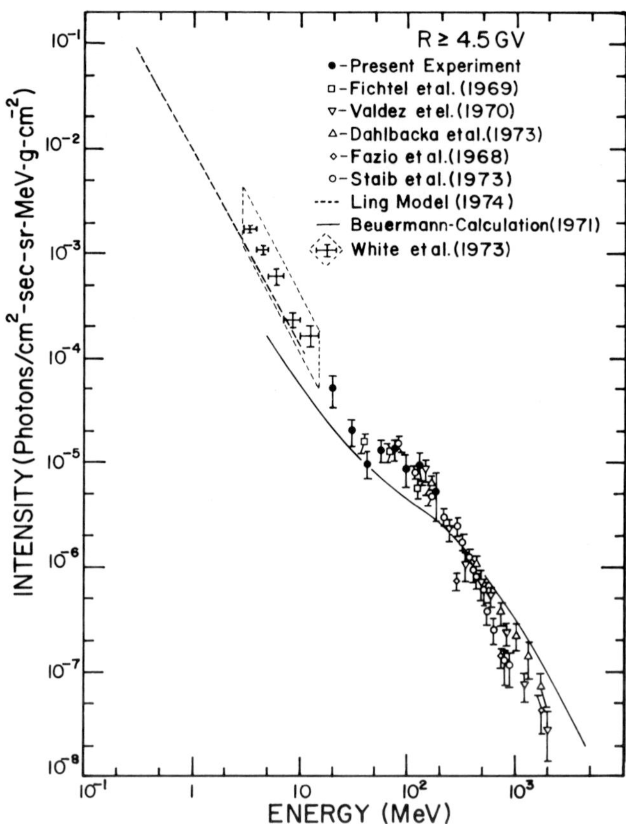

Fig. VI-5. The growth rate of the atmospheric γ-ray flux near the top of the atmosphere at cut-offs $R \sim 4.5$ GV vs. γ-ray energy. (From R.L. Kinzer *et al., J. Geophys. Res.* **79,** 4567, 1974, copyrighted by American Geophysical Union.)

In general, the natural activities ^{40}K, Ra and Th are found in most materials. The isotope ^{40}K undergoes β^- decay to ^{40}Ca or β^+/EC decay to ^{40}Ar giving a γ-ray line at 1.46 MeV from the second mode of decay. The first decay mode accounts for 89% of the activity and the second, 11%, and the total half-life of 1.26×10^9 yr corresponds to a decay constant of $\lambda(^{40}\text{K}) = 1.74 \times 10^{-17} \text{s}^{-1}$. There are also numerous emissions from radium and thorium parent and daughter nuclides, such as α-particles, β-particles and numerous discrete energy γ-rays. The γ-ray spectrum of either naturally-occurring thorium and radium, is a continuous spectrum with several peaks and typical examples of such spectra as measured by a NaI(Tl) spectrometer may be found in Heath (1964). The presence of thorium in materials is usually identified by the detection of a 2.62 MeV γ-ray line from the first excited state of ^{208}Pb (ThD) which is populated by β^- decays of ^{208}Tl(ThC″). ThC″ is a short half-life (3.1 min) daughter nuclide in the decay chain of ^{232}Th whose half-life is 1.4×10^{10} yr.

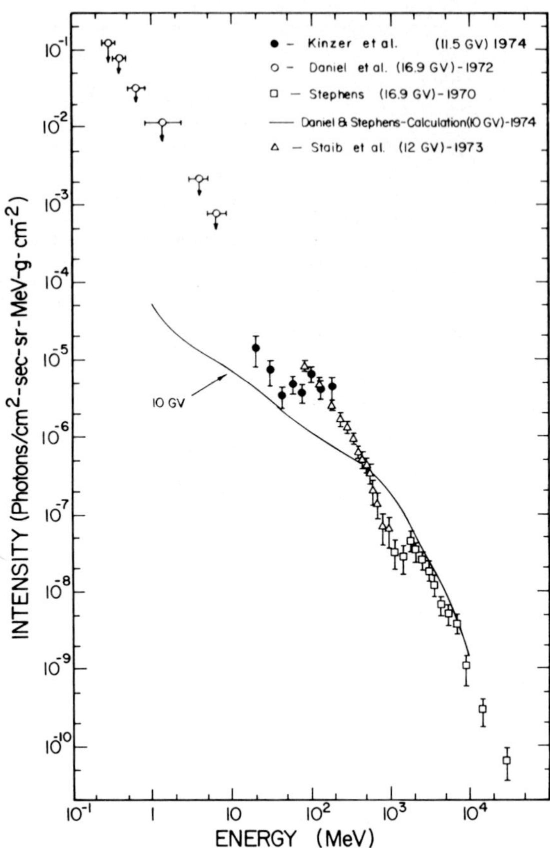

Fig. VI-6. The growth rate of the atmospheric γ-ray flux near the top of the atmosphere at cutoffs $R \geqslant 11.5$ GV vs. γ-ray energy. (From R.L. Kinzer *et al., J. Geophys. Res.* **79**, 4567, 1974, copyrighted by American Geophysical Union.)

The relative magnitude of a natural background γ-ray spectrum may be seen in Figure VI-22 compared to the atmospheric γ-ray background. This figure shows the energy loss spectrum in a 3″ x 3″ NaI(Tl) γ-ray spectrometer in Peterson's laboratory at UCSD and during a balloon flight at 3.6 g cm^{-2} over Palestine, Texas. The sea level spectrum shows the 1.46 MeV line from the parent ^{40}K and the 2.62 MeV line from ^{208}Pb(^{208}Tl daughter) and the continuous spectrum; ^{228}Th levels also give several other γ-ray lines whenever ^{232}Th is present (see Lederer *et al.,* 1968). The background spectrum in Figure VI-22 is the result of the total intrinsic activity in the detector and associated materials in the instrument and the surrounding environment. As can be seen by comparing the ground spectrum with the atmospheric spectrum at 3.6 g cm^{-2}, the room background is the major contributing factor, since there is no 1.46 MeV line in the latter spectrum. Also, the natural background spectrum falls steeply above 2.62 MeV and is about two orders of magnitude below the atmospheric spectrum above this energy. It should be remembered that the γ-ray lines from any radioactivity in the NaI(Tl) crystal itself will not, in general, produce a photopeak. This is because the beta decay electron-energy-loss spectrum will be continuous, and a full energy loss count from a time-coincident γ-ray line still gives a continuous spectrum rather than a photopeak, which would be the case when the discrete energy γ-ray enters the crystal from the outside. Decay schemes and β-decay energetics should be used to evaluate the expected background spectrum for each case (cf. Lederer *et al.*, 1968).

In order to evaluate the magnitude of the intrinsic activity of various materials, a few guidelines are useful. According to Kreger and Mather (1967), the current purification process for spectrometer quality NaI gives about 1 to 2 parts per million (ppm) of ^{40}K. Let P_m be the ppm of ^{40}K present in a cylindrical NaI(Tl) crystal of diameter D_{in} and of length L_{in} where both dimensions are in inches. By using an empirical result of Marinelli *et al.* (1962), it can easily be shown that the β activity in such a crystal is $A(^{40}\mathrm{K}) \cong 1.5 \times 10^{-3}\, P_m D_{in}^2 L_{in} (\mathrm{s}^{-1})$. For the 3″ x 3″ crystal used by Peterson *et al.* (1973b) in obtaining the spectrum shown in Figure VI-22, the maximum contribution to the count rate will be $\sim$0.04 counts s^{-1} for $P_m = 1$. These counts will be distributed over a spectrum extending downward from 1.35 MeV which is the end point energy of the β^- decay to ^{40}Ca. The decay rate of ^{40}K which gives the 1.46 MeV γ-ray line is about 10% of the above estimate. The work of Peterson *et al.* (1973b) indicates that the total counting rate of such a crystal at an atmospheric depth of 3.6g cm^{-2} is $\sim$250 counts s^{-1} between 0.2 and 10 MeV for a geomagnetic latitude of 40°. This high integral counting rate is due predominantly to atmospheric photons. Therefore intrinsic ^{40}K activity would be negligible compared to atmospheric background, but could become important from other local sources in and around the detector.

More details on the intrinsic activities found in various materials used in NaI(Tl) spectrometers may be found in Kreger and Mather (1967) and Van Lieshout *et al.* (1966). Normally, manufacturers of scintillation crystals, NaI(Tl) and CsI(Tl), and photomultipliers take the necessary precautions to keep all naturally-occurring activities as low as possible in the raw materials. A problem of special concern, however, is the highly variable contamination found in some of the other materials used in fabricating balloon and satellite components. For example, Kreger and Mather (1967) have pointed out that reworked Al usually has more radioactive contamination than virgin Al since the Ra in instrument dials is melted down with the Al scrap. Also, stainless steels produced before 1956–57 were relatively clean, but since then various steels have shown a variable contamination.

Finally, γ-ray telescopes utilizing a coincidence technique, such as a Compton telescope (Section VI-6.3.2), usually can circumvent problems associated with the intrinsic activity of detector materials. However, the simpler NaI(Tl) or CsI(Tl) spectrometers could experience background limitations from this source. A prudent approach would be to check the intrinsic activity of detector components in a low-level counting facility. With such a precaution, the limiting source of background in many γ-ray astronomy experiments will be a combination of atmospheric γ-rays and neutrons, the diffuse radiation, and activation, but which background source is predominant will, of course, depend on the experiment design.

d. *Atmospheric Neutrons*

We wish to give here a brief description of the atmospheric neutron spectrum expected at intermediate latitudes near the top of the atmosphere. The energy range of neutrons which can produce important background effects in γ-ray telescopes goes from thermal energies to several hundred MeV. Most of the work on the atmospheric neutron flux has been directed toward obtaining the 'albedo' neutron flux in space in order to determine the CRAND (cosmic ray albedo neutron decay) source function for the Van Allen belt protons (White, 1973; Williams, 1972), and a large number of measurements have been made for this purpose.

Recently, a series of papers has been published (Merker *et al.*, 1973; Light *et al.*, 1973; Mendell *et al.*, 1973) reviewing the atmospheric neutron measurements at balloon altitudes (and below) for various latitudes, but these concern primarily the omnidirectional flux below 10 MeV. Complete angular distributions of the neutron flux at all energies of importance for γ-ray experiments at different latitudes near the top of the atmosphere are, however, not available. Even if it were available, the effect of a neutron flux seen by a γ-ray telescope would be, in general, modified because of local production (see Section VI-6.1.2). This same remark applies, of course, to the atmospheric γ-ray spectrum.

We content ourselves, therefore, with giving an indication of the total omnidirectional atmospheric neutron flux for neutrons below 10 MeV and some recent significant results on the directional neutron flux above 10 MeV. Light *et al.* (1973) have calculated the global distribution of atmospheric neutrons using a Monte Carlo program taking into account neutron production and neutron transport in the atmosphere. Figure VI-7 shows the altitude and latitude distributions of the 1 to 10 MeV neutron flux during solar minimum when the primary galactic cosmic ray flux is a maximum. The results of the calculations are shown by the solid curves labeled A-G. The units on the ordinate give the *total* neutron flux in the indicated energy interval without regard to direction. Often the neutron leakage rate (flux) is reported in many papers on atmospheric neutrons, but this refers specifically to the upward moving neutron flux ($cm^{-2}\ s^{-1}$). Figure VI-7 also shows, by dashed lines, the experimentally determined total neutron fluxes of Merker *et al.* (1973) made at geomagnetic cutoffs of ~17 GV, ~4.5 GV, and 0.2 GV, indicated respectively by G′, C′, and A′. Near the top of the atmosphere the agreement between the calculations and the measurements is sufficiently good that calculated curves can be used, except where there is a significant discrepancy as at the highest cutoff of 17 GV.

A similar curve for the period of maximum solar activity is also given by Light *et al.* (1973), and it shows that the fast neutron flux is down by less than a factor of 2 in all cases as compared to the solar minimum flux. Neutrons below 1 MeV can also be an important contributor to background in γ-ray experiments, especially since the differential neutron spectrum continues to rise at lower energies and the neutron capture cross section is often $\propto 1/v$ (where v is the neutron velocity) in many materials. Therefore, it is useful to have a good measure of the shape of the neutron energy spectrum. An

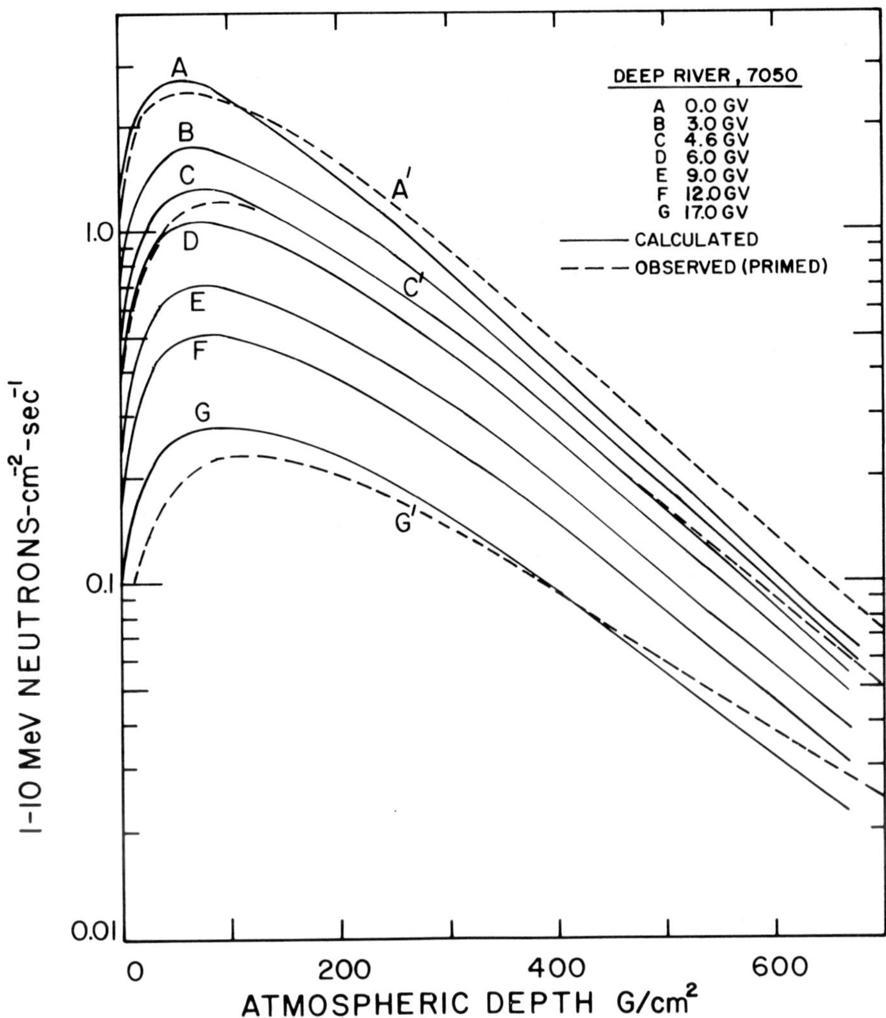

Fig. VI-7. Fast neutron flux vs. atmospheric depth at various cutoff rigidities; balloon observations (dashed line) compared with Monte Carlo calculations (solid line). (From E.S. Light *et al.*, *J. Geophys. Res.* **78**, 2741, 1973, copyrighted by American Geophysical Union.)

adequate representation is given by the Monte Carlo calculations of Armstrong *et al.* (1973) near the top of the atmosphere for an intermediate latitude, $\lambda = 42°$, corresponding to a cutoff of ~4.5 GV. This differential neutron flux is shown in Figure VI-8, plotted vs. neutron energy for several atmospheric depths. It is important to note the broken ordinate on the left. Also, the flux units (cm^{-2} s^{-1} MeV^{-1}) give the total omnidirectional neutron flux at all atmospheric depths except for 0 g cm^{-2} where the ordinate value gives the leakage rate or current of neutrons moving vertically out of the atmosphere (dotted histogram) or the neutron flux (solid histogram). Armstrong *et al.* (1973) should be consulted for mathematical expressions defining neutron flux and current.

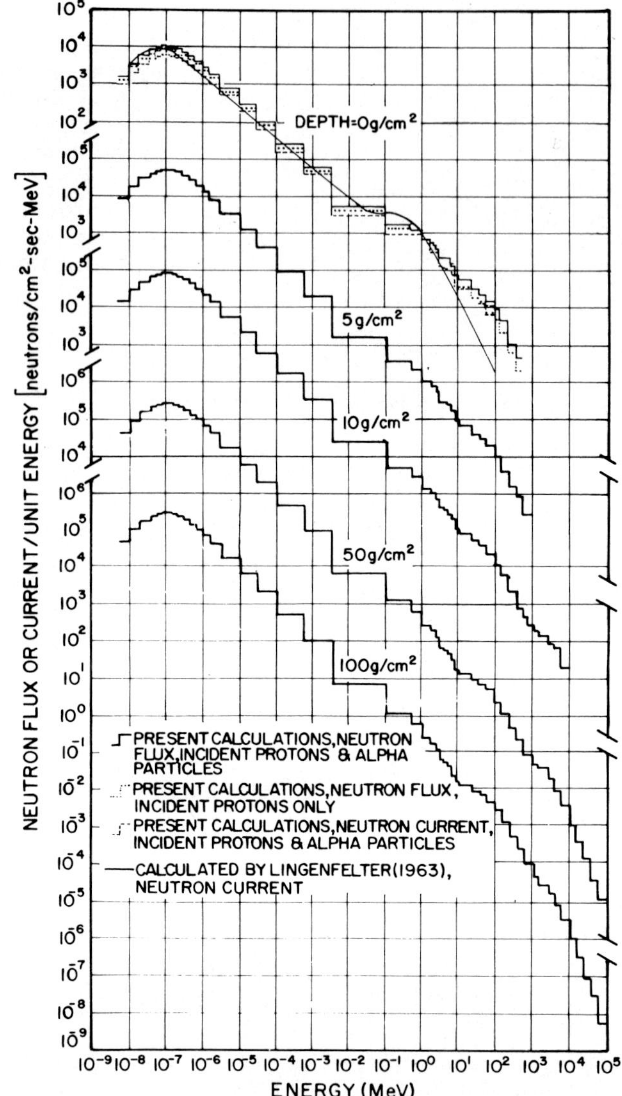

Fig. VI-8. Calculated neutron spectra at various depths from top of the atmosphere for solar minimum and $\lambda = 42°$ N. (From T.W. Armstrong *et al.*, *J. Geophys. Res.* **78**, 2715, 1973, copyrighted by American Geophysical Union.)

At the higher neutron energies (>10 MeV), measurements of the angular distributions of the neutron flux near the top of the atmosphere are now available. These have been made by Preszler *et al.* (1973), using a large scintillator recoil proton telescope, and

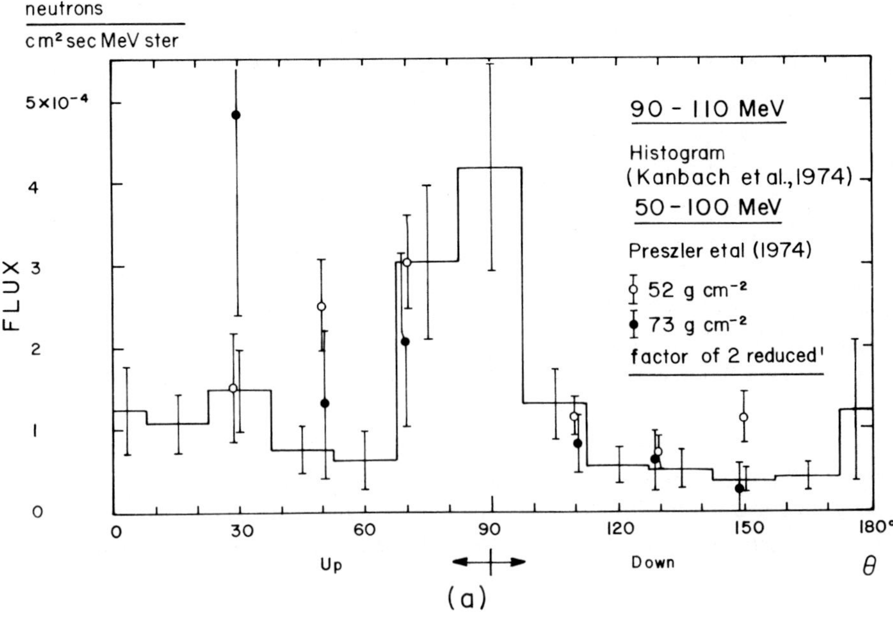

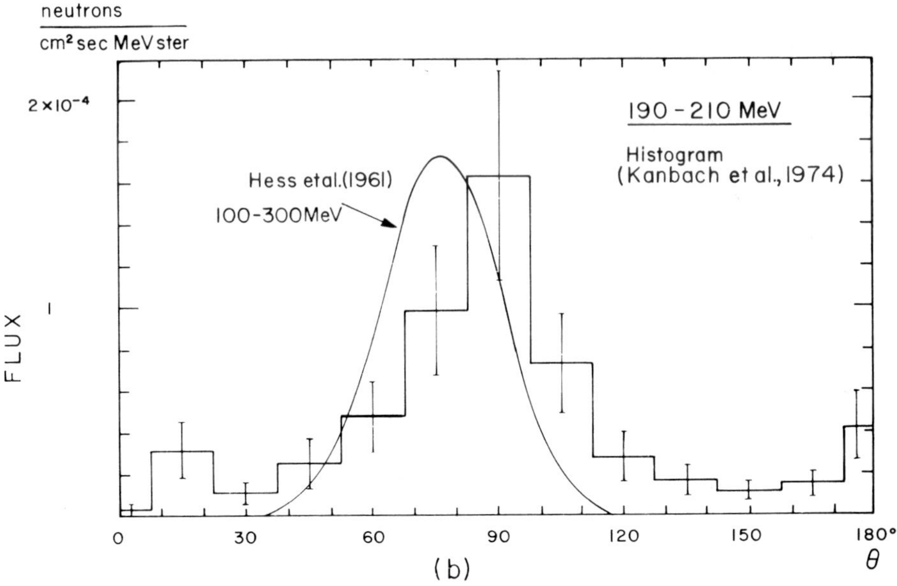

Fig. VI-9. Histogram of the angular distribution of atmospheric neutrons measured at an atmo spheric depth between 4.7 and 8.6 g cm^{-2} in two energy intervals. The results of Preszler *et al.* (1974) (9a) have been reduced by a factor of 2 in order to take into account the different energy intervals. The theoretical curve of Hess *et al.* (1961) (9b) has been folded with angular resolution of 15°. (From G. Kanbach *et al. J. Geophys. Res.* **79**, 5159, 1974, copyrighted by American Geophysical Union.)

by Kanbach *et al.* (1974) using a spark chamber which recorded the double elastic scattering of neutrons on protons. The former measurements covered the energy range from 10 to 100 MeV and the latter from 70 to 250 MeV. In Figure VI-9(a, b) the angular distribution of the atmospheric neutrons near the top of the atmosphere is shown in three energy groups. In the upper figure, the directional flux measured by Kanbach *et al.* (1974) is shown for the neutron energy interval 90 to 110 MeV and for atmospheric depths between 4.7 and 8.6 g cm^{-2}. These are compared with the results of Preszler *et al.* (1973) at a lower energy range (50 to 100 MeV) by reducing the latter by a factor of 2. The latter measurements were also made at similar atmospheric depths of 5.2 g cm^{-2} and 7.3 g cm^{-2}. No measurements were made at 90° because the spark chamber was flown vertically. Upward moving neutrons have a zenith angle $< 90°$ and are indicated as 'up' in the figure; similarly, downward moving neutrons have a zenith angle $> 90°$. According to the comparison shown in Figure VI-9a, the two experimental angular distributions are in reasonable agreement where measurements overlap, except for neutrons moving upward in the zenith angle interval ~30° to 60°. Kanbach *et al.* (1974) suggest that this discrepancy may reflect a different angular flux distribution at the lower energies. In the lower part of Figure VI-9, the angular distribution of the higher energy neutrons (190 to 210 MeV) is shown. As is the case for the lower energy neutrons (90 to 110 MeV) the angular distribution is strongly peaked at 90°. This was expected from the work of Hess *et al.* (1961), who predicted the angular distribution for neutrons *leaking out* of the atmosphere as shown by the smooth curve in the lower part of Figure VI-9. The experimental and theoretical curves cannot be compared directly since the former includes, in the 90° bin, neutrons moving up and down near the horizontal, whereas the latter included only upward moving neutrons. However, the strong angular asymmetry can cause a variable background source in some γ-ray telescopes (cf. White and Schönfelder, 1975).

6.1.2. INSTRUMENT ACTIVATION (LOCAL PRODUCTION)

Even though the local production of γ-ray background was evident in early balloon experiments (Jones, 1961) and in the first γ-ray satellite experiments (Peterson, 1965), no comprehensive effort was made to quantitatively understand the physical mechanisms involved until the recent work of Dyer and Morfill (1971) and Fishman (1972a). They studied energy spectra from the radioactive decay of spallation products using theoretical predictions based on the formula given by Rudstam (1966) and, in addition, calculated the activation expected from cosmic rays and from protons in the South Atlantic anomaly.

There are several ways in which activation of the anticoincidence shield or the detector itself can lead to γ-ray counts which are not eliminated by the anticoincidence shield. γ-rays produced in the spacecraft or any inert material inside or outside an anticoincidence shield can also, in principle, produce a background which cannot be simply determined. The spacecraft background activation can be reduced by increasing the thickness of the active shield or by placing the detector on a boom, but background from activation of inert material inside the anticoincidence shield cannot be easily eliminated. Activation of the anticoincidence shield by secondary spacecraft neutrons can also produce a background, in which case the effects are the same as for activation through charged particles. γ-Rays from inelastic neutron scattering in the active shield can also produce an unwanted background. Particularly troublesome is the possibility that activation of an active anticoincidence shield can produce radioactive nuclei which decay by electron capture and give a γ-ray background which cannot be electronically eliminated.

We will now describe the important aspects of the Dyer and Morfill (1971) approach, later work along the same lines by Fishman (1972a), and related calculations by Shima and Alsmiller (1970) and Silberberg and Tsao (1973a, b). In the original Dyer-Morfill method, the basis of the theory is the use of the semi-empirical formula of Rudstam (1966) for the production cross section of spallation products. The Rudstam formula was developed using the available experimental data on proton, neutron and α-particle spallation yields for targets from V ($Z = 23$) to U ($Z = 92$), and a complex five parameter formula was developed to give the cross section, $\sigma(A_t, E, Z, A)$, for producing a nuclide (Z, A) in its ground state when a target nucleus (Z_t, A_t) is bombarded by protons of energy E. The cross section formula is considered good to a factor of 2 or 3 primarily for proton energies from 50 MeV to 30 GeV for targets from V to Bi. However, large inaccuracies could be expected for estimates from this formula for local production in much of the material of detectors and spacecraft, which has typically $\bar{Z} \sim 13$. In addition, in some cases corresponding to a single or a few nucleon removal from the target, Funk and Rowe (1967) have shown the formula to be greatly in error. Therefore, Dyer and Morfill (1971) have assumed that the total spallation cross section is normalized so that

$$\sum_{Z,A} \sigma(A_t, Z, A, E) \leqslant 0.92\ \sigma_i\ (A_t, E) \tag{VI.4}$$

where σ_i is the total 'measured' inelastic interaction cross section for a proton of energy E on a target A_t. The factor 0.92 comes from the assumption by Waddington (1969) that the cross section for removal of a single proton or neutron is 0.08 σ_i. The above sum, then, does not include products with $A = A_t - 1$. The Rudstam formula also only gives the production of residual nuclei and not the evaporation products which tend to be light fragments. Also, production in excited states and metastable states is not included.

Of interest here is the application of this formula to detector activation by inner belt protons or cosmic rays. Dyer and Morfill (1971) have done this for a CsI(Tl) crystal with dimensions close to those of a UK-5 X-ray experiment. For both Cs and I, σ_i was taken to be the geometric cross section (1260 mb for both Cs and I), so the interaction length in CsI is 38 cm (171 g cm^{-2}).

TABLE VI-2

CsI spallation products with short half-lives (< 18h) for: (A) 10 min exposure to inner belt protons, (B) exposure to 155 MeV protons, and (C) exposure to cosmic rays > 2 GV. Each case is normalized to 3×10^5 interactions. (From C.S. Dyer and G.E. Morfill: 1971, *Astrophys. Space Sci.* **14**, 243. Used by permission D. Reidel Publishing Company, Dordrecht, Holland.) Note: These authors have used in some cases out-dated values for nuclear properties. See Lederer *et al.* (1968) for verification.

Isotope	Decay mode and energy (MeV)		Half-life $\tau_{1/2}$ (min)	Predicted nos. produced (A) Inner belt	(B) 155 MeV	(C) Cosmic rays
^{130}Cs	β^+	1.97	30	2132	2122	475
	β^-	0.442				
^{128}Cs	β^+	3.0(70%)				
		2.5(30%)	3	5744	5632	2525
	E.C.	(25%)				
	γ	0.460(20%)				
		0.285(20%)				
^{127}Cs	E.C.					
	γ	0.406	360	7965	7460	4734
^{126}Cs	β^+	3.8(82%)	1.6	9287	8105	7282
	E.C.	(18%)				
	γ	0.386(38%)				
^{125}Xe	E.C.	0.187	1080	2878	2287	2910
	γ	0.056				
		0.243				
^{123}Xe	β^+	1.7	120	4379	2712	6913
	γ	0.148				
^{123}I	E.C.					
	γ	0.160	780	8333	7929	3999
^{122}I	β^+	3.0	4	11923	10918	7796
^{121}I	β^+	1.2	96	14376	12345	12479
	γ	0.21				
^{118}Sb^m	β^+	3.1	3.5	1573	967	2775
^{117}Sb	E.C.					
	γ	0.161	168	2601	1376	5586
^{116}Sb	β^+	2.4	15	3579	1601	9204
	γ	1.3				
		0.900				
^{111}Sn	E.C.	(71%)	35	653	105	3427

E.C. Electron Capture.

TABLE VI-3

CsI spallation products with long half-lives (1.9h to 154 days) for cases identical to those of the previous table. (From C.S. Dyer and G.E. Morfill: 1971, *Astrophys. Space Sci.* **14,** 243. Used by permission D. Reidel Publishing Company, Dordrecht, Holland.) Note: These authors have used in some cases outdated values for nuclear properties. See Lederer *et al.* (1968) for verification.

Isotope	Energies of γ rays (MeV) and branching ratio		Half-life $\tau_{1/2}$ (days)	Predicted nos. produced (A) Inner belt	(B) 155 MeV	(C) Cosmic rays
^{132}Cs [a]	E.C.	0.670	6.2	6000	6000	6000*
^{131}Cs	E.C.		10	1145	1106	175
^{129}Cs	E.C.	0.380	1.3	3669	3672	1163
^{127}Cs	E.C.	0.406(80%)	0.25	7965	7460	4734
^{131}Xe^{m}		0.163	12	46	45	7
^{129}Xe^{m}	0.196; 0.040		8	244	244	77
^{127}Xe	E.C.0.370(40%); 0.203(60%)		34	1002	938	596
^{125}Xe	E.C.0.187; 0.243		0.7	2878	2287	2909
^{123}Xe	E.C.	0.148	0.08	4379	2712	6913
^{126}I [a]	E.C.(55%) 0.386(34%); 0.650(33%)		13	6000	6000	6000*
^{125}I	E.C.	0.035	60	2898	2750	692
^{124}I	E.C.(70%) 0.605(95%)		4	5157	4964	1769
^{123}I	E.C.0.160		0.5	8333	7929	3999
^{125}Te^{m}	0.110; 0.035		58	136	129	32
^{123}Te^{m}	0.089; 0.159		104	685	652	328
^{121}Te^{m}	0.082; 0.214		154	2560	2199	2222
^{121}Te [b]	E.C.0.570(87%); 0.506(13%)		17	14376	12345	12479
^{119}Te	E.C.		4.5	6070	4264	8638
^{122}Sb	β^-0.566(66%)		2.8	71	65	46
^{120}Sb	E.C.	0.089; 0.199	6	397	312	446
^{119}Sb	E.C.		1.66	832	585	1184
^{117}Sn^{m}	0.159; 0.162		14	283	150	608
^{113}Sn	E.C.	0.392	120	2249	562	9162
^{111}In	E.C.0.172; 0.247		2.8	1318	212	6920

[a]Single nucleon removal.

[b]Secondary product.

E.C. Electron Capture.

* Corrections via personal communication from C.S. Dyer.

The yield (s^{-1}) of a particular isotope was taken approximately to be

$$\frac{N_B(Z, A)}{600} = \int_{100}^{2500} \frac{\sigma(A_t, E, Z, A)}{\sigma_i(A_t)} (1 - e^{-t/\lambda}) G_0 \times j'(E) \, dE \qquad \text{(VI.5)}$$

where $[1 - \exp(-t/\lambda)]$ is the fraction of protons interacting in a crystal of thickness t(cm), $G_0(\text{cm}^2)$ is the omnidirectional geometrical factor of the crystal, and $j'(E)$ is the radiation belt spectrum $j(E)$ modified by energy loss in 18 g cm^{-2} of CsI equivalent corresponding to the spacecraft and collimator material. The peak radiation belt flux used was

$$j(E) = 2 \times 10^6 E^{-2.54} \text{ protons cm}^{-2}\text{s}^{-1}\text{MeV}^{-1} \qquad \text{(VI.6)}$$

for the deepest penetration into the South Atlantic anomaly. The lower limit of the integration at 100 MeV is due to the fact that below this energy the majority of particles lose energy through ionization without undergoing nuclear interactions.

Table VI-2, column (A), gives the number of some predominant spallation nuclei with half-lives shorter than 18h, produced from CsI during a 10 min exposure to the peak radiation belt flux given above, for the UK-5 crystal of approximate dimensions 5 cm x 3.4 cm diameter ($G_0 \sim 23 \text{ cm}^2$). Table VI-3, column (A), gives the corresponding results for nuclides with half-lives in the range from 1.9h to 154 days. It should be emphasized that the yield of radioactive nuclei given is for a single exposure as specified. In the real situation, there are about four orbits per day for an orbit of $\sim 30°$ inclination which would give a radiation dose of 6×10^5 protons cm^{-2} for a 10 min segment of time in each orbit. Therefore, the instantaneous number of radioactive nuclei will be a function of intermittent bombardments, building up a given activity, each followed by decay with an eventual buildup in the isotopes with longer half-lives. The second columns in both Tables VI-2 and VI-3 give the decay mode of the specific nuclide and the resulting γ-rays from the daughter nuclide.

For the case of activation of the CsI by primary cosmic rays, Dyer and Morfill (1971) have assumed that the production cross section is independent of energy for $E_p \geqslant 2.1$ GeV and used 950 protons $\text{m}^{-2}\ \text{s}^{-1}\ \text{sr}^{-1}$ for a 2 GV cutoff rigidity (Webber, 1967). Columns (C) in Tables VI-2 and VI-3 give the nuclide production for a 6.3 day exposure to these cosmic rays. This gives the same total number of interactions in the crystal (3×10^5) as for the radiation belt dosage shown in columns (A). The cosmic ray interaction rate was 0.55 interactions s^{-1}. Since the number of radioactive nuclides present with a given decay constant, λ, at a given time under steady bombardment, reaches a saturation value equal to the production rate divided by the decay rate, λ, the abundance of those nuclides with half-lives less than ~ 3 days will not build up under further cosmic ray bombardment; however, nuclides with half-lives longer than 6 days will continue to build up. For comparison with accelerator measurements using 155 MeV protons, columns (B) in Tables VI-2 and VI-3 show the nuclide production for an exposure giving 3×10^5 interactions.

Carpenter and Dyer (1973) have published the decay spectra observed in CsI crystals following irradiation by 155 MeV protons. One crystal of dimensions 3.4 cm diameter by

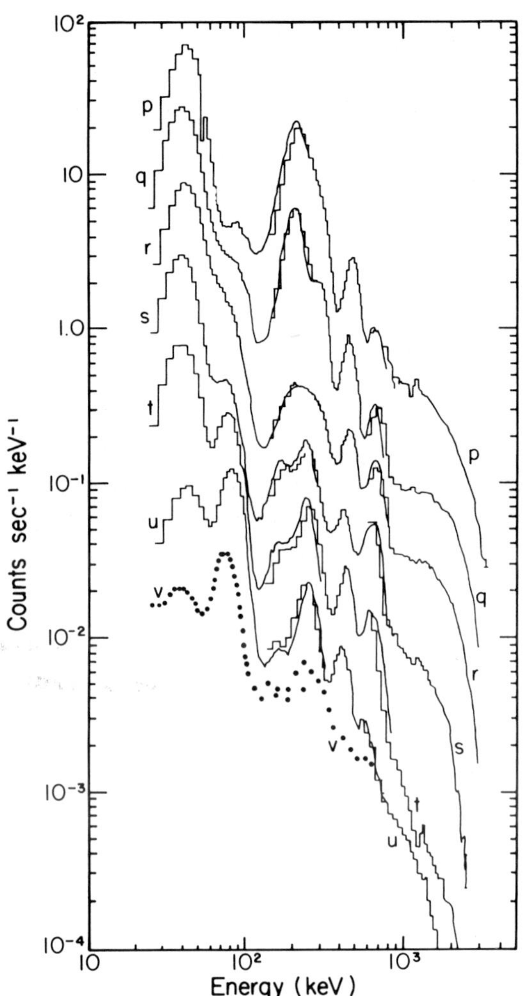

Fig. VI-10. The activation γ-ray spectra in CsI following an exposure to 10^{10} 155 MeV protons. (From G. Carpenter and C.S. Dyer: 1973, *Astrophys. and Space Sci.* **14**, 95. Used by permission D. Reidel Publishing Company, Dordrecht, Holland.)

5.0 cm length was irradiated for 1 min by $\sim 10^8$ protons, and a second crystal of 2.6 cm diameter by 3.7 cm length was exposed to $\sim 10^{10}$ protons. The light pulses resulting from radioactive decays inside the irradiated crystals were detected by a photomultiplier, and the pulse height spectra were recorded at regular intervals for the study of the decay characteristics of the spectra. It should be noted here that this scheme would record only a continuous spectrum predominantly for γ-rays following $\beta^{\pm}$ decay in the crystal as discussed above, and photopeaks in the spectrum would only generally occur if the decay resulted from electron capture (EC).

In Figure VI-10 we show the resulting spectra obtained by Carpenter and Dyer (1973) for a proton dosage corresponding to $\sim 10^9$ nuclear interactions in the crystal, which is $\sim 10^4$ times the dose for a single South Atlantic anomaly exposure as discussed by Dyer and Morfill (1971). Each of the separate spectra shown were taken from approximately 7 h after irradiation for the spectrum denoted as (p) to 224 days after irradiation for the spectrum shown as (v). Above 600 keV this spectrum has decayed to background after 94 days but at lower energies broad decay peaks at $\sim$35 keV, $\sim$80 keV and $\sim$200 keV are still evident above the background after 225 days. (The background is not shown in this figure.) Each of the broad features evident at earlier times has been considered by Dyer and Morfill (1971), who conclude that these are due to the superposition of several isotopes which contribute lines at slightly different energies. The likely sources of the various features are interpreted by Carpenter and Dyer (1973) as follows:

The peak at $\sim$35 keV is a superposition of K-shell X-rays emitted following K-capture (EC) decay by a number of spallation nuclides with (Z, A) slightly less than ^{133}Cs. Tables VI-2 and VI-3 list several possible contributing isotopes. The broad feature at $\sim$200 keV could have contributions from several EC decays ^{125}Xe, ^{123}I, ^{117}Sb, or decays of some isomeric states such as ^{121m}Te or ^{123m}Te with half-lives of 154 days and 117 days respectively. Possible contributing EC decays giving the broad feature at around 400 keV are ^{127}Cs, ^{129}Cs, ^{127}Xe, and ^{126}I, with the feature around 600 to 700 keV due to ^{124}I, ^{126}I, and ^{132}Cs.

Carpenter and Dyer (1973) have pointed out the relatively rapid decay of the spectrum beyond 600 keV and suggest this is due to the large number of short-lived β^+ decays, several of which are given in Table VI-2. Notice also the peak in Figure VI-10 which appears at about 80 to 90 keV and is likely due to ^{123m}Te and ^{121m}Te, which have half-lives $\gtrsim$ 100d.

The nuclide yields given in Tables VI-2 and VI-3 were also used to predict approximate energy loss spectra in CsI and NaI crystals by the following procedure: X-ray source response functions and Monte Carlo estimates of the photopeak efficiency and Compton plateau were used to arrive at the spectra assuming the β^+ decays gave an energy loss at $1/2 \times$ (maximum β^+ energy + annihilation energy). Since the agreement between the experimentally produced activation and that predicted was moderately close, Dyer and Morfill (1971) have scaled the experimental results at 155 MeV to correspond to the 3×10^5 interaction dosage used for the spallation yields for the inner belt given in Tables VI-2 and VI-3. The estimated activation spectrum for inner belt protons and cosmic rays is shown for the OSO-3 NaI detector in Figure VI-11 (Dyer, 1973) and compared with early measurements of the diffuse X-rays on Ranger 3 up to 1 MeV and on ERS-18. The last two measurements were made in space outside the radiation belts. The activation predictions shown in the figure are given in terms of an equivalent flux of external photons as are the experimental values shown in Figure VI-11. It can be seen that the estimated contribution from spallation activity is comparable to the measured flux on OSO-3 between 50 keV and 200 keV. Note that the OSO-3 data have been corrected by Schwartz (see Dyer, 1973), and that above 200 keV the Ranger 3 and ERS-18 flux values are actually in units of counts cm^{-2} s^{-1} sr^{-1} keV^{-1}. The estimated spallation

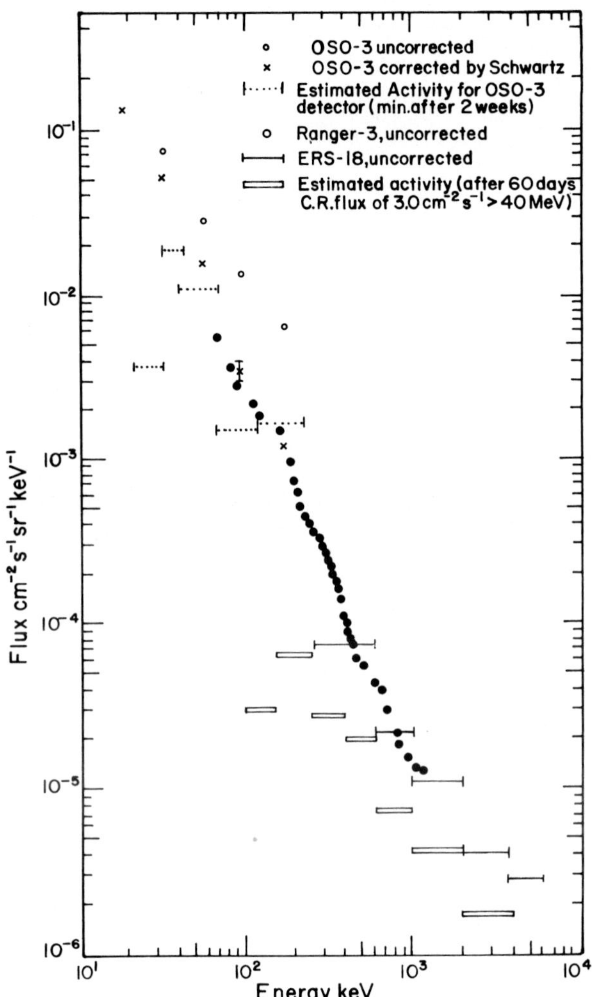

Fig. VI-11. The uncorrected measurements of the diffuse X-ray spectrum are compared with estimates of proton activation. (From C. Dyer: 1973, *NASA SP-339*.)

contribution in these NaI detectors is also shown for a cosmic ray flux of 3.0 protons cm^{-2} s^{-1} above 40 MeV. If the long-lived spallation contribution from the radiation belt passage of the Ranger 3 and ERS-18 is included, the total contributions from spallation could be comparable to the measured fluxes above 200 keV (See Selzer (1975) for recent calculations on the response of scintillation detectors to internally induced radioactivity.)

Fishman (1972a) has also used the Rudstam formula to predict the activation or spallation yields in NaI for cosmic ray primary proton energies above 100 MeV. Modifications to the Rudstam formula made by Silberberg and Tsao (1973a, b) (see below) were also incorporated later by Fishman (1973), who concludes that the majority of spallation

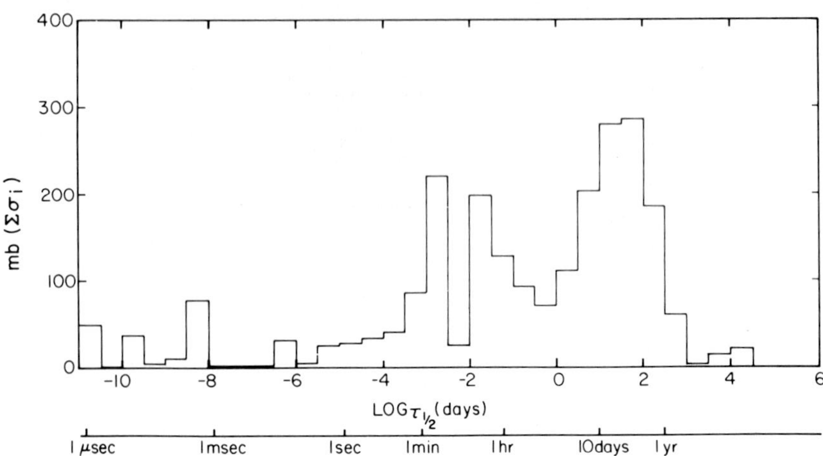

Fig. VI-12. Half-life distribution of iodine spallation products. (From G.J. Fishman: 1972, *Proton Induced Radioactivity in* NaI(Tl) *Scintillation Detectors,* Teledyne Brown Engineering Summary Report SE-SSL-1497.)

products (~80%) are due to interactions in I. It would seem that the spallation yields for CsI and NaI should be nearly the same, because of the closeness in (Z_t, A_t) of ^{133}Cs and ^{127}I. Fishman (1973) gives the cross sections used for predicting the major radioactive spallation products of ^{127}I, which show that significant spallation yields are expected from all energies, and that above 3000 MeV there is no detectable change of the cross sections with energy. The distribution by half-life of the calculated iodine spallation products is given by Fishman (1972b) and is shown in Figure VI-12. The ordinate gives the total cross section for 1 GeV protons to produce radioactive spallation nuclides with half-lives in the time intervals as indicated on the abscissa.

Experimental measurements were also made using a 600 MeV proton beam on a 17 g cm^{-2} thick NaI(Tl) crystal (Fishman, 1972b). The resulting counting rate spectrum seen by a phototube attached to this crystal is shown in Figure VI-13 taken ~7 h after an exposure to 7×10^{10} – 600 MeV protons. Several features are apparent in the spectrum and the features likely due to EC decays of ^{123}I and ^{126}I are indicated. Figure VI-14 gives the measured energy loss spectrum with broad resolution compared with an empirical prediction of what the general shape of the continuum from activation should be. Fishman (1973) has assumed this continuum to have an exponential shape, given by $dF/dE \propto \exp(-E/E_e)$, in analogy with the γ-ray decay spectrum of a large number of mixed fission products, since these have Z and energy level spacings similar to ^{127}I spallation products. The e-folding energy, E_e, which determines the hardness of the spectrum, was taken to have values $E_e = 0.9$ MeV and $E_e = 1.4$ MeV as shown in Figure VI-14. Fishman (1972b) has found that the long-lived NaI activation spectrum is exponential with $E_e = 0.6$ MeV up to ~3 MeV. This was derived from a spectrum taken several hours after irradiation in which no correction was made for the effect of short-

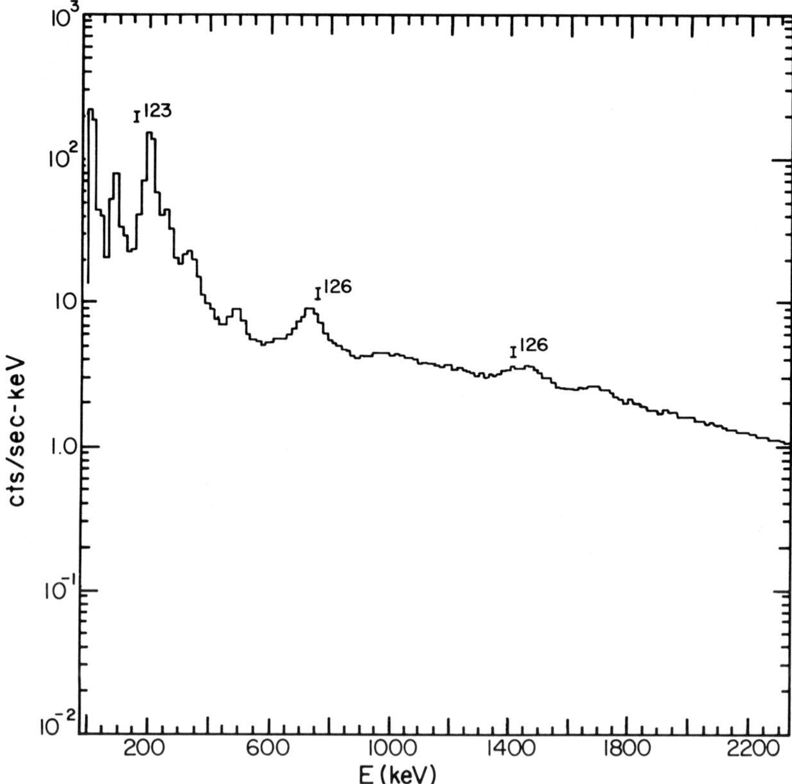

Fig. VI-13. The activation spectrum in NaI ~ 7 h after an exposure to 7×10^{10} 600 MeV protons. (From G.J. Fishman: 1972, *Proton Induced Radioactive in* NaI(Tl) *Scintillation Detectors,* Teledyne Brown Engineering Summary Report SE-SSL-1497.)

lived decays. Fishman (1973) suggests that a harder spectrum, $E_e \cong 1.4\,\mathrm{MeV}$, may be closer to the truth for a spallation background such as may be seen by Apollos 15 and 17 since shorter half-lives tend to have a harder spectrum. This is supported by the spectra of Carpenter and Dyer (1973) in Figure VI-10, which clearly show the fall-off, or steepening, of the spectra with time after 7 h. The shorter half-lives cannot be isolated, of course, if one is dealing with cosmic ray activation, but after anomaly passages this effect should be seen.

The Rudstam formula (1966) cannot be used to calculate the yield of every spallation nuclide since it represents an empirical fit for targets $Z > 20$ and was intended to apply only to particular cases. Specifically, if the spallation product corresponds to $3 < \Delta A < 30$ for $E_p < 200\,\mathrm{MeV}$, then the formula has reasonable accuracy; however, for spallation nuclides corresponding to removal of one or two nucleons, the formula is highly inaccurate. Silberberg and Tsao (1973a, b) have extended the Rudstam formula to cover targets from Li to Bi. Basically, they have tabulated all available experimental

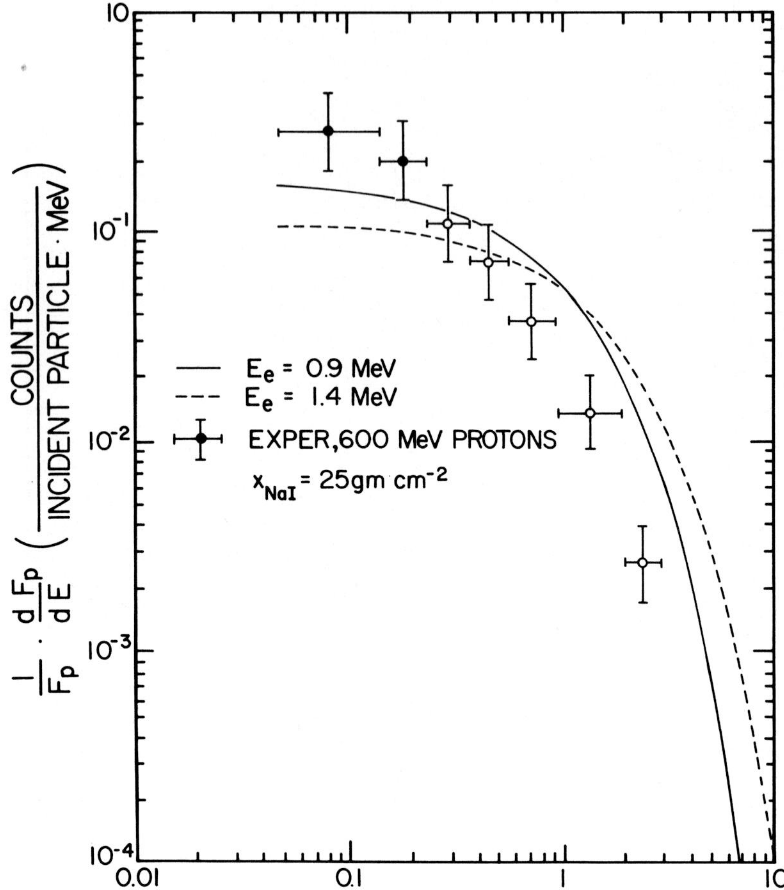

Fig. VI-14. The energy loss spectrum of induced radioactivity. The data points are from direct measurements of 600 MeV proton-induced radioactivity in NaI(Tl) corrected for the expected contribution of unmeasured, short-lived products (Fishman, 1973). Also shown are two exponential spectra described in the text, normalized to a total rate of 0.15 cts per incident high energy proton. (From G.J. Fishman: 1973, *NASA SP-339*.)

cross sections (Silberberg and Tsao, 1973b) and derived from these new empirical cross sections. Figure VI-15 is a comparison of these new calculations with experiments and with the Rudstam formula for the case of 150 MeV protons on Fe. First of all, notice that the new calculations of Silberberg and Tsao (1973b) fit the few experimental points shown very well, while the Rudstam formula underestimates the cross section by one to three orders of magnitude. This example should suffice to demonstrate the large inaccuracies that can be introduced in calculated activation effects. Nevertheless, the new empirical cross sections of Silberberg and Tsao (1973b) have only recently become

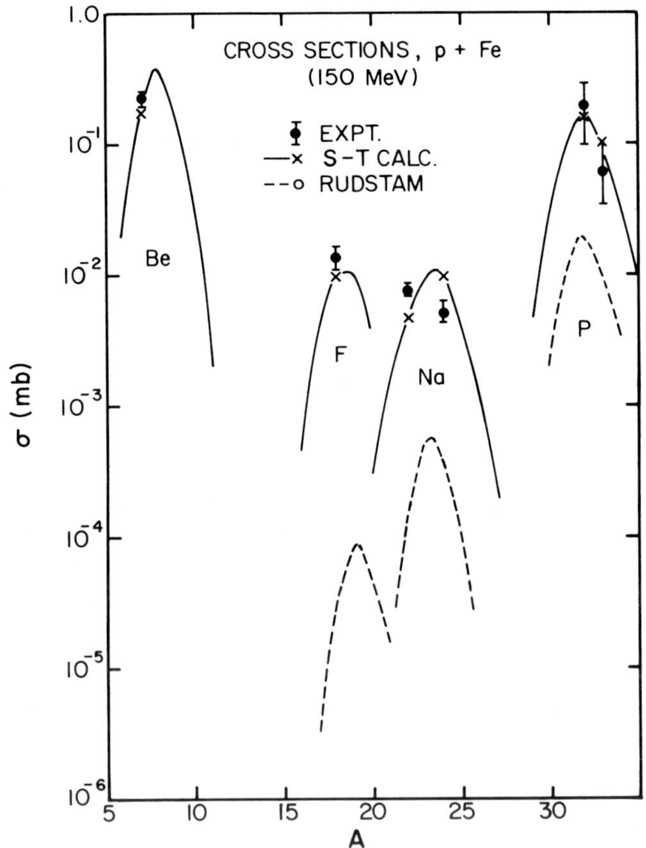

Fig. VI-15. Comparison of experimental, Rudstam, and Silberberg-Tsao cross sections from 150 MeV protons on Fe. (From R. Silberberg and C.H. Tsao, *Astrophys. J. Suppl.* **25**, 315. Copyright 1973, The American Astronomical Society. Used by permission of the University of Chicago Press.)

available and should be very valuable in improving the calculations of Dyer and Morfill (1971). It is relevant to point out here that all the calculations discussed so far give the yield of spallation products in the ground state of the product, except as mentioned below.

Another method of determining the yield of γ-rays from proton-nucleus interactions is to use the method described by Shima and Alsmiller (1970), known as the intranuclear-cascade-evaporation model, which is based on early work by Bertini (1963, 1965, 1966) with modifications of Guthrie (1969). These Monte Carlo calculations assume that the reactions take place in three stages and that γ-ray emission does not compete with particle emission until the latter is no longer energetically possible. In the *first stage,* a cascade takes place within the nucleus and fast nucleons are emitted leaving the residual nucleus in an excited state. In the *second stage*, the high temperature nucleus evaporates nucleons, deuterons, tritons, α-particles, etc. This evaporation process continues until particle emission is no longer energetically possible, and the final nucleus may still be in an excited state, in which case γ-ray emission occurs as a *third stage*. Each Monte Carlo history gives a third stage nucleus with specific values of (A, Z) and excitation energy. Thus, the photon emission is calculated for each Monte Carlo history, if the energy levels of the final nucleus and its γ-ray branching ratios are known. These calculations were mentioned previously (see Section II-2.4.3) in connection with the γ-ray spectrum measurements of Zobel *et al.* (1968). There it was seen that the detailed comparison of these calculations with experimental γ-ray yields for protons on light elements (C, O, etc.) *was not good*. The calculated *total* photon production cross sections were, however, within a factor of 2 of the experimental values for the energy range 16 to 160 MeV for all elements considered by Shima and Alsmiller (1970). There were serious discrepencies, however, between the calculated and experimental intensities of specific γ-ray lines. This may be due to the fact that Shima and Alsmiller (1970) did not allow for the spins and parities of the excited nuclei since many are unknown; hence, they did not take into account the appropriate selection rules for γ-ray emission. Numerous corrections have been made to these calculations which could be made available for background estimates for γ-ray astronomy.

The activation of detector and spacecraft materials by neutrons is also of interest, since prompt and delayed γ-rays are emitted in most neutron reactions and these γ-rays cannot always be eliminated by an anticoincidence method. For γ-ray astronomy experiments carried out in the atmosphere an estimate of the yield of prompt and activation γ-rays from neutrons can be carried out by using the neutron flux given in Section VI-6.1.1 and the cross sections given in Section II-2.4.3. Neutrons, however, are also a secondary product of the interactions of primary cosmic rays, solar cosmic rays, and trapped radiation in spacecraft detector materials. Background effects from this local neutron production have been reduced on some space experiments by placing the γ-ray detector on a boom, such as was the case for the Ranger and Apollo experiments (Metzger *et al.,* 1964; Trombka *et al.,* 1973). For some γ-ray experiments carried out near the Earth, the neutrons leaking into space could be an important source of background. A recent paper by Lockwood (1973) reviews the present knowledge of the leakage neutron flux as measured near the Earth.

In order to further assess the problem that neutrons present for space experiments, Fishman (1974) has measured directly the neutron flux inside Skylab, which was in orbit from 1973, November 16 until 1974, February 5. The spacecraft was in an orbit of inclination $\sim 33°$, for which about 4 or 5 orbits per day penetrate the South Atlantic anomaly deeply. The neutron flux at several locations inside the Skylab was measured by standard neutron activation techniques which, in this case, consisted of exposure of samples during the 84 day mission and later counting the activity in a low-level γ-ray spectrometer facility on the ground. Results on both the total neutron and proton fluxes were obtained, and some typical values are of interest. The fast neutron flux (3 to 15 MeV) as measured from the activation-reaction ^{58}Ni (n, p) ^{58}Co was ~ 0.96 neutrons cm^{-2}s^{-1}. The slow neutron flux ($E_n \leqslant 0.3$ eV), measured by ^{181}Ta samples with and without Cd shields, was typically Φ_n (<0.3 eV) $\lesssim$ 0.06 neutrons cm^{-2} s^{-1}. This latter flux is comparable with the atmospheric neutron flux (<1 eV) at a depth of ~5g cm^{-2}. An approximate estimate of the proton flux was also made using the reactions $^{48}Ti(p, n)$ ^{48}V and $^{58}Ni(p, 2pn)$ ^{56}Co. This gave Φ_p(8 to 15 MeV) ≅ 0.24 protons cm^{-2} s^{-1} and Φ_p (30 to 100 MeV) ≅ 2.3 protons cm^{-2} s^{-1}. It should be noted that the flux estimates given above refer to average values over the full 84 day mission lifetimes, but, of course, during radiation belt passage the instantaneous fluxes of neutrons and protons are several orders of magnitude higher, so activation of short half-life nuclides by neutrons and protons can be a serious problem. The neutron and proton fluxes, of course, vary with location in Skylab (see Fishman, 1974). That neutron effects can produce a significant background in γ-ray spectrometers is evident from experiments discussed in Sections VI-6.1.3 and VI-6.1.5.

6.1.3. SATELLITE ORBITS

The nature of the actual backgrounds seen by γ-ray line detectors in different satellite orbits is of great interest. In general, all the potential sources of background given in Table VI-1 can contribute, and, because of the Van Allen radiation belts, there is often a very substantial contribution from interactions of the strong charged particle flux in the spacecraft and the detector that overrides backgrounds from the atmosphere and from the diffuse radiation. In principle, these latter contributions could be calculated separately from the basic fluxes we have given in Section VI-6.1.1. At the present time, however, the induced background from charged particles represents a complex unresolved problem, and the diffuse flux itself is uncertain in the energy range of interest here.

a. *Intermediate Latitudes*

The first satellite γ-ray experiment in the nuclear line region was carried out on OSO-1 by Peterson (1965). The γ-ray detector consisted of an array of NaI(Tl) scintillation counters operated in various logic and shield configurations to provide directional properties and reject charged particle effects. The principal element in the array was a NaI crystal (5.1 cm diameter × 5.4 cm length) surrounded by a 0.32 cm layer of plastic scintillator and operated as a phoswich detector. The satellite orbit had an inclination of 33° and a nominal altitude of 550 km. No results on extraterrestrial γ-rays have been reported from this experiment, and no large solar flare events occurred during the satellite's lifetime. Peterson (1965) has, however, reported on the induced activation of the NaI detector from trapped charged particle flux in the South Atlantic anomaly. It was expected that high backgrounds would be experienced when the detector was in the trapped particle region of the South Atlantic anomaly. The observations indicated, however, that the detector counting rate was abnormally high for a long period of time after the satellite had passed through the anomaly. In Figure VI-16 the counting rate vs. time after a radiation belt anomaly passage is shown for γ-rays in two energy regions: 0.3 → 1.0 MeV and 1.0 → 3.0 MeV. It is seen in the figure that the counting rates in both channels decay sharply from a maximum rate in about 5 min, followed by a slower

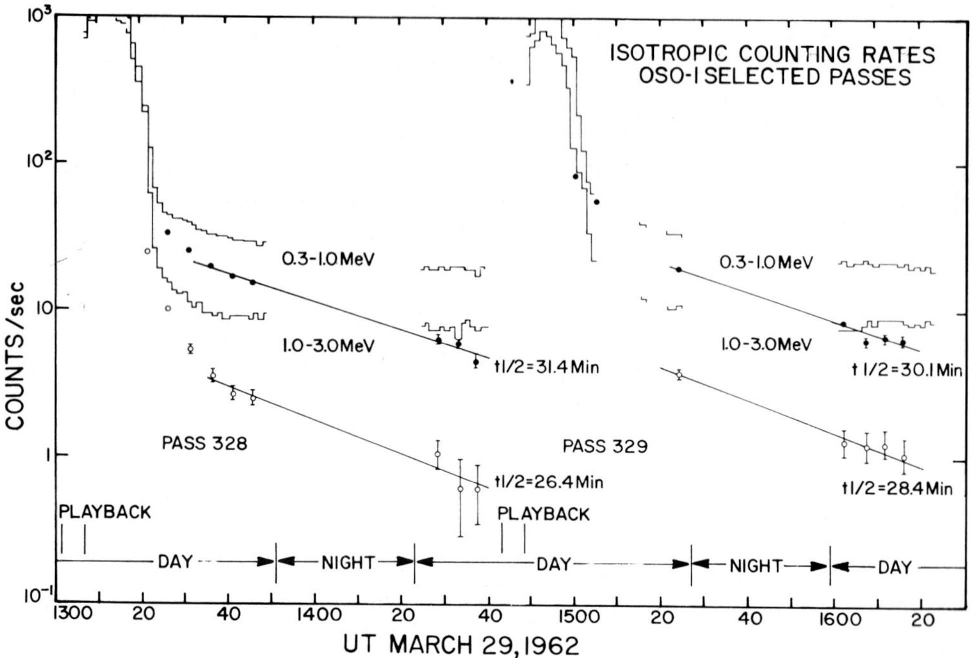

Fig. VI-16. The counting rate vs. time in the OSO-1 NaI(Tl) γ-ray spectrometer in two energy bands after two passages through the radiation anomaly. The results of subtracting out cosmic ray background and fitting the excess rates to a radioactive decay law are also shown. (From L.E. Peterson, *J. Geophys. Res.* **70,** 1762, 1965, copyrighted by American Geophysical Union.)

decay, until another anomaly passage in the next orbit. In order to understand the origin of the slower decay background, Peterson (1965) assumed that the counting rate was made up of contributions from a latitude-dependent background due to cosmic rays and any radioactivity that might be induced by the trapped radiation.

The procedure to evaluate this latitude dependent background can be summarized as follows: Since Lin *et al.* (1963) have shown that the invariant latitude Λ obtained from the magnetic shell parameter L gives a good description of the geomagnetic cutoff effects of cosmic rays, equal values of Λ imply equal geomagnetic cutoffs and, therefore, equal cosmic-ray-produced background. The relation between Λ and L is $\Lambda = \cos^{-1} (L^{-1/2})$ (Lin *et al.*, 1963). The counting rate curves observed for southbound orbits passing through the geographic meridian 60° E, well away from the radiation anomaly, were plotted vs. the geographic latitude of OSO-1 (see Peterson, 1965). This was then transformed to a 'universal curve' of counting rate versus Λ which was used for background correction on a worldwide basis. Therefore, for each point of time as shown in Figure VI-16, a value for Λ was computed from the knowledge of the satellite location and a cosmic ray background counting rate was found from the universal curve described above. This rate was subtracted from the measured rates shown. The resulting excess counting rate curves for the two energy bands are shown as points fitted by the solid line in Figure VI-16. Since the plot is semi-logarithmic in counting rate vs. time, the apparent radioactive decay half-life is readily obtained. For the four curves shown in Figure VI-16, the half-lives range from 26.4 to 31.4 min. Peterson (1965) has carried out this procedure for twenty additional passes and has found that the excess rate is proportional to the penetration of the satellite into the anomaly and that the single half-life decay values range from about 20 to 30 min. On the assumption, which is undoubtedly oversimplified, that the background is due to the decay of a single radioactive species, Peterson (1965) concluded that ^{128}I ($t_{1/2} = 25$ min) must have been produced in the detector by the reaction ^{127}I(n, γ) ^{128}I where the neutrons are locally produced when the satellite is in the anomaly. The beta decay of ^{128}I has an end point energy of 2.12 MeV and populates levels in ^{128}Xe whose decay gives γ-rays at ~0.45, 0.54, 0.75, and 0.99 MeV. The steep energy loss spectrum suggested by the relative rates in the two energy bands in Figure VI-6 is consistent with this decay.

That this single nuclide is solely responsible for the data in Figure VI-16 seems highly unlikely in the face of our earlier discussion on activation by protons. Even though secondary fast neutron production is copious in a satellite, neutrons must be moderated to increase the capture probability. (Moderation could have been effected by a large block of polyethylene aboard the OSO-1 used in a University of California neutron experiment; see OSO-1 Report, *NASA SP*-57, 1965.) If the ^{128}I interpretation is correct, Peterson concludes that the activation of the OSO-1 crystal upon leaving the trapped region was typically 8×10^{-2} disintegration g^{-1} s^{-1}. From the thermal neutron capture cross section for ^{127}I of 6.3b, this implies an average thermal neutron flux of 7 cm^{-2} s^{-1} on the NaI crystal during each 10 min passage through the anomaly. It is interesting to note that this thermal neutron flux is about two orders of magnitude higher than Fishman found in Skylab (see Section VI-6.1.2).

Peterson's X-ray telescope onboard the OSO-3 satellite (Schwartz *et al.*, 1970) was in essentially the same orbit as OSO-1. The instrument, in this case, consisted of a 3.5 × 0.5 cm NaI(Tl) crystal in a CsI anticoincidence shield, constituting a telescope with a 23° FWHM conical field of view. The counting rate data were telemetered in six logarithmically spaced channels between 7.5 keV and 210 keV. Schwartz and Gursky (1973) have used the data from this experiment to study the diffuse cosmic X-ray background. They have argued that the most serious background contribution in this low energy experiment was from sporadic charged particles. The selection criteria used for good data were a limit on the upper threshold integral rate, a requirement that the magnetic shell parameter was $L \leqslant 1.2$, and acceptance of data only when not pointed within the local magnetic loss cone. These conditions rejected most of the data (80%). Interestingly, these authors argued that the remaining source of background was radioactivity of the satellite but *not* spallation products of CsI and NaI as proposed by Dyer and Morfill (1971). In particular, they suggested that secondary neutrons produced in the anomaly interact with Al of the spacecraft according to ^{27}Al (n, α) ^{24}Na $\xrightarrow{15h}$ ^{24}Mg. ^{24}Na undergoes β decay with an end-point energy of 1.39 MeV, accompanied by γ-rays at 2.75 MeV and 1.37 MeV. It is presumed that the γ-rays Compton scatter throughout the space craft before interacting in the detector. Apparently, the major reason for identifying the activity with ^{24}Na is that a 15h half-life gives a good fit to their monitor counting rates over the time interval of 30 min to 12h after penetrating the anomaly. The spectrum of radioactivity background in the OSO-3 central X-ray detector immediately after emerging from the radiation anomaly, therefore, has been interpreted as due to ^{24}Mg γ-rays. It is interesting to note that an isomeric state of ^{24}Na with an energy of 473 keV and half-life 0.020 s could be produced in this reaction and may be observable if this scheme is a dominant one. Schwartz and Gursky (1973) concluded that the spallation mechanism is not effective on this time scale ($\sim$ 15h). They have also studied the buildup in counting rates in the 38 to 65 keV channel over a period of weeks after launch of the OSO-3 and compared this with the spallation buildup curve predicted by Dyer *et al.* (1972). They concluded that the proton dose which the latter used should be decreased by about a factor of 3.

The OSO-7 γ-ray spectrometer, which was discussed in Section V-5.1.1, also provides detailed spectra that may be compared with the Dyer and Morfill (1971) and Fishman (1972a) calculations. Again, this satellite was in an orbit of about the same inclination as OSO-1 and OSO-3; however, due to a launch error, OSO-7 was placed in a highly eccentric orbit, which at apogee ($\sim$ 575 km) penetrated the anomaly deeply and at perigee ($\sim$ 375 km) was well below the anomaly. Figure VI-17 (Suri *et al.*, 1974) shows summed spectra obtained when the wide angle OSO-7 γ-ray spectrometer viewed two different regions of the sky. The basic spectra used in the summation were 3-min real-time integrations, with a 25% duty cycle. The spectra were selected for the lowest background conditions, that is, when the detector was well out of the South Atlantic anomaly and when no solar activity was evident. The upper curve shows the total spectrum obtained for a live time of 1.33×10^4 s (which includes $\sim$ 300 separate spectra) when the Galactic Center region of the sky was in the field of view. The lower spectrum shows a similar summation for a live time of 1.06×10^4 s when the detector was viewing the opposite region of the sky including the Sun and the Crab Nebula.

This second spectrum is shifted down by a factor of 2 at all energies so a visual comparison may be made more easily. When the spectra are directly overlaid, there are no statistically significant differences. A complete understanding of the origin of these OSO-7 spectra is not yet available; however, at energies below 1 MeV, long-lived activation of the spacecraft and the instrument is believed to be the major contribution, and above 1 MeV it is possible that diffuse cosmic γ-radiation could make a significant contribution. The features at low energies indicated by the numbers 1 to 4 are undoubtedly due to activation of the CsI (Na) and NaI (Tl) portions of the spectrometer by the radiation belt protons (see Table VI-3). The peaks at 1.17, 1.33 and 2.50 MeV are due to an onboard ^{60}Co calibration source. This spectrum can be considered as the lower limit to the γ-background for this particular instrument. A further discussion of the OSO-7 background with particular reference to the activation problem is given by Dunphy *et al.* (1975) and Dyer *et al.* (1975).

b. *Polar Orbits*

The only γ-ray background data available for polar-orbiting near-Earth satellites are from

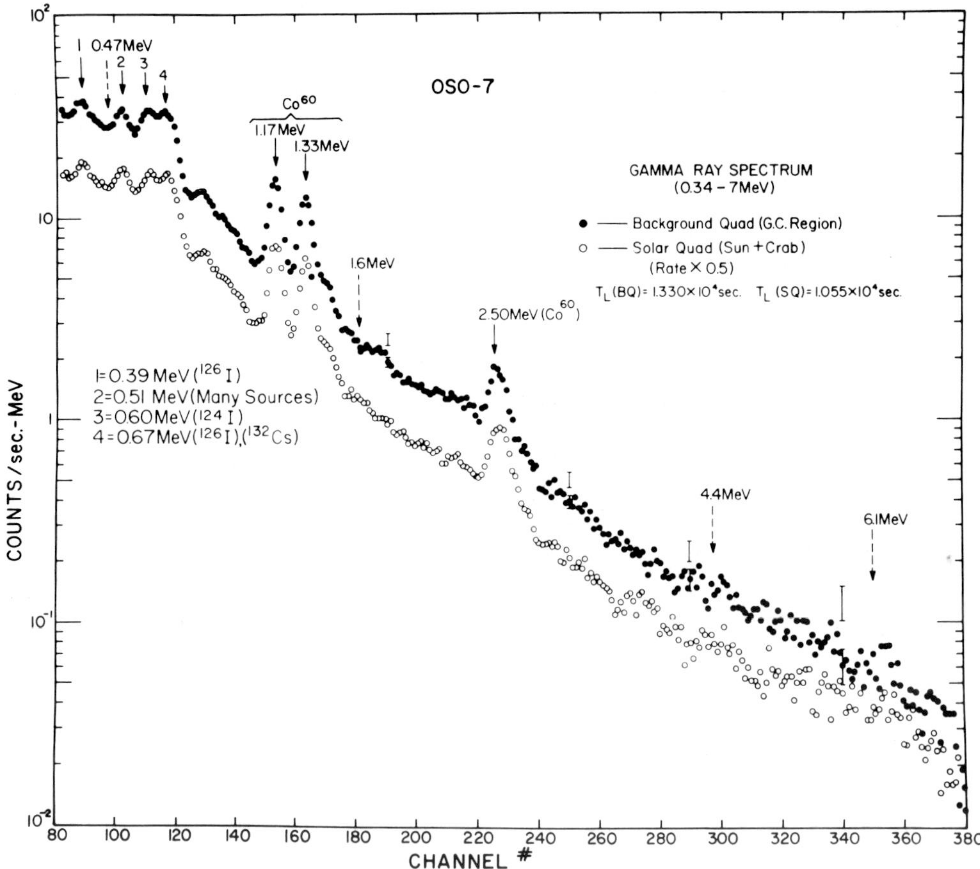

Fig. VI-17. The energy loss spectrum is shown for the OSO-7 γ-ray spectrometer. The features indicated by 1, 3, and 4 are a result of long term activation due to repeated passage through the South Atlantic anomaly.

the recent solid state detector experiment described by Nakano *et al.* (1973) and Imhof *et al.* (1973). Figure VI-18 shows a cross sectional view of the detector, which was on the spin stabilized satellite (1972-076B) launched into a noon-midnight, Sun-synchronous orbit of inclination 98.4° with apogee and perigee of 761 km and 736 km, respectively. The spin period was 5 s and the detector viewed the Sun once every revolution of the satellite in a manner similar to the OSO-7 γ-ray experiment mentioned previously. The basic detector was a 50 cm^3 Ge(Li) spectrometer with an active area of 15 cm^2 cooled by solid CO_2 as shown in Figure VI-18. The detector was surrounded by a W collimator and a stainless steel tungsten rear shield, which in turn was surrounded by a 4π anticoincidence plastic scintillator. γ-Radiation entering from outside the spacecraft was restricted effectively to a cone of 45° half-angle. The minimum shielding in other directions was 30 g cm^{-2}. The energy loss spectrum in the Ge(Li) detector was recorded over the energy

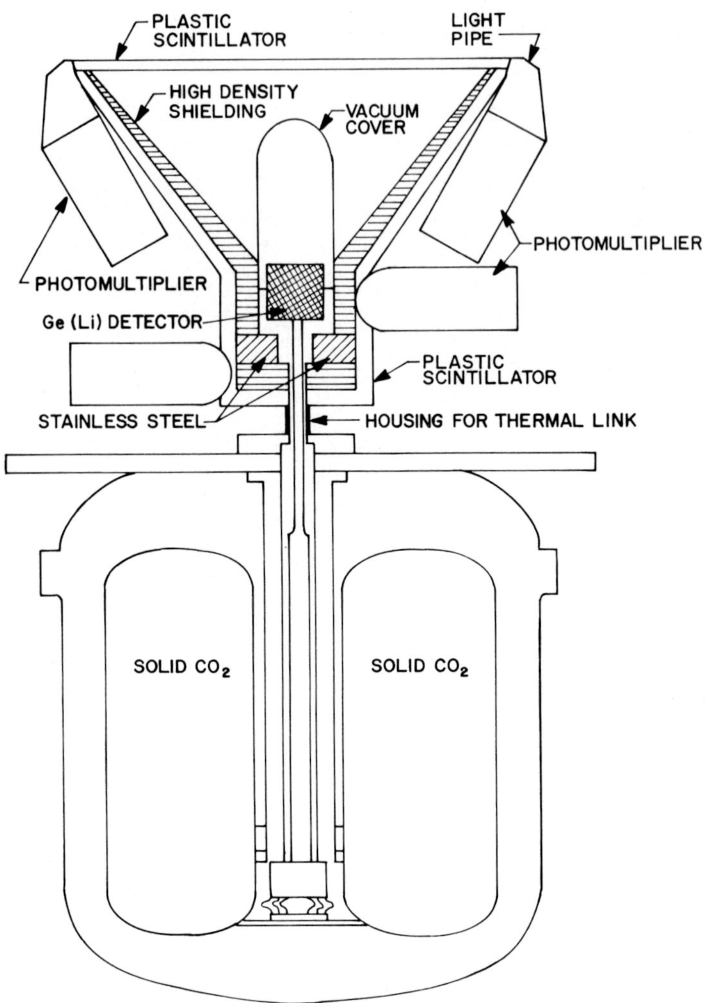

Fig. VI-18. The cross sectional view of the important features of the Ge(Li) spectrometer in polar orbiting satellite (1972 – 076B). (From G. H. Nakano *et al.*: 1973, *NASA SP-339.*)

range 40 keV to 2.8 MeV with a 4096-channel linear pulse height analyzer. The best energy resolution obtained was 3.5 to 4.0 keV over the full energy range, which, incidentally, corresponds to a resolution improvement over the OSO-7 instrument by a factor of 20 at 1.5 MeV and a factor of 30 at 3 MeV. The energy loss counting rate spectrum obtained by averaging over all spin angles is shown in Figure VI-19. From the discussion of Nakano *et al.* (1973), there was no strong evidence for a modulation of this spectrum in different viewing directions, so it is reasonable to take this spectrum as a lower limit to the locally produced γ-ray line spectrum for this instrument at low to mid-latitudes. This background could then, as in the case of OSO-7, be a result of all effects mentioned previously: spallation activation in the satellite and in the detector,

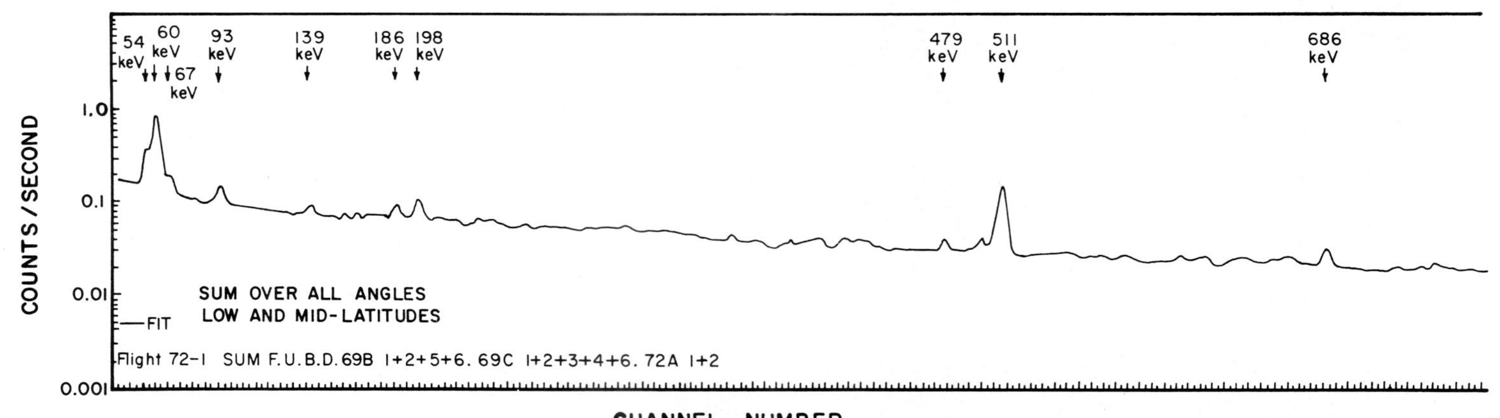

Fig. VI-19. The background counting rate spectrum in a Ge(Li) spectrometer in a polar orbiting satellite. (From G.H. Nakano *et al.*: 1973, *NASA SP-339*.)

neutron activation of spacecraft materials as well as the detector, and similar phenomena in the heavy W collimator (20 kg) near the detector.

It is not easy to compare this background spectrum with the OSO-7 background spectrum just mentioned, which was observed in a NaI(Tl) detector with a CsI shield. The solid state detector background shown in Figure VI-19 must be considered as representative for such a detector in a particular shielded configuration and orbit. As may be seen in the figure, the spectrum from 40 keV to 700 keV is basically a slowly falling continuum with several peaks. The preliminary identification and interpretation of the origin of the peaks is given in Table VI-4, which gives their approximate energies in the first column. The strongest peak, at 59.6 keV, is due to the onboard ^{241}Am calibration source. The peaks at 54 keV, 67 keV, 139 keV, 198 keV, and 511 keV were also seen by Womack and Overbeck (1970), using a balloon-borne Ge(Li) detector (see Section VI-6.1.5). In this experiment and the Lockheed experiment, these lines (except for the one at 511 keV) are clearly identified with γ-ray decays in the Ge(Li) crystal itself, and represent excitation of isomeric levels in Ge isotopes produced by neutron reactions as indicated in column 2 of Table VI-4. As discussed earlier, only decays through electron capture or isomeric states can give a line spectrum if the detector material itself is activated, whereas, with β^- decay in the detector, the subsequent γ-ray transition does not give a single energy peak. In column 4 of the table is given the counting rate measured for the background lines in the balloon experiment of Womack and Overbeck (1970). Column 5 indicates the rates for the background lines recently recorded by Jacobson *et al.* (1975) in a balloon experiment using a Ge(Li) spectrometer (see Section VI-6.1.5). Nakano *et al.* (1973), Womack and Overbeck (1970), and Jacobson *et al.* (1975) have attributed several observed peaks to (n, γ) reactions in Ge as shown in Table VI-4. Neutron capture is not necessarily the only production mode for the isomeric states shown but it seems most probable. No calculations have been carried out to determine the consistency of observed line intensities with the neutron cross sections and the fluxes and the moderation of local neutrons.

Support for the neutron activation mechanism is also given by the presence of the lines at 479 keV and 686 keV shown in Table VI-4. They are attributed by Nakano *et al.* (1973) to (n, γ) reactions in the heavy W shield producing ^{187}W with a half-life of 23.9 h. It is interesting, and perhaps coincidental, that Womack and Overbeck (1970) see lines at nearly these two energies, which they relate to neutron interactions in Ge (see a later discussion on γ-ray backgrounds in the atmosphere in Section VI-6.1.5). Finally, there is the omnipresent 511 keV line which can originate from a multitude of sources. This line is always enhanced in satellite or balloon experiments when additional inert material surrounds the detector. Nevertheless, the atmosphere also seems to be a natural source for this line as discussed by Chupp *et al.* (1970) and Peterson *et al.* (1973b).

The continuous spectrum shown in Figure VI-19 is of crucial importance for γ-ray astronomy experiments in satellite orbits. Since the Lockheed detector passes through the outer radiation belts, as well as the inner belt, in its polar orbit, electron bremsstrahlung becomes a source of background. Imhof *et al.* (1973) discuss two classes of bremstrahlung background. One class arises from radiation belt electrons stopping in the

TABLE VI-4

The origin of the peaks in the Ge(Li) spectrum shown in Figure VI-19 by Nakano *et al.* (1973) compared with those seen by Womack and Overbeck (1970) and Jacobson *et al.* (1975). (Rates from E.A. Womack and J.W. Overbeck, *J. Geophys. Res.* **75**, 1811, 1970, copyrighted by American Geophysical Union; and A.S. Jacobson *et al.*: 1975, *Nucl. Instrum. and Methods* **127**, 115. Used by permission of the North Holland Publishing Company.)

Energy, keV	Reaction producing precursor	Isomer half-life	Rate c s^{-1} (1)	(2)
54 (3)	^{72}Ge (n, γ) ^{73m}Ge	0.53 s	$1.30 \pm 0.03 \times 10^{-1}$	1.57×10^{-1}
59.6 (3)	^{241}Am in-flight calibration source			
67 (3)	^{72}Ge (n, γ) ^{73m}Ge	4.6 μs	$7.6 \pm 0.2 \times 10^{-2}$	6.1×10^{-2}
74 (3)	^{72}Ge (n, γ) ^{73m}Ge		$4.4 \pm 0.2 \times 10^{-2}$	
93 (3)				
94 (1)			$1.8 \pm 0.1 \times 10^{-2}$	
139 (3)	^{74}Ge (n, γ) ^{75m}Ge	48 s	$3.4 \pm 0.1 \times 10^{-2}$	6.5×10^{-2}
159 (1)	^{76}Ge (n, γ) ^{77m}Ge	54 s	$7.1 \pm 0.6 \times 10^{-3}$	6×10^{-3}
175 (1)	^{70}Ge (n, γ) ^{71m}Ge	70 ms	$1.8 \pm 0.1 \times 10^{-2}$	9×10^{-2}
186 (3)			(Observed)	(Observed)
198 (3)	^{70}Ge (n, γ) ^{71m}Ge	20 ms	$5.6 \pm 0.2 \times 10^{-2}$	1.23×10^{-1}
473 (1)	^{77}Ge (β^-) ^{77m}As	1.3 h	$2.2 \pm 0.3 \times 10^{-3}$	
479 (3)	^{187}Re from ^{186}W (n, γ) ^{186}W $\xrightarrow[1d]{\beta^-}$			
511 (3)	Positron/electron annihilation		(Observed)	3.2×10^{-2}
686 (3)	^{187}Re from ^{186}W (n, γ) ^{187}W $\xrightarrow[1d]{\beta^-}$			
691 (1)	^{72}Ge (n, n′) ^{72}Ge	(totally converted)	$4.9 \pm 0.2 \times 10^{-3}$	
1911 (1)			$1.0 \pm 0.3 \times 10^{-3}$	
4367 (1)			$5.0 \pm 2.0 \times 10^{-4}$	

(1) Womack and Overbeck (1970) – Balloon (see Section VI-6.1.5).
(2) Jacobson *et al.* (1975) – Balloon (see Section VI-6.1.5).
(3) Nakano *et al.* (1973) – Satellite (see Figure VI-19).

satellite near the detector, and the other arises from electrons precipitating towards the atmosphere. A large component of the continuum background may also be due to activation of both the spacecraft and the detector during their repeated passage though the proton belts and from the normal cosmic ray flux at the pole. Unfortunately, no data exist for the background for a solid state detector in an orbit of $\sim 33°$ inclination or in shuttle-type orbits, but, if it is similar to polar orbit background, it will constitute a serious problem for nuclear line experiments (see Section VI-6.3.1 for a discussion of the high resolution experiment planned for HEAO-C).

6.1.4. SPACE PROBES

The first γ-ray spectrum measurements made outside the terrestrial magnetosphere on space probes were reported by Arnold *et al.* (1962) and Metzger *et al.* (1964). These results from Ranger 3 in *cis*-lunar space were discussed in Section V-5.3, along with later results on the ERS-18 satellite reported by Vette *et al.* (1970), and give an estimate of the upper limit cosmic diffuse gamma ray spectrum valid to ~ 1 MeV. More extensive deep space background measurements covering the energy range from 0.3 to 27 MeV are now available from the Apollo 15 and 16 missions, and have been reported by Trombka *et al.* (1973). We presented these results earlier (Section V-5.3) in terms of a diffuse γ-ray continuum; here we discuss the counting rate data and spectral features. The measurements were made with a 7.0 cm diam × 7.0 cm long NaI(Tl) γ-ray spectrometer surrounded by a 1 cm thick plastic scintillator anticoincidence shield on all sides of the crystal except that facing the photomultiplier. The spectrometer and associated electronics were enclosed in a thermal shield and mounted on a boom which could be extended as desired up to a distance of 7.6 m from the Apollo service module. The material on the boom external to the γ-ray spectrometer was 5 g cm^{-2} averaged over all directions

The basic energy loss spectra obtained under various conditions are shown in Figure VI-20. The actual counting rates can be obtained by multiplying the ordinate values by the geometrical factor $G_0 = 57.5\ \text{cm}^2$. Calibration of the spectrometer was performed while the detector was inboard using a ^{203}Hg source, the intense spacecraft annihilation line at 0.511 MeV, and other identifiable spacecraft background lines. The count rates were summed over a number of channels corresponding to the detector's energy resolution which was 8.6% at 662 keV. When the detector was extended on the boom the spacecraft subtended a solid angle of ~0.28 sr, and many of the line features seen when the detector was inboard disappeared or were reduced in intensity. The lower curve in Figure VI-20 shows this energy loss spectrum extending up to ~27 MeV. It is suggested by Trombka *et al.* (1973) and Peterson *et al.* (1973a) that any contribution from the spacecraft (service module) background to this spectrum is typically less than ~20% at all energies. This is based on the fact that the intensity changed by a factor of 5 while the spacecraft solid angle changed by a factor of 20 when the detector was extended. This means that the resulting energy loss spectrum shown in Figure VI-20 is due to a combination of local production effects and a possible cosmic γ-ray flux (diffuse or point sources). Trombka *et al.* (1973) have also pointed out that the energy loss spectra measured with the NaI(Tl) detector on Apollo 15 are very similar, in overlapping energy regions, with that measured with the Ranger 3 CsI(Tl) spectrometer (Metzger *et al.*, 1964) and the ERS-18 NaI(Tl) spectrometer (Vette *et al.*, 1970), even though there are slight differences in the detector's response and the material near the extended detectors.

It is interesting to compare the background experienced in similar detectors in other environments. For example, at an energy of 1 MeV in a balloon experiment at $\lambda \sim 40°$ Peterson *et al.* (1973b) find (Figure VI-22) that the energy loss rate is ~0.6 counts cm^{-2} s^{-1} MeV^{-1}, while the corresponding rate in deep space is ~0.2 counts cm^{-2} s^{-1} MeV^{-1}. Since this factor of 3 is also about the background reduction expected for an equatorial balloon experiment, there appears to be no essential advantage to performing γ-ray astronomy experiments with omnidirectional γ-ray spectrometers in deep space from the point of view of background alone. On the other hand, in a balloon experiment background production in the few grams of overlying atmosphere will impose the ultimate limitation. Trombka *et al.* (1973) have also attempted to determine the residual contribution of discrete γ-ray lines to the energy loss spectrum shown in Figure VI-20. This was done using an iterative procedure described by Trombka *et al.* (1970)

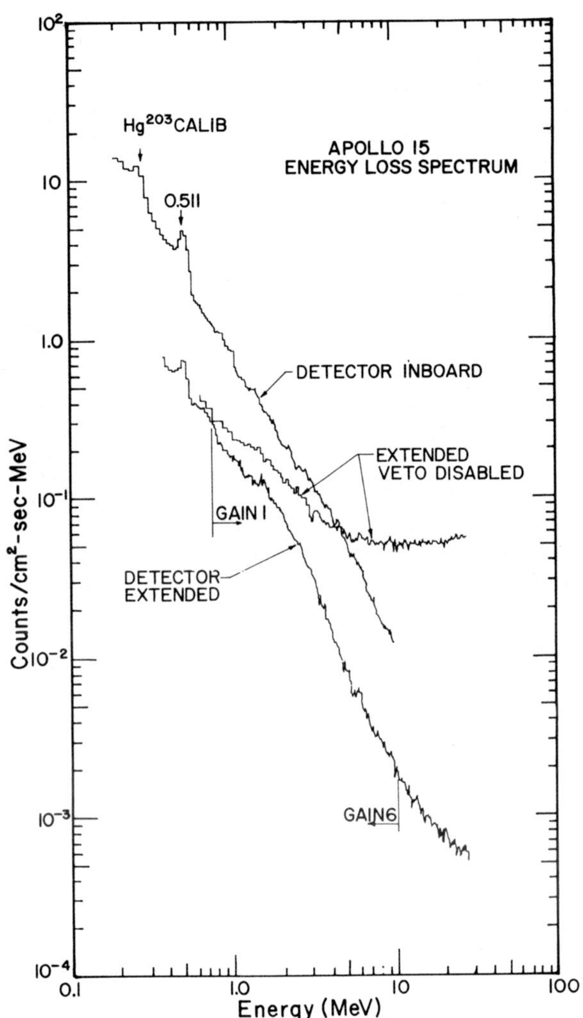

Fig. VI-20. Energy loss spectra in the 7 cm dia. by 7 cm long NaI(Tl) scintillation counter, measured on Apollo 15 during trans-Earth coast. The rates decreased a factor of about five when the detector was extended to 7.6 m. The spectrum with the anticoincidence disabled agrees with that expected from cosmic rays passing through the crystal edges. (From J.E. Trombka *et al.*, *Astrophys. J.* **181**, 737. Copyright 1973, The American Astronomical Society, by permission of the University of Chicago Press.)

and Reedy *et al.* (1973). Line features are apparent at several energies with the 0.51 MeV line due to positron annihilation being the strongest. Many of the discrete lines seen are the same as those shown in Figure VI-17 for the OSO-7 spectrometer. In Table VI-5 the composition of the Apollo 15 energy loss spectrum is broken down into several components for 2 energy ranges.

TABLE VI-5

The composition of the Apollo 15 energy loss spectrum (trans-Earth coast, detector deployed). From J.E. Trombka *et al., Astrophys. J.* **181**, 737. Copyright 1973, The American Astronomical Society. Used by permission of the University of Chicago Press.)

Component	Energy range 0.6 to 3.5 MeV	3.5 to 9.0 MeV
γ-Ray lines (%)	15.9	3.7
Spallation in NaI crystal (%)	15.8	0.5
Spacecraft continuum (%)	10.2	21.7
Cosmic upper limit (%)	58.1	74.1
Total (%)	100.0	100.0

A direct attempt at measuring the effect of proton activation has been carried out by Trombka and collaborators on Apollo 17 (see Peterson *et al.,* 1973a) using a NaI(Tl) crystal which was identical to that used on Apollo 15 and 16. The crystal was in the command module of Apollo 17, and therefore passed through the radiation belts twice. An identical central crystal was kept on the Earth but the K and Th content of the flight crystal was slightly higher. The activated crystal was attached to a photomultiplier and spectra were obtained within $1\frac{1}{2}$ h after reentry. In Figure VI-21 is shown the net energy loss spectrum of the activated crystal after the spectrum of the control crystal was subtracted. The peak energies for line identification are shown in the figure. Most of the lines in the recovered crystal are due to nuclides with half-lives greater than ~12 h; however, the line at ~0.44 MeV is due to ^{128}I which has a half-life of 25 min. It is important to note that this experiment does not give the important data on the activation of nuclides with half-lives shorter than ~10 min. In the case of the peak positions indicated by ^{123}I, ^{124}I, ^{126}I, and ^{128}I, there is a shift upward by 27 keV from the associated γ-ray line energy, since these lines follow predominantly electron-capture decay of the above nuclides. The atomic electron vacancy is filled, giving the characteristic Te Kα X-ray, which is readily absorbed in the crystal. Of course, much of the continuum is due to a mixture of β^- decay energy and line emission energy. Unfortunately, the proton dose received was not readily available, so direct comparison with the predicted activation spectra is not possible. (See Notes Added in Proof.)

6.1.5. OBSERVED BACKGROUNDS IN SPECIFIC BALLOON EXPERIMENTS

In Sections VI-6.1.1 to VI-6.1.3 we have discussed some general considerations concerning background contributions and have given information on the current knowledge of the diffuse γ-ray radiation and the atmospheric γ-ray and neutron spectra, as measured both in the atmosphere and above the atmosphere. In principle, this information can be used to estimate backgrounds in any experiment, whether it is carried out in a balloon or in a satellite. In this section, we wish to describe the actual backgrounds observed in various balloon experiments at different altitudes. Here we wish to deal with the observed

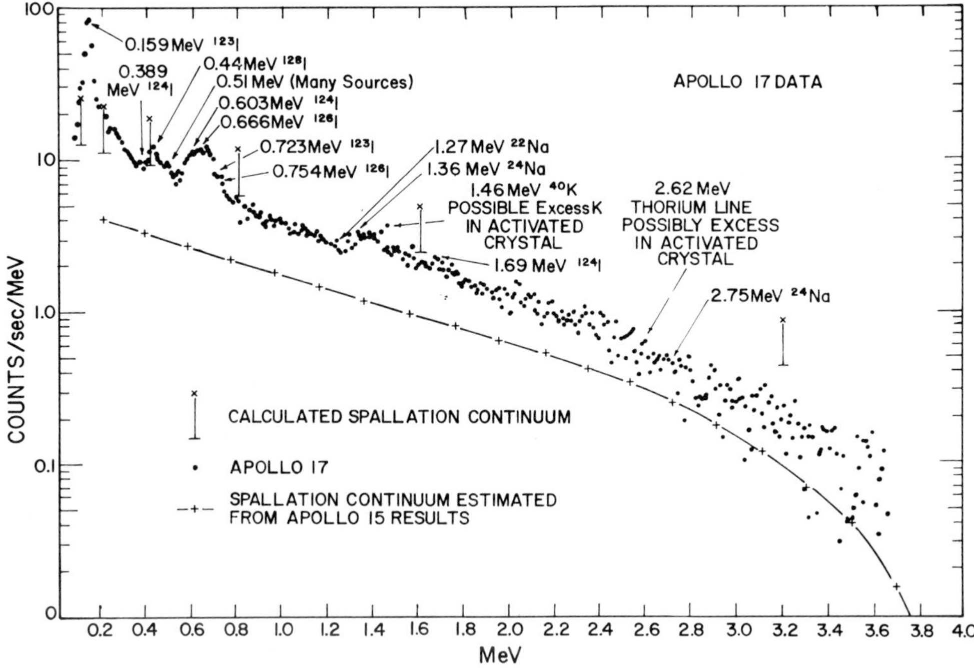

Fig. VI-21. Proton-induced activity in 7 cm by 7 cm NaI(Tl) crystal $1\frac{1}{2}$ h after reentry. The background has been subtracted. Counting time was 1800 s. The spectrum was obtained by direct internal counting of the activated crystal. (From L.E. Peterson *et al.*: 1973, *NASA SP-339.*)

counting rate spectra, since it is often more straightforward to extrapolate from the raw data of one experimental situation to another without the uncertainties introduced by spectrum-unfolding techniques and other assumptions often used when deriving true atmospheric or diffuse γ-ray spectra. We choose as representative experiments: detectors using shielded but uncollimated NaI(Tl) crystals, collimated NaI(Tl) and Ge(Li) crystals, and Compton telescopes.

a. *Alkali Halide Spectrometers*

The primary data available on the atmospheric γ-ray backgrounds measured by omnidirectional NaI(Tl) spectrometers with 4π plastic scintillation anticoincidence shields have been obtained by groups at the University of California, San Diego, and at the University of New Hampshire. In Figure VI-22 is shown the energy loss spectrum obtained by Peterson *et al.* (1973b) with a 3″ × 3″ NaI(Tl) spectrometer flown beneath a balloon from Palestine, Texas to an atmospheric depth of 3.6 g cm^{-2} and at a nominal geomagnetic cutoff of 4.5 GV. The detector was surrounded by a 3/4″ thick plastic anticoincidence shield on all sides except where the photomultiplier was placed against the crystal. The open side faced downward during the balloon flight and corresponded to

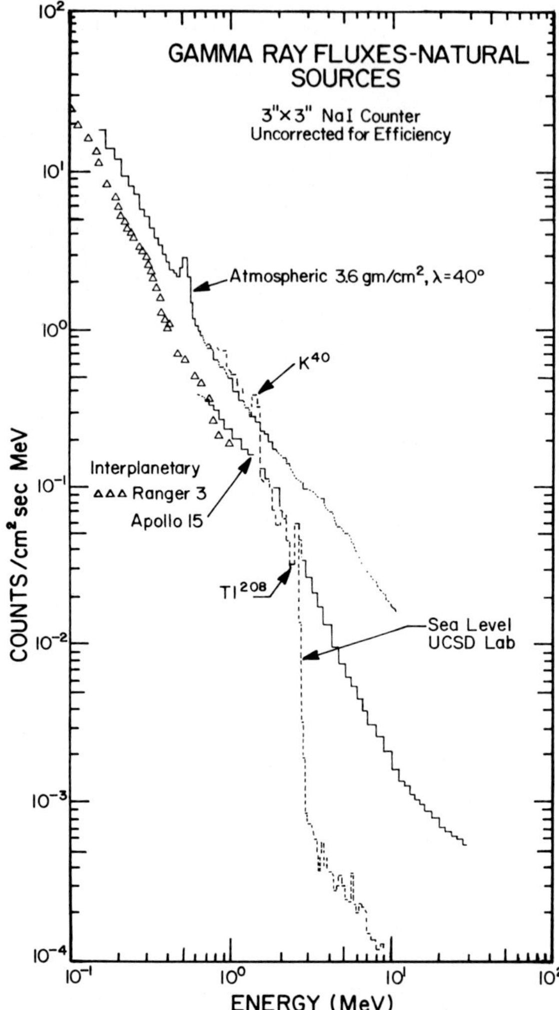

Fig. VI-22. The energy loss spectrum for a standard active shielded 3″ × 3″ NaI(Tl) spectrometer at balloon altitude above Palestine, Texas is compared with the corresponding spectra at sea level and on Ranger 3 and Apollo 15. (From L.E. Peterson *et al.*, *J. Geophys. Res.* **78**, 7942, 1973, copyrighted by American Geophysical Union.)

~15% of the full solid angle. The actual detector counting rate spectrum (counts s^{-1} MeV^{-1}) may be found simply by multiplying the ordinate by the omnidirectional geometrical factor $G_0 = 67$ cm². Also shown in Figure VI-22 for comparison is the observed counting rate spectra for the same size NaI(Tl) detector on Ranger 3 below 1 MeV from Metzger *et al.* (1964) and on the Apollo 15 from Trombka *et al.* (1973). Also, of fundamental interest are the sea level spectra observed in the UCSD laboratory. Two features dominate this spectrum at 1.46 MeV and 2.61 MeV. The lines are due to radioactive decay of ^{40}K and a ^{232}Th granddaughter [^{208}Tl (ThC″)], which are prevalent in soils and building materials, activities which vary considerably on the Earth's surface. These lines are generally not seen in balloon experiments, but should be considered when selecting photomultipliers and other detector materials in order to avoid possible excessive contamination of spectra (see Section VI-6.1.1c).

The high altitude γ-ray energy loss spectrum for an alkali halide crystal as shown in Figure VI-22 is generally featureless except for the line at 0.51 MeV. As discussed previously, this line is seen in all balloon-borne detectors with sufficient energy resolution. The intensity of this line varies with the amount of matter surrounding the detector, since many reactions can produce β^+ emitters. Peterson (1963), Rocchia *et al.* (1965), Chupp *et al.* (1970), and Kasturirangan *et al.* (1972) have also carried out measurements with detectors with a minimum of exterior inert mass at different latitudes and obtained spectra similar to the one shown here. It is currently believed that detectors using omnidirectional NaI(Tl) spectrometers are measuring a true atmospheric component at 0.51 MeV.

This is consistent with calculations that show that the 0.51 MeV line should have its origin predominately in the electromagnetic component of the secondary cosmic rays in the atmosphere. The apparent latitude dependence of the 0.51 MeV line (Kasturirangan *et al.*, 1972) also fits the electromagnetic origin. Additional support that the omnidirectional detectors are measuring an atmospheric 0.51 MeV flux, and not a locally produced component, comes from an experiment by Peterson *et al.* (1973b) at 40 g cm^{-2} atmospheric depth. Spectra were obtained with and without an Al cylinder of 7.5 g cm^{-2} average thickness surrounding a 2.75 inch × 2.75 inch NaI counter. A reduction of the counting rate in the 0.2 to 10 MeV range was observed with the Al mass in place, and they have concluded that absorption exceeds production in a low-Z inert shield of thickness <1 mean-free-path. We note, however, that a small increase in the intensity of the 0.51 MeV line could be missed in this experiment unless one studied the spectrum with sufficient energy resolution.

For consideration of backgrounds in balloon experiments, the flux of the *external* atmospheric 0.51 MeV line is typically (0.20 ± 0.021) photons cm^{-2} s^{-1} at 3.9 g cm^{-2} and $P_c = 4.5$ GV (Chupp *et al.*, 1970) and (0.09 ± 0.01) photons cm^{-2} s^{-1} at 6 g cm^{-2} and $P_c \sim 15$ GV (Kasturirangan *et al.*, 1972). These are equivalent 4π fluxes; no information is available on the angular distribution. Nevertheless, it is likely that a highly directional detector would see a significantly larger counting rate from large zenith angles near the top of the atmosphere (see Section VI-6.1.1b).

Much larger versions of the omnidirectional active-shielded NaI(Tl) spectrometer have been flown from Palestine by Kurfess (1971) and Orwig (1971). In Figure VI-23 we show the counting rate spectrum obtained by Orwig (1971) at 3.5 g cm^{-2} and $P_c = 4.5$ GV (40° N). Notice that the units for this spectrum are the same as in Figure VI-22. Raw counting rate spectra in both cases have been normalized by dividing the counting rate in each channel by the omnidirectional geometrical factor G_0 (cm^2), which in this case is 570 cm^2. Several of the spectral features need explanation. First, there is the line at

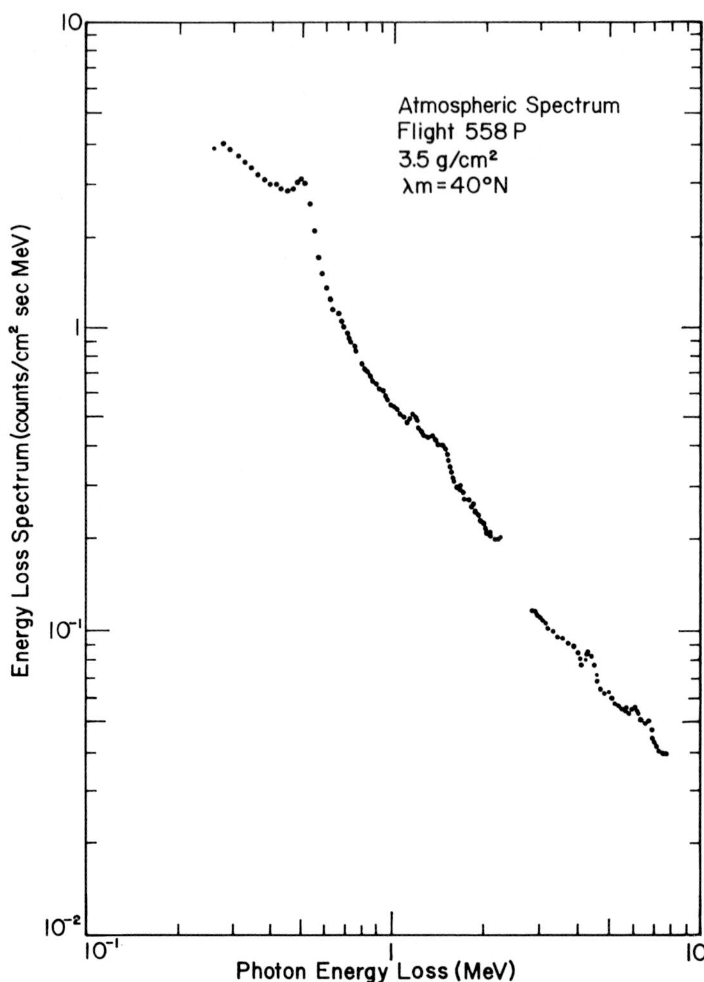

Fig. VI-23. The energy loss spectrum for a large active-shielded 11″ × 4″ NaI(Tl) spectrometer at balloon altitude above Palestine Texas. (From L. E. Orwig, Ph.D Thesis, University of New Hampshire, Durham, 1971.)

0.5 MeV seen in all γ-ray spectral measurements in the atmosphere. The equivalent omnidirectional flux for this line was obtained by fitting a Gaussian and two exponentials to the shape of the counting rate spectrum in the energy region of interest. After correction for dead time loss and attenuation in the plastic shield and Al case and using a photopeak efficiency of 0.79 at 0.5 MeV, the resulting flux found by Orwig (1971) is 0.17 photons cm^{-2} s^{-1}. This is in excellent agreement with the measurements mentioned above made with smaller omnidirectional NaI(Tl) spectrometers under very similar experimental conditions. This observation gives support to the view that this is a reasonable value for the true flux of atmospheric annihilation photons. The features at 1.17 and

1.33 MeV are due to a small leakage from a gated ^{60}Co calibration source (Forrest *et al.*, 1972). The faint bump at $\sim$1.46 MeV is due to ^{40}K intrinsic to the detector, probably the photomultiplier glass. The enhancements at $\sim$4.4, 6.1, and 6.8 MeV are definitely not seen in the background of this detector at sea level. Whether these features are true lines from the atmosphere or due to some unknown background produced in the apparatus at balloon altitude is not certain at this time.

Ling (1974), however, has recently made a comparison of the measured fluxes of atmospheric γ-ray lines with his calculations, which used a semi-empirical model specified for a latitude of 40° and an atmospheric depth of 3.5 g cm^{-2}. Lines at the above nominal energies are expected as a result of neutron interaction in atmospheric O_2 and N_2 and are reportedly seen by Kurfess as well as Orwig (see Ling, 1974). Although we believe the question of the true flux values of these possible atmospheric lines is uncertain at the present time, Ling (1974) reports the measured flux to be $\sim 1.4 \times 10^{-2}$ photons cm^{-2} s^{-1} for the 4.4 MeV feature compared to his calculated fluxes of $\sim 10^{-4}$ cm^{-2} s^{-1} for a line at 4.49 MeV from $^{14}N(n, \gamma)^{15}N$ and $\sim 3 \times 10^{-3}$ cm^{-2} s^{-1} for a line at 4.44 MeV from $^{14}N(n, \alpha)^{11}B$. Several lines could contribute to the 6.1 MeV feature, and the apparent flux observed by Kurfess is reported as $\sim 6 \times 10^{-3}$ cm^{-2} s^{-1} compared with a calculated flux of $\sim 7 \times 10^{-3}$ cm^{-2} s^{-1} for a line at 6.13 MeV from the reaction ^{16}O $(n, n'\gamma)^{16}O$. Orwig's flux for this line is reported as about 2.5×10^{-2} cm^{-2} s^{-1} or four times higher than the above calculated and measured values. The feature at $\sim$6.8 MeV is observed at the level of $\sim 2 \times 10^{-2}$ cm^{-2} s^{-1}, while Ling's prediction is over an order of magnitude lower.

It is clear that there are significant disparities between predicted line strengths and the equivalent fluxes inferred from measurements. Variations in the materials surrounding the detectors may cause significantly different spectral features due to activation effects and a comparison with calculations may be extremely difficult. Finally, it is interesting to note by comparing Figures VI-22 and VI-23 that the larger detector gives a flatter energy loss spectrum at higher energies than the results of Peterson *et al.* (1973b) indicate. This is expected, since the larger spectrometer has a relatively greater sensitivity for high energy photons than the 3″ × 3″ spectrometer.

Very few measurements are available on the latitude dependence of the γ-ray line spectrum near the top of the atmosphere. For γ-ray experiments carried out at equatorial latitudes, the results obtained by the Indian group at Hyderabad (Kasturirangan *et al.*, 1972) are of interest. The instrument used for these measurement consisted of a 5.1 cm × 5.1 cm NaI(Tl) spectrometer in a 4π plastic anticoincidence shield flown to $\sim$6 g cm^{-2} at a geomagnetic latitude of 7.6° N. The energy bins used in this experiment did not allow a clear resolution of the line at 0.5 MeV; however, the authors were able to arrive at a flux of the 0.5 MeV line of 0.079 ± 0.010 photons cm^{-2} s^{-1} at 6 g cm^{-2} over Hyderabad. This is about a factor of 2 to 3 lower than the Palestine value given by Chupp *et al.* (1970). Kasturirangan *et al.* (1972) have also summarized all the measurements made of the 0.5 MeV line flux at various latitudes; their summary is shown in Figure VI-24, normalized to 6 g cm^{-2}. Even though all of the flux values shown were made by several investigations using different techniques the latitude dependence ob-

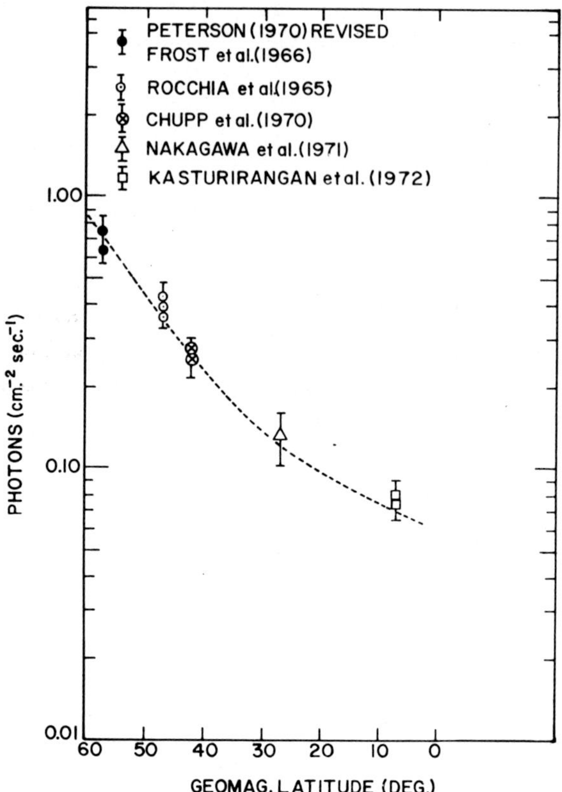

Fig. VI-24. The latitude dependence of the 0.5 MeV line flux normalized to 6 g cm^{-2} (From K. Kasturirangan *et al.*: 1972, *Planet Space Sci.* **20**, 1961. Used by permission.)

served agrees with that expected for an electromagnetic origin. The flux value is to be interpreted as an equivalent omnidirectional flux, since the measurements were all made with detectors that have an isotropic response. The conversion from counting rate to flux was made essentially by dividing by $\bar{\epsilon}G_0$ where $\bar{\epsilon}$ is the average photopeak efficiency at 0.5 MeV and G_0 is the usual 4π projected area for an omnidirectional detector. This interpretation implies that the counting rate contribution to the detector is the same from all directions in a given solid angle. However, within a photon absorption mean free path of the top of the atmosphere, the counting rate contribution measured with a highly directional detector should vary with zenith angle if the local background contribution and leakage through the shield is negligible.

The Rice University group (Haymes *et al.*, 1969) have flown a larger NaI(Tl) spectrometer collimated with a large well-type NaI(Tl) collimator in balloons. This instrument was a 4″ diam. x 2″ thick central crystal in a well of thickness equivalent to ~2 absorption lengths at 500 keV and of such length as to provide an opening half-angle of ~9°. The opening aperture was covered by a 1/4″ thick plastic anticoincidence scintillator

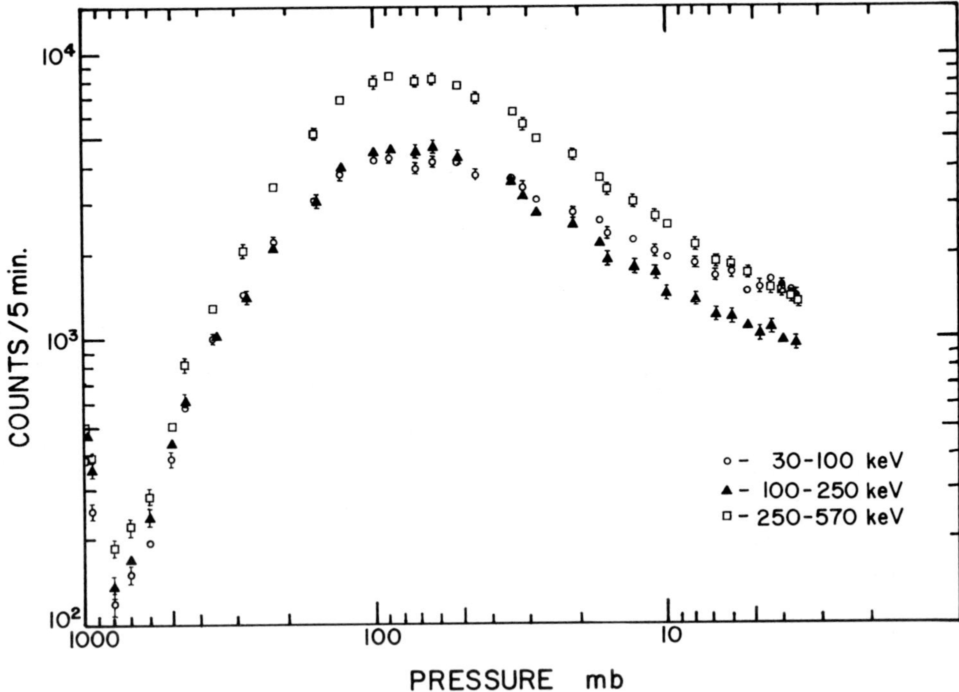

Fig. VI-25. The counting rate vs. atmospheric depth measured over Texas on 1967, August 29 with the detector of Haymes *et al.* (From R.C. Haymes *et al., J. Geophys. Res.* **74**, 5792, 1969, copyrighted by American Geophysical Union.)

above which was the phototube which viewed the NaI(Tl) spectrometer. In Figure VI-25, the counting rates for several different energy bands are shown vs. atmospheric depth for this spectrometer for mid-latitude flights from Palestine, Texas in 1967 and 1968. This basic data can be used to scale background counting rates for larger detectors shielded in a similar manner and used at mid-latitudes. In this connection, Forrest *et al.* (1975b) have developed a large partially-collimated γ-ray spectrometer using an 8″ diameter by 4″ thick NaI(Tl) spectrometer. This instrument was test flown from Palestine, Texas in July, 1974. The counting rate in this spectrometer agrees with that scaled from the data in Figure VI-25 for the same photon enrgies when account is taken of the different instrument volumes and efficiencies. This gives one confidence in using the method of scaling backgrounds, even if a detailed understanding of the major sources of background is not yet available (see Section VI-6.3.1).

Figure VI-25 also shows several features characteristic of atmospheric γ-ray measurements which are useful in checking the overall instrument behavior. First, the counting rate at an atmospheric depth of 800 mb is about a factor of 5 lower than at sea level because of the intense radioactivity in the Earth's crust. The general atmospheric growth curve of the soft component of secondary cosmic rays is then followed until the transition maximum is reached at ~70 mb. Above this maximum the γ-ray detector counting rate falls in a manner determined by several factors which are not easily separated. As we have previously discussed, the total detector background rate will be due to true atmospheric γ-rays, any diffuse cosmic flux penetrating to the detector, as well as any flux from other cosmic sources, and also activation effects due to charged cosmic ray particles and ambient and locally-produced neutrons. The latter effects depend strongly on the specific type of detector used.

b. *Solid State Detector Measurements*

The real future of γ-ray line astronomy will depend on use of the highest resolution detectors available. Even though the liquid-N_2-cooled lithium-drifted germanium [Ge (Li)] detector has been extensively adopted in laboratory nuclear physics, only a few balloon experiments have used this device. The first work reported was that of Jacobson (1968), who searched for *r*-process lines from the Crab Nebula using a 15 cm^3 Ge(Li) spectrometer in an active NaI(Tl) shield/collimator (see Section V-5.2.1a). Other Ge(Li) balloon experiments have been carried out by Nakagawa *et al.* (1971), Womack and Overbeck (1970), and Jacobson *et al.* (1975).

The instrument used by Womack and Overbeck is shown schematically in Figure VI-26 and contained a 27.6 cm^3 Ge(Li) spectrometer. The active NaI(Tl) collimator defined a solid angle of 0.09 sr (~10° half-angle). The instrument was used to search for solar γ-ray lines in May, 1968 from a balloon above Palestine, Texas at 4.7 g cm^{-2} atmospheric depth. No solar lines were observed; however, an excellent spectrum was obtained which shows the nature of the background involved with this type of instrument. In Figure VI-27 the raw counting rate spectrum is shown as obtained in this experiment. There are several line features seen when the instrument is in its normal operating mode. One of the strongest lines is at 511 keV. When the anticoincidence is disabled, the spectrum increases strongly and becomes featureless except at 511 keV, where the intensity of this line is greatly increased. Womack and Overbeck (1970) have found that 0.80 ± 0.01 counts s^{-1} occur at 511 keV in coincidence with shield events while 0.018 ± 0.001 cts s^{-1} are unaccompanied by shield events. The typical 'equivalent flux' found by Womack and Overbeck (1970) for the 511 keV line was $\sim 5 \times 10^{-2}$ cts cm^{-2} s^{-1}. It is not clear from this work whether a correction has been made for photopeak efficiency of the Ge(Li) cell. This 511 keV line intensity does refer to a parallel flux at the top of the atmosphere (Womack and Overbeck, 1970) but cannot be directly compared with measurements with isotropic detectors. The other background lines seen in Figure VI-27 are summarized in Table VI-4, column 4, with line identifications, counting rates, and possible sources for these lines as given by Womack and Overbeck (1970).

Very recently, a large volume Ge(Li) spectrometer has been flown on a balloon from Palestine, Texas (Jacobson *et al.*, 1975). This is a smaller version of a high resolution γ-ray telescope planned for the HEAO-C mission which is described in Section VI-6.3.1d. The construction of the new spectrometer of Jacobson *et al.* (1975) is somewhat similar to that of Womack and Overbeck (1970) except that the shield-collimator of the new

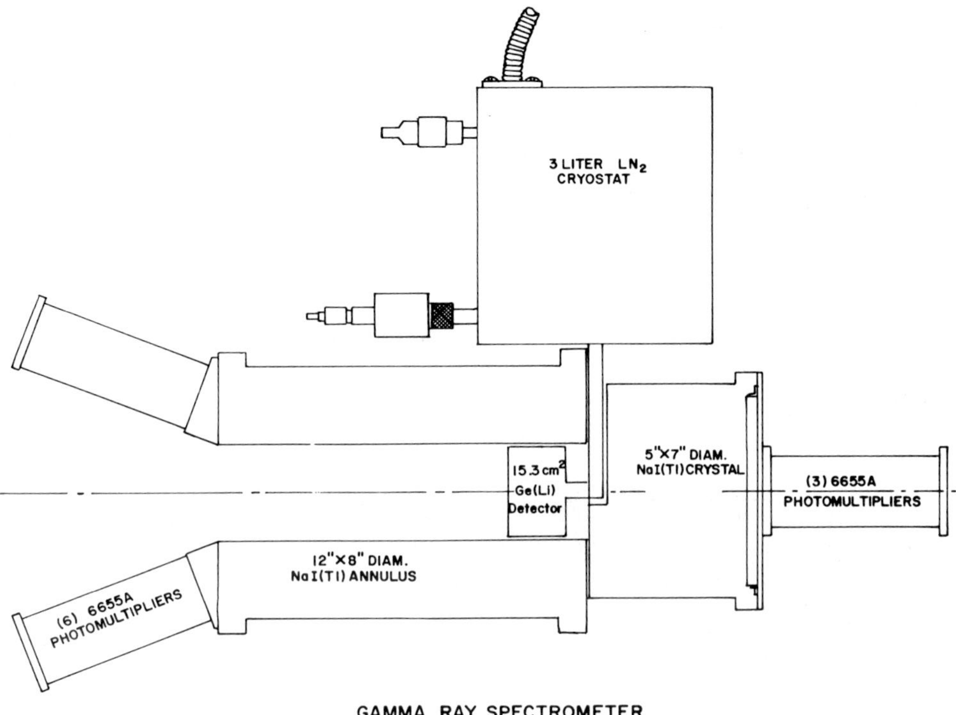

Fig. VI-26. The first balloon-borne Ge(Li) spectrometer telescope is shown in a NaI(Tl) collimator. (From E.A. Womack and J.W. Overbeck, *J. Geophys. Res.* **75**, 1811, 1970, copyrighted by American Geophysical Union.)

telescope uses CsI(Na) and contains four 40 cm^3 Ge(Li) cells. In Figure VI-28 we show the background counting rate spectrum obtained with a single 40 cm^3 Ge(Li) cell at an atmospheric depth of 2.9 g cm^{-2} to compare with the spectrum in Figure VI-27. In Table VI-4, column 5, are given the counting rates for several lines observed by Jacobson *et al.* (1975). It is interesting that, even though there are detailed differences in the construction of the two Ge(Li) spectrometers, the counting rates are quite comparable for the same lines. The main source of the neutrons giving rise to the majority of the lines is, most likely, production in the high-Z shield and surrounding materials and probably cannot be eliminated by anticoincidence. These background lines can easily be separated from any extraterrestrial γ-ray lines, however, because of the high resolving power of the Ge(Li) spectrometer.

c. *Compton Telescope*

Recently, Compton telescopes using plastic scintillators have been developed by Schönfelder *et al.* (1973), Zych *et al.* (1975), and Herzo *et al.* (1975) for use in γ-ray astronomy. The principle of operation of this type of instrument is discussed in Section

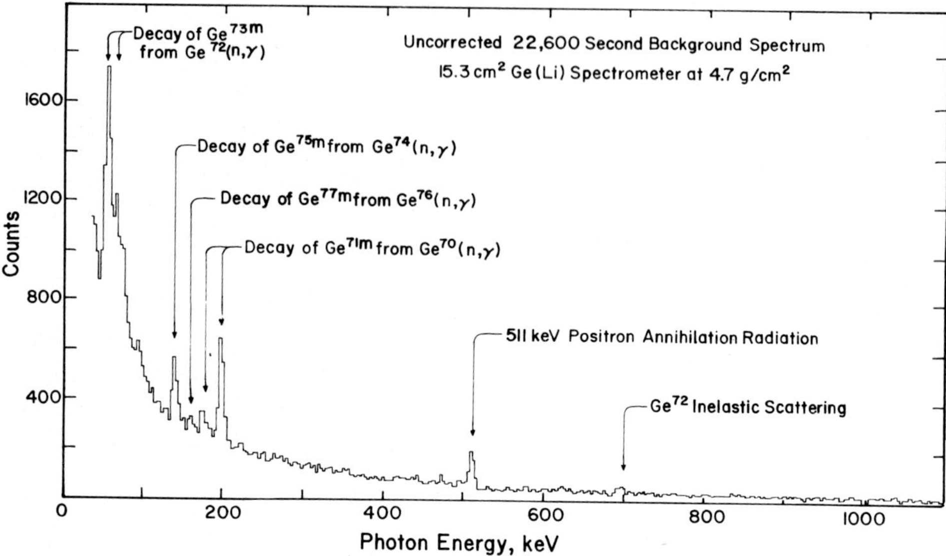

Fig. VI-27. The raw counting rate spectrum obtained with a Ge(Li) telescope at balloon altitude over Palestine, Texas. Line identifications are given. (From E.A. Womack and J.W. Overbeck, *J. Geophys. Res.* **75**, 1811, 1970, copyrighted by American Geophysical Union.)

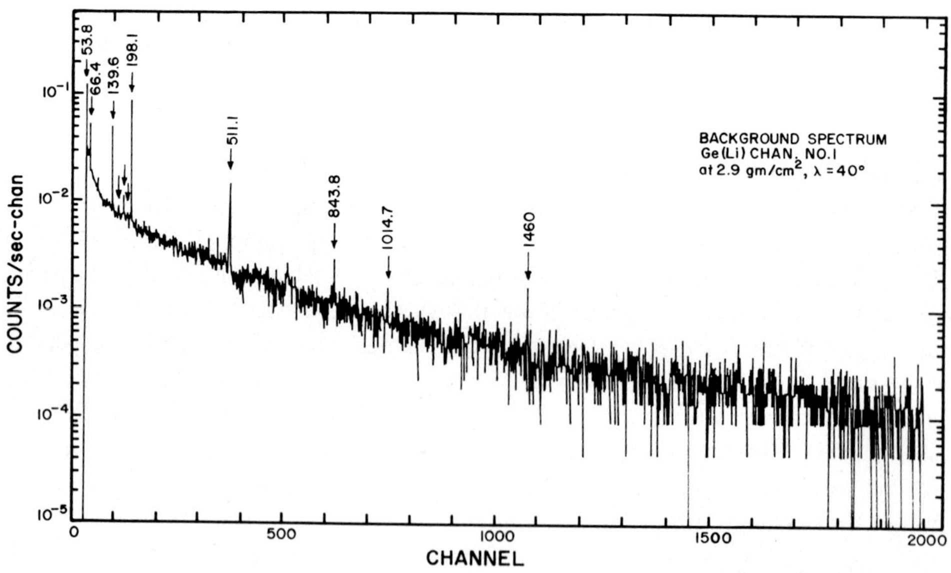

Fig. VI-28. The background counting rate spectrum obtained in a large (40 cm^3) Ge(Li) detector at balloon altitude over Palestine, Texas. (From A.S. Jacobson *et al.*: 1975, *Nucl. Instrum. and Methods* **127**, 115, by permission of the North Holland Publishing Company.)

VI-6.3.2 and will not be repeated here. We are interested, however, in the background expected with this type of instrument, since the technique offers promise for attacking several problems in γ-ray astronomy. White *et al.* (1973) have obtained preliminary data from a Compton telescope flown on a balloon from Palestine, Texas in May, 1972. The result of this experiment, expressed as a vertical atmospheric flux, was given in Figure VI-5, which summarized several other experimental results results. The points of White *et al.* (1973), however, should only be considered as upper limits to the vertical atmospheric γ-ray flux, due to an interesting neutron background effect that can masquerade as a γ-ray in a Compton telescope (see discussion below). Also, no correction was made for a possible diffuse γ-ray flux contribution.

As discussed in Section VI-6.3.2, Compton telescopes are usually operated to separate downward-moving γ-rays from neutrons by time of flight. White and Schönfelder (1975) and Herzo *et al.* (1975) have considered the effect of $(n, n'\gamma)$ and $(n, x\gamma)$ reactions with the C in the upper plastic scintillator element of a Compton telescope. For example, in the first case neutrons coming from below the telescope can produce a start pulse in the upper element when the inelastically scattered neutron n′ produces a recoil proton. The 4.43 MeV γ-ray from the deexcitation of C goes directly to the lower Compton scintillator, where it interacts, and the coincident event is recorded as a true γ-ray. In the second case, the 'x' proton or α-particle initiates a start signal and the γ-ray from an excited product nucleus (^{12}B or ^{9}Be) completes the event. White and Schönfelder (1975) also discuss two other possible background effects for Compton telescopes: First, 0.511 MeV annihilation γ-rays from the positron emitter ^{11}C produced by cosmic ray spallation of ^{12}C can appear as a low energy γ-ray (~0.5 MeV) if one annihilation photon interacts in the upper Compton element and the second in the lower Compton element. Secondly, neutrons captured by the protons in the plastic scintillator would give a deuteron recoil and the 2.22 MeV capture γ-ray. In this latter case, the deuteron recoil energy for capture at rest is very near 1 keV, so unless the threshold is very low in the upper Compton element, this effect would not be seen.

In terms of assessing the flux sensitivities of Compton telescopes for γ-ray astronomy, it is most conservative to use the counting rates inferred from Figure VI-5 and the sensitivities of the instrument given by Herzo *et al.* (1975). These authors report that γ-rays in the energy range 5 to 30 MeV can be detected down to a flux level of $\sim 10^{-4}$ photon cm^{-2} s^{-1} MeV^{-1} (see also Section VI-6.3.2).

6.2. Design Limitations

For a realistic appraisal of instrument requirements, we will assume that the next generation of experiments will be carried out in the HEAO program in the late 1970's and in the US Space Shuttle Program in the 1980's or by scientific balloon experiments. It is clear that improvements in the ability to detect cosmic photons in the nuclear line region are possible at present. Instrument parameters of importance are: spatial resolution, energy resolution, minimum detectable flux (or highest sensitivity), and time resolution. Improvement of all these parameters simultaneously in one instrument is usually not possible; however, we will discuss here primarily some statistical limitations to the sensitivity attainable. For simplicity, we will consider the measurement of γ-ray lines from a point source and explore the instrument requirements by setting down formulas appropriate to this case. First, let us consider the flux sensitivity, the signal-to-noise ratio, and the experiment stability required for the measurement. Similar considerations may be made for other source distributions (see, e.g., Peterson, 1975).

The practical goal of measuring weak γ-ray line fluxes is ultimately constrained by the actual instrument backgrounds. Consequently, a flux measurement is limited by statistics if all background counts occur at random in time. Therefore, let us assume that a particular instrument has a measured average background counting rate, C_b, integrated over an energy interval equal to the instrument resolution width, and that this same background is applicable when viewing a point source. If it is assumed that the average background rate is constant in time and the detector views a source plus the background, and the background, for the same time, T_{obs}, then the 1σ error in the counting rate difference between source-on and source-off measurements is $\sqrt{2C_B/T_{obs}}$ (Beers, 1957) for the case of the signal counting rate $C_s \ll C_B$. If there are no detectable source counts, then the limiting flux for the measurement may be specified as

$$F_{min} \leqslant \frac{n}{S(E)}\sqrt{\frac{2C_B}{T_{obs}}} \quad (\text{photons cm}^{-2}\,\text{s}^{-1}) \tag{VI.7}$$

where $S(E)$ is the photopeak sensitivity (cm^2) or area-efficiency factor for an energy E and n determines the confidence level of the null measurement. It is best to use $n = 5$ to avoid an overly optimistic judgement of instrument capability. It is important to keep in mind that F_{min} is a measure of the smallest γ-ray line flux detectable with the instrument.

Now consider an observable flux from the source F (photons cm^{-2} s^{-1}) where $F \geqslant F_{min}$, then the signal-to-noise ratio of the measurement is the observed signal counts, $(F \cdot S(E) \cdot T_{obs})$, divided by the fluctuation in the background counts $\sqrt{C_B \cdot T_{obs}}$ giving

$$\text{S/N} = \frac{F \cdot S \cdot T_{obs}}{\sqrt{C_B \cdot T_{obs}}} = F \cdot S\sqrt{\frac{T_{obs}}{C_B}}. \tag{VI.8}$$

We can also specify the magnitude of the effect as the signal counts divided by the total background counts, giving

$$M(\%) = \frac{10^2 \cdot F \cdot S \cdot T}{C_B \cdot T} = \frac{10^2 \cdot F \cdot S}{C_B}. \tag{VI.9}$$

This ratio, expressed as a fraction, is sometimes called the signal-to-noise ratio, but Equation (VI.8) is more appropriate when one is limited by counting statistics. If the background flux is constant, then the ultimate statistical precision obtained in a counting experiment is

$$P_s(\%) = 10^2 \frac{\sqrt{C_B \cdot T_{obs}}}{C_B \cdot T_{obs}} = \frac{10^2}{\sqrt{C_B T_{obs}}} \quad \text{(VI.10)}$$

This quantity also defines a precision criterion for the experiment since C_B must be constant to a precision $P_s(\%)$, otherwise real changes in the background may mimic a γ-ray source. In other words, one cannot expect to do better than instrument stability allows. Several factors can cause a change in background, such as altitude and latitude changes in a balloon experiment, or gain change or sporadic noise in a detector.

A very desirable goal in this field would be the measurement of a γ-ray line flux of $F_{min} = 10^{-4}$ photons cm^{-2} s^{-1} at, for example, 1 MeV. We consider three general types of feasible instruments in a high altitude balloon environment: (1) a large, high resolution γ-ray spectrometer with $S = 300$ cm^2 and a background counting rate of $C_B \cong$ 18 cts s^{-1} in the energy resolution width at 1 MeV, (2) a large Compton telescope with $S = 100$ cm^2 and with $C_B \cong 2$ cts s^{-1}, and (3) a large volume Ge(Li) spectrometer with $S = 6.4$ cm^2 and with $C_B \cong 1.6 \times 10^{-2}$ cts s^{-1}. These parameters are based, respectively, on an instrument of Orwig (1971), on one described by Herzo *et al.* (1975), and on an instrument planned for HEAO-C but using the balloon background observed by Womack and Overbeck (1970) scaled to the larger spectrometer (Jacobson *et al.*, 1975).

In the first case, the Expression (VI.7) requires that the source and background each be observed for a time of $\sim 10^6$ s ($\sim$24 days total). From Equation (VI.8) it follows that S/N $\sim$ 7, and the precision criterion that the experiment must meet is $<0.02\%$ over a time period of $\sim$24 days! Such a detector with the above experimental conditions would record $\sim 3 \times 10^4$ source counts and $\sim 2 \times 10^7$ background counts in the instrument resolution width. In the case of the Compton telescope, the time for the observation is also $\sim 10^6$ s. The signal-to-noise ratio is $\sim$7 and the stability requirement is $P_s(\%) \leqslant 0.07\%$. In this case the recorded source counts would be 1×10^4 and the background counts would be 2×10^6. Finally, for the corresponding Ge(Li) spectrometer, the observation time is 2×10^6 s, the signal to noise ratio is also $\sim$7 and the stability requirement is $P_s(\%) < 0.6\%$. The recorded source counts would be $\sim$1300 and the background counts $\sim$30 000.

The purpose of this illustration is not to compare the three types of γ-ray instruments, but rather to illustrate how crucial the sensitivity, background and stability factors are for low flux γ-ray line experiments using instruments currently available. The necessity of insuring that the background is stable for several days is probably the most severe requirement if one is searching for a weak but steady source, since all of the 'integrated source counts' could be due to a fluctuation in the background which occurred while viewing the source, but which did not occur in the background measurement. The best method of guarding against this possibility is to measure source and background simultaneously using two detectors. Knowledge of the exact arrival time of events (counts) in both systems is also desirable, since the flaring of a source could also be determined.

The finest spatial or angular resolution attainable at energies of $\sim$1 MeV will only be $\sim 1°$; this is limited by the instrument techniques currently available, some of which are described in Section VI-6.3.1. At the lower energies ($\lesssim$ 100 keV) collimator techniques are being developed for the NASA Solar Maximum Mission (SMM) which will achieve a spatial resolution of $\lesssim 30''$. At energies above 100 keV and below 30 MeV, there does not now appear to be a practical approach for attaining spatial accuracies $\ll 1°$, although renewed efforts in this direction are highly desirable. The angular resolution of an instrument also directly effects the measured background of a telescope. This is especially true in a track-imaging device such as a spark chamber observing electron pairs. Here background pairs may be restricted to a small solid angle defined by multiple scattering effects and very low background rates are found as compared to those used above in comparing possible γ-ray line instruments. Unfortunately, in the cases discussed, backgrounds are accepted from very large solid angles, effectively 4π sr, even though all the devices could achieve angular resolutions of $\sim 1°$ to $10°$.

High energy resolution for γ-ray spectroscopy is possible with both alkalai halide scintillators and solid state detectors. The former, if NaI is used, are capable of resolutions (FWHM) of $\Delta E_\gamma / E_\gamma \sim 7\%$ at the standard energy, 662 keV, varying with photon energy $\propto E_\gamma^{-1/2}$. The highest energy resolution measurement that can be achieved with cooled Ge(Li) solid state detectors is $\gtrsim 1$ keV and is limited by electronic noise. Therefore, the fractional energy resolution FWHM $\propto E_\gamma^{-1} + K' E_\gamma^{-1/2}$ and is a considerable improvement over the alkalai halide scintillators at all energies. In many cases of astrophysical importance, γ-ray line widths may be much wider than the resolution achievable by solid state detectors. For example, if inelastic proton scattering on ^{12}C, in a low density gas, gives the 4.43 MeV line, then it is possible that the line width due to the distribution of carbon recoil velocities will be ~ 100 keV (see Section II-2.5.1), and it would be sufficient to use NaI(Tl) which has a line width of ~ 130 keV at 4.43 MeV, whereas a typical solid state detector would have a line width of ~ 3 keV at the same energy.

6.3. Current Detection Methods

Several instruments are now in operation, or under development, which are suitable for γ-ray astronomy investigations in the next few years and which operate effectively in the energy region below 20 MeV. Only a preliminary assessment of the capabilities of these instruments is possible based on design specifications given in various publications. We will briefly review active-shielded detectors which include a well collimated NaI(Tl) spectrometer for HEAO-A, a partially collimated NaI(Tl) spectrometer with an occulting disk or shutter, a rotating array of NaI(Tl) spectrometers which successively occult one another, and a collimated solid state spectrometer. Two types of Compton telescope designs will also be discussed.

6.3.1. SHIELDED DETECTORS

a. *HEAO-A Collimated X- and γ-Ray Spectrometer*

The γ-ray instrument designed for the HEAO-A mission (ca. 1977) is a multi-element NaI(Tl) spectrometer with coarse and fine collimators which will scan the sky for sources in the energy range 10 keV to ~ 10 MeV. The instrument concept planned for this mission

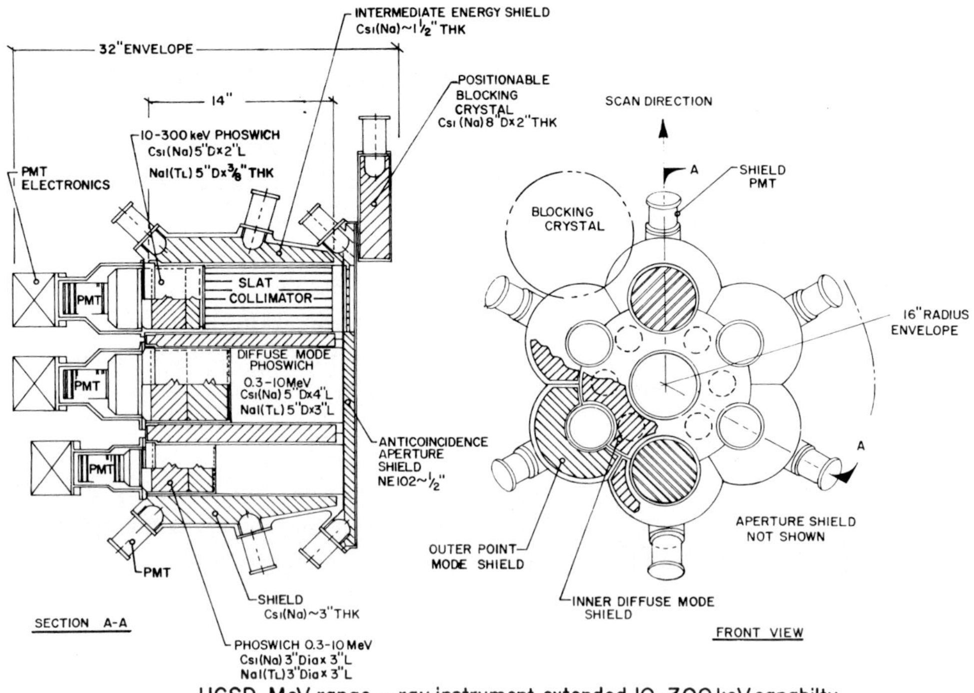

Fig. VI-29. UCSD/MIT HEAO-A X-ray and γ-ray spectrometer for the energy range 10 keV – 10 MeV. (From A. Metzger: 1973, *NASA SP-339.*)

evolved from earlier satellite and balloon experiments, some of which were discussed in Section VI-6.1. The present design is a result of a collaboration between L. Peterson and co-workers at the University of California, San Diego and W. Lewin at the Massachusetts Institute of Technology. A schematic view of the detector is shown in Figure VI-29 (Metzger, 1973; Matteson, 1974). It consists of seven separate NaI(Tl) – CsI(Na) phoswich spectrometers, each viewed by a separate photomultiplier. The central NaI(Tl) detector is 5″ in diameter and 3″ in length backed by a CsI(Na) crystal of the same diameter and of 4″ length which acts as the rear anticoincidence shield. This central element, which covers the energy range from 300 keV to 10 MeV, is surrounded by an annular anticoincidence shield of CsI(Na) split in two halves which can also operate in coincidence with the central detector as a pair spectrometer. The FWHM field of view provided by the collimator is $\sim 40°$. Surrounding the central detector is a circular group of 6 NaI(Tl) – CsI(Na) spectrometers. Two of these have 5″ diameter by 0.38 cm thick NaI(Tl) crystals and passive slat collimators giving a field of view of 1.5° by 20° for energies ~ 10 to 300 keV. The other four spectrometers have 3″ diameter by 3″ length NaI(Tl) crystals and are collimated to a field of view of $\sim 20°$ for the same energy range as the central detector. An external CsI(Na) anticoincidence shield surrounds the full array of seven spectrometers, and a thin (~1/2″ thick) plastic scintillator covers the aperture of all detectors.

The reported flux sensitivity of this instrument (Metzger, 1973) is ~1/30 of the Crab flux at 300 keV and 1/3 of the Crab flux at 3 MeV. A plot of the Crab differential spectrum is shown in Figure VI-32, from which it can be inferred that at 3 MeV a γ-ray line flux of $\sim 10^{-4}$ photons cm^{-2} s^{-1} should be measurable if background is as low as expected when the HEAO-A is in orbit (see Section VI-6.2 for general considerations).

The ability of this experiment and other HEAO-A instruments to study the Vela type cosmic γ-ray bursts has also been discussed recently by Matteson (1974). Depending on whether this phenomena is of extragalactic or galactic origin, as many as 1600 or as few as 60 bursts could be detected, and the possibility exists of determining source location to $\lesssim 1'$, using collimated X-ray telescopes on the HEAO-A.

b. *Collimator with Shutter-Occulter (Large γ-Ray Telescope)*

Larger detecting areas are needed in order to improve the γ-ray flux sensitivity; however, active anticoincidence shielding must be used in order to reduce the background counting rate, and some method must be used to give the instrument angular resolution. For γ-rays with energies above 0.1 MeV, the angular resolution is usually obtained with a well- or slat-type extension of the main anticoincidence shield (see Sections VI-6.1.5 and VI-6.3.1a). However, this approach greatly increases the instrument weight, so other approaches are desirable.

In Figure VI-30, we show a schematic of a large balloon γ-ray telescope (LGT) designed at the University of New Hampshire (Forrest *et al.*, 1975b) and recently flown from Palestine, Texas. The gain-stabilized central detector is an 8″ × 4″ NaI(Tl) scintillator viewed by a single 5″ photomultiplier tube. The resolution of the unit at 0.662 MeV is ~9% and the ^{60}Co peak-to-valley ratio is ~3. It is completely surrounded by an anticoincidence shield. An interesting feature of this detector is the occulting disk, which is movable on rails as indicated. The shutter-occulter and all of the shielding portions of the detector are constructed of small (1 to 2 in^3) pieces of NaI immersed in silicon oil. Optical transmission of the light produced in the NaI is provided by the clear viscous oil into which the photomultiplier tubes are also immersed. This scheme provides a significant cost saving and also considerable versatility in shield fabrication over a large single NaI crystal. The geometry of the shielding can be changed by merely changing the container and placing the NaI pieces and oil into the new container. In terms of volume of NaI, the cost of the present shield is ~1/3 of the cost of an equivalent solid NaI shield. When the occulter is closest to the main detector, the background can be measured, and when extended (up to 2 ft in the current design), it acts as an occulter for a celestial source of γ-rays. When fully extended, the occulter gives an angular response of ~20° FWHM.

The first balloon flight of the Large Gamma-Ray Telescope (LGT) was carried out on 1974, July 31. Detailed scientific studies of the data are now (1975) being performed; however, final results are not yet complete. The prominent 511 keV line and an interesting feature at ~7 MeV were observed. Initial estimates of the expected detector back-

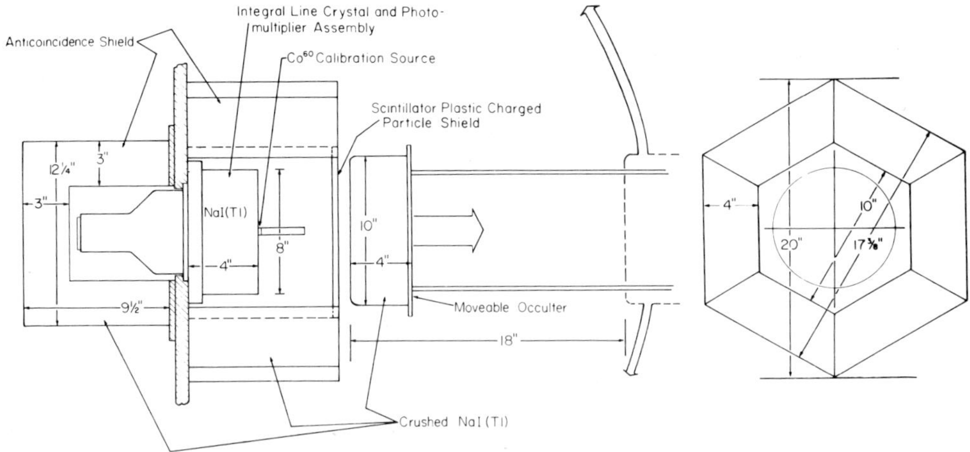

Fig. VI-30. University of New Hampshire – Large γ-Ray (shutter) Telescope. (From D.J. Forrest *et al.*, 1975b.)

ground indicated that a flux as low as 1.4×10^{-3} photons cm^{-2} s^{-1} could be measured with this detector at 0.5 MeV in balloon flights of ~4 h duration. Future flights of this instrument from the Southern Hemisphere should permit the verification of the existence of the Galactic Center line reported at ~480 to 530 keV.

c. *Anti-Collimator Spectrometer*

Use of the type of collimated detectors discussed above requires that a heavy price be paid in terms of how much of the instrument weight is actually put into the main γ-ray detecting element. Since the weight of a shield of CsI(Na) a few inches thick obviously far exceeds the weight of the central NaI(Tl) or Ge(Li) γ-ray detector, it is important to investigate the advantage that might accrue if all of the allowed weight were put into the detector. The sensitivity of a γ-ray detector $S(E, \theta, \phi)$, given in Equation (VI.1), actually increases approximately as the volume and, hence, as the weight of the γ-ray detector neglecting any shield weight. The background also increases essentially as the volume, so the overall flux sensitivity given by Equations (VI.1) or (VI.7) improves as $\sim(\text{weight})^{-1/2}$. One needs, however, a method of giving such an instrument some angular resolution capability. The shutter-occulter technique discussed above is a step in this direction; however, the ultimate application of this approach has recently been investigated by Morfill and Pieper (1974).

This scheme utilizes a battery of parallel cylindrical NaI(Tl) spectrometers which rotate about an axis parallel to that of the detectors. Figure VI-31 shows a schematic diagram of this assembly applied to the case of a point source of γ-rays. Here the axis of rotation is perpendicular to the plane of the paper, and a given detector in position A occults the radiation (from the source) reaching the detector at position D. The signals

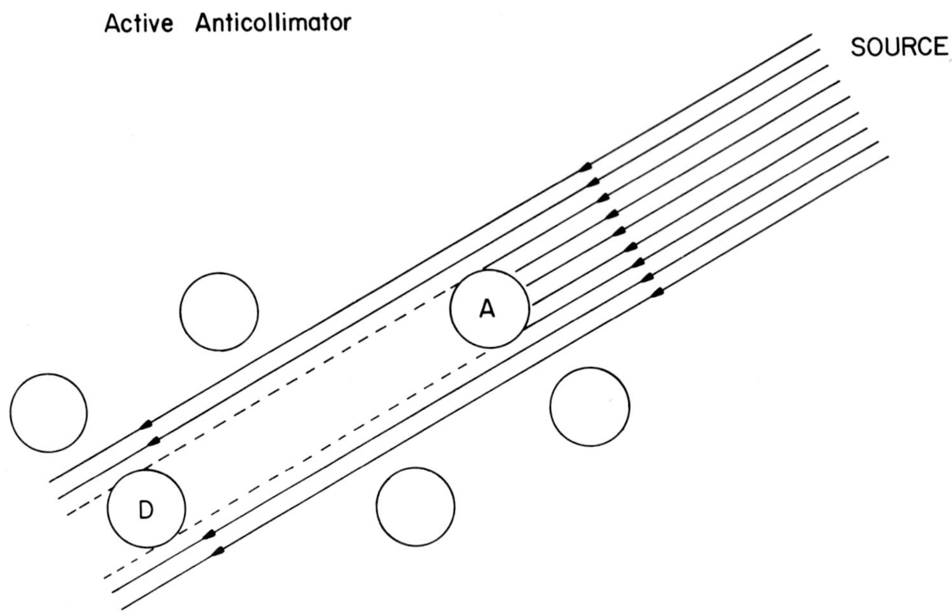

Fig. VI-31. Schematic diagram of the active anticollimator detector assembly, viewed from an angle above the rotation plane. A is the absorber and D is the detector crystal for the source position shown. (From G. Morfill and G.F. Pieper: 1974, in I.B. Strong (ed.), *Proceedings of Conference on Transient Cosmic Gamma- and X-Ray Sources,* LA-5505-C, p. 206.)

for such detectors at D are therefore modulated in intensity, but in only one detector at a time.

This arrangement has been termed an 'Active Anticollimator' (AAC) by Morfill and Pieper (1974), and they have compared the performance of this system with an active collimated system of approximately equal mass. Their analysis has assumed that both types of detectors see an isotropic background F_B (photons cm^{-2} s^{-1} sr^{-1} MeV^{-1}) such as measured by the Ranger 3 and Apollo spacecraft (see Section V-5.3) and a source strength F_S (photons cm^{-2}s^{-1}MeV^{-1}). For a signal to be detectable in the anticollimator, they require that the modulation be significant to 3 standard deviations of the combined signal and background fluctuations. The modulation is the difference in the occulted number of counts in a given detector, or $N_S + N_B - N_{S'} - N_{B'}$, where the S and B refer to the number of unocculted counts due to the source flux F_S and the background F_B, respectively, and the primes refer to the corresponding occulted counts all in some energy interval ΔE. The modulation ratio is then

$$R_a = \frac{N_S + N_B - N_{S'} - N_B}{[N_S + N_B + N_{S'} + N_{B'}]^{1/2}} \geqslant 3. \qquad \text{(VI.11)}$$

In the case of n identical detectors arranged on a ring as shown in Figure VI-31, and where the occulting detector subtends a mean angle α (FWHM) in degrees at the occulted detector, then for a total viewing time T (s) (many spin periods), Morfill and Pieper (1974) show that

$$R_a(AAC) = F_S(8F_B)^{-1/2} (\bar{\alpha}T/360)^{1/2} (n/n-1)^{1/2} R(1-e^{-2\mu R})^{3/2} \quad \text{(VI.12)}$$

where R is the radius of the cylindrical NaI(Tl) spectrometers and μ is the total absorption coefficient of the NaI(Tl) at the gamma ray energy of interest. The length of each spectrometer is $2R$. In order to compare this signal/noise ratio with that from a normal active-collimated spectrometer of overall cylindrical radius R and with a shield thickness X, Morfill and Pieper (1974) give

$$R_a(AC) = F_S(8F_B)^{-1/2} T^{1/2}(R-X)(1-e^{-2\mu(R-X)})^{1/2} \times (1-e^{-\mu X})e^{\mu X/2} \quad \text{(VI.13)}$$

where T is the time for observing the source or the background and all other symbols have been defined previously. These expressions have been used to compare two such systems of approximately equal total weight in measuring the γ-ray continuum flux from the Crab Nebula whose source strength is $F_S(E) = 9 \times 10^{-3} E^{-2}$ photons cm^{-2} s^{-1} MeV^{-1}, and using a background flux $F_B = 0.026E^{-2}$ photons cm^{-2} s^{-1} sr^{-1} MeV^{-1}.

In Figure VI-32 the minimum flux sensitivity for $R_a \geqslant 3$ in Equation (VI.12) is shown for six 4 cm radius by 8 cm length spectrometers in the anticollimator geometry with both CsI and NaI as the detector material. The detector separation is such that the angular resolution $\bar{\alpha} = 7°$ (FWHM) and the observing time $T = 8.6 \times 10^4$ s (1 day). The corresponding flux sensitivity for a normal active collimated system of total cylindrical radius 5 cm and optimum shield thickness is shown in Figure VI-32. The Crab Nebula γ-ray spectrum is also given in the figure. It can be seen that, on a weight-for-weight basis, the active anticollimator scheme can measure a flux more than an order of magnitude lower than the conventional collimated detector. The AAC and AC systems compared have nearly the same volume, or 1608 cm^3 and 1586 cm^3, respectively, and the weight of the detecting and shield material alone in each case would be ~12 lb of NaI. It should be noted, however, that active collimated detectors in use now are much larger than that used in this comparison; for example, the weight of the NaI in the collimator-shutter arrangement discussed in Section VI-6.3.1b is ~300 lbs. Also, it is not known if an active anticollimator experiment has been flown and tested, but it certainly seems that the approach is worth exploring further to see if the relative advantage shown in Figure VI-32 can be achieved.

d. *HEAO-C Collimated Solid State Spectrometer*

An important γ-ray line experiment is planned for the HEAO-C mission, anticipated for launch in 1979, which will provide the most advanced capability for searching for γ-ray lines from cosmic sources. HEAO-C will be the first long-term space mission using the highest energy-resolution γ-ray spectrometer currently available, a cooled solid state detector. The HEAO-C instrument utilizes four cooled Ge(Li) spectrometers, each of

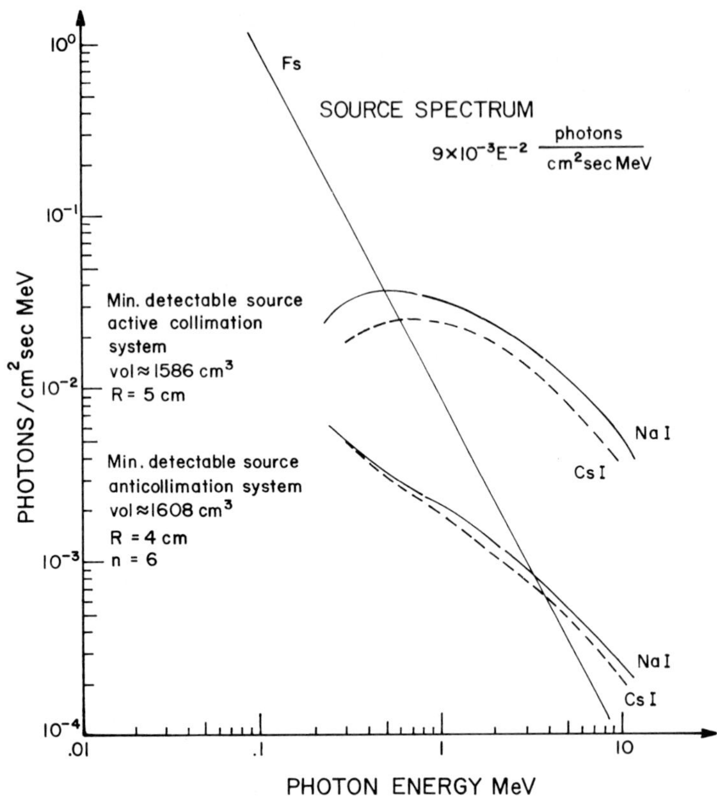

Fig. VI-32. Comparison of the sensitivities of an optimum design active collimator with an active anticollimator system consisting of six cylindrical detectors of diameter and length = 8 cm. The minimum detectable source strength at 3σ is shown for each case, as well as the extrapolated Crab X-ray spectrum for comparison. (From G. Morfill and G.F. Pieper: 1974, in I.B. Strong (ed.), *Proceedings of Conference on Transient Gamma- and X-Ray Sources,* LA-5505-C, p. 206.)

volume $60\,cm^3$, with an overall effective area of $64\,cm^2$. These elements are contained in a CsI(Na) anticoincidence shield-collimator shown in Figure VI-33. The CsI(Na) well-collimator gives a field of view of $\sim 30°$ FWHM at ~ 1 MeV and the design goal for the instrument resolution is $\lesssim 2.5$ keV FWHM.

In order to maintain the resolution capability of the Ge(Li) unit, cryogenic cooling is required. This is accomplished with a two-stage sublimation system using solid methane and ammonia during flight. Liquid N_2 is used to stabilize the coolants prior to launch. In the present design, there are some deviations from the configuration shown in Figure VI-33. In particular, the 'cold finger' from the refrigerator enters the spectrometer from the rear; also, intrinsic Ge cells could be used instead of Ge(Li) cells (Jacobson *et al.*, 1975). Intrinsic Ge needs to be cooled during operation, but not before launch, therefore

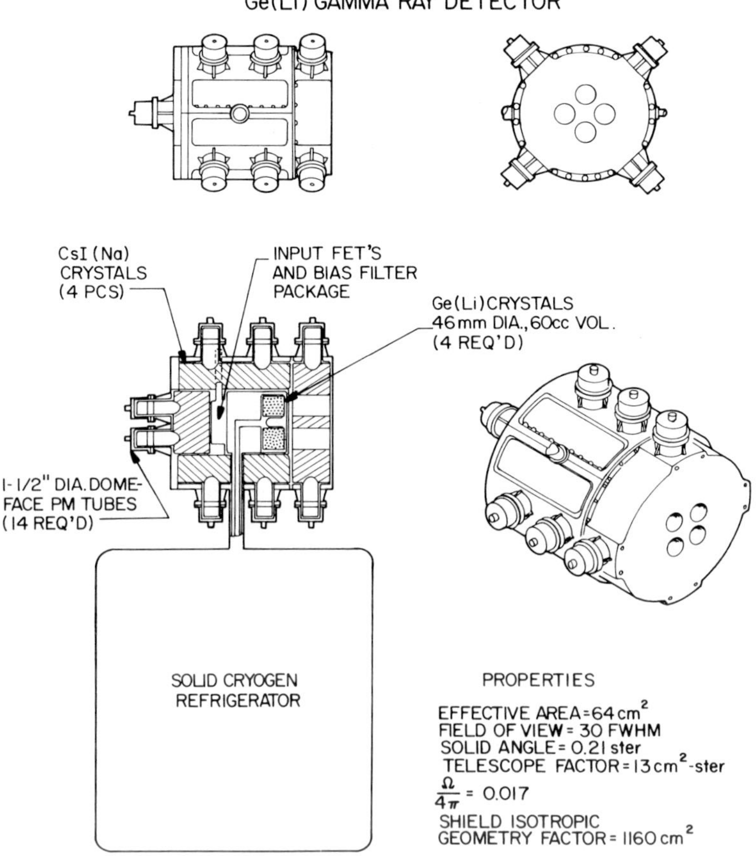

Fig. VI-33. The high resolution HEAO-C Ge(Li) spectrometer consisting of four 60 cm³ crystals in a CsI(Na) anticoincidence shield. (From A. Metzger: 1973, *NASA SP-339*. See also Peterson, 1975, p. 505 for a more recent schematic.)

some simplification of pre-launch activities would result. The electronic and command capability will also permit the four Ge(Li) cells to operate as a single large total absorption spectrometer, as a pair spectrometer, and as a sum-coincidence spectrometer. These various ways of operating a multiple array of γ-ray detectors are standard techniques used in laboratory γ-ray spectroscopy and provide a means of unraveling complicated spectra.

In order to determine the expected flux sensitivity of this telescope in the planned HEAO orbit, it was necessary to estimate the background contributions from terrestrial radiation, diffuse cosmic γ-radiation, and local production in the spacecraft due to passage through the South Atlantic anomaly. This is a difficult problem since there have not been any orbital flights of a solid state spectrometer except for the Lockheed experiment discussed in Section VI-6.1.3b for which the shielding configuration was quite

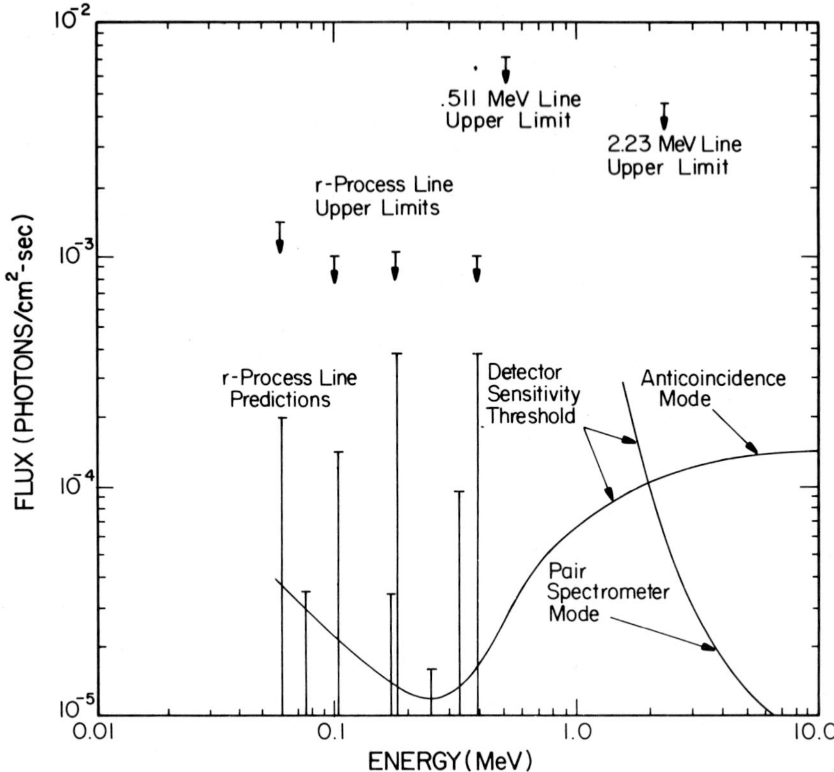

Fig. VI-34. The estimated line flux sensitivity for the HEAO-C Ge(Li) spectrometer compared to the predicted *r*-process line limits. (From A. Metzger: 1973, *NASA SP-339*.)

different from that shown in Figure VI-33. Nevertheless, an estimate of the expected flux sensitivity for the HEAO-C spectrometer is available and is shown in Figure VI-34. The solid curve gives the expected 3σ flux sensitivity for two operating modes of the instrument based on an expression similar to Equation (VI.1) (see Jacobson *et al.*, 1975). The predicted fluxes of several *r*-process lines from the Crab Nebula are also shown for comparison, as well as the current upper limit to some *r*-process lines as determined by the experiment of Jacobson (1968).

Jacobson *et al.* (1975) have recently (June 1974) flown a preliminary version of this solid-state detector telescope on a balloon from Palestine, Texas to study the γ-ray emissions from the direction of the Crab Nebula. The instrument was also provided with a shutter so the detector background could be determined. The data are currently being analyzed so, unfortunately, there are as yet no results to report on possible Crab line emissions. The experiment, however, should be able to detect fluxes at least an order of magnitude lower than the earlier version of the Ge(Li) telescope flown in 1967 (Jacobson, 1968). One valuable result can be mentioned concerning the observed background, which consisted of a continuum and approximately 20 lines. The counting rate from the background lines alone contributed about 10% of the observed detector background rate. In addition, the majority of background lines observed can be attributed to neutron interactions in the Ge(Li) crystal itself. The background observed in this balloon instrument was discussed in more detail in Section VI-6.1.5b.

6.3.2. COMPTON TELESCOPES

The shielded γ-ray detectors described in the previous section have directional properties due to active anticoincidence collimation or occulting elements. Another method of achieving directionality in a γ-ray detector is to make use of the kinematics of the Compton scattering process. This approach appears to be very promising since it is the dominant γ-ray interaction mechanism in the nuclear transition energy region as shown in Figure IV-3. γ-Ray telescopes which require that two Compton scattering interactions occur in pairs of separated detectors have been developed by Schönfelder *et al.* (1973) and Herzo *et al.* (1975). Alvarez *et al.* (1973) have also proposed development of a Compton telescope using liquid Xe, a new type of scintillator.

The basic principle of operation of a Compton telescope can be understood by reference to Figure VI-35, which illustrates schematically the instrument which was designed and tested by Herzo *et al.* (1975). At the right of the figure a γ-ray, γ_0, is shown undergoing a Compton scattering in scintillator S_1, and continuing with degraded energy as γ_1 to the lower scintillator S_2, where another Compton scattering takes place. The first scattered electron, e_1, loses energy in the upper scintillator and gives a start pulse for a delayed coincidence ($\lesssim$ 7 ns) with the electron pulse from the scattered electron, e_2, in the lower scintillator. As shown at the left of the figure a delayed coincidence can also be obtained if a neutron undergoes an elastic scattering with protons in the hydrogenous scintillators; however, the transit time for neutrons is typically much greater (7 to 128 ns) than for photons so the two types of events can be separated. The coincidence and time of flight requirement on events also insures that upward moving γ-rays and any neutrons undergoing elastic scattering are completely eliminated and the direction of the downward moving γ-rays is determined from the kinematics of Compton scattering using the electron pulse heights in the separate cells in S_1 and S_2. Notice that the upper and lower scintillators in Figure VI-35 actually consist of a large number of cells (28 each) so the direction of the scattered γ-ray is approximately defined from the knowledge of the specific cells in S_1 and S_2 which record the delayed coincidence signals. In Figure VI-35 the incident γ-ray direction undergoes a scattering through an angle θ, which also defines the half angle of the cone for all allowed directions of γ_0 for a particular γ_1 trajectory. That is, the direction of the incident γ-ray can only be defined to lie on a cone of half angle θ. The indeterminacy in the photon's arrival direction is a result of the lack of knowledge of the direction of the first recoil electron (see Section IV-4.2.2).

The scattering angle, θ, can be found from the experimental data and Compton kinematics as follows (see Section IV-4.2.2 and Herzo *et al.*, 1975). The energy of the incident γ-ray is given exactly by

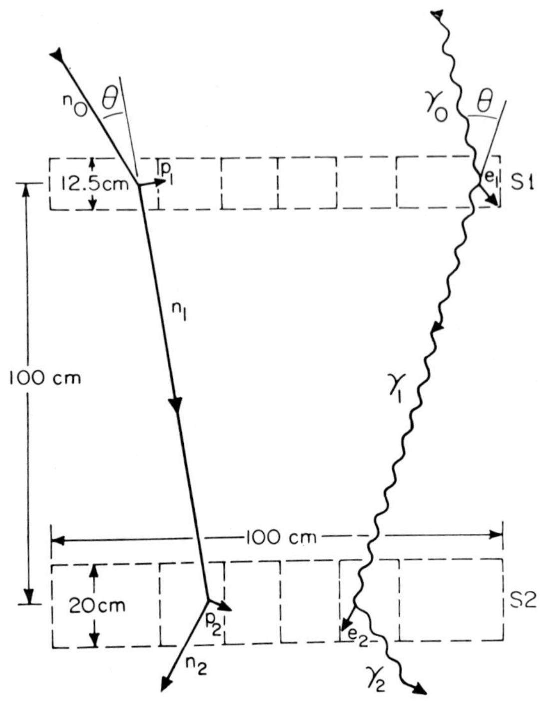

Fig. VI-35. Illustration of the experimental set-up for a double Compton scattering telescope (From D. Herzo *et al.*: 1975, *Nucl. Inst. and Methods* **123**, 583. Used by permission.)

$$E_{\gamma_0} = E_{e_1} + E_{\gamma_1} \tag{VI.14}$$

where E_{e_1} and E_{γ_1} refer to the energies of the scattered particles defined above.

From Equation (IV.4) it directly follows that:

$$\cos\theta = 1 + m_0c^2 \left(\frac{1}{E_{\gamma_0}} - \frac{1}{E_{\gamma_1}}\right) \tag{VI.15}$$

where the value for E_{γ_1} is generally

$$E_{\gamma_1} = E_{e_2} + E_{\gamma_2}\,. \tag{VI.16}$$

The quantities E_{e_1} and E_{e_2} are the measured pulse heights (in energy units) in the scintillator cells in S_1 and S_2. If γ_1 loses all of its energy in S_2 because of multiple Compton scattering and/or photoelectric absorption, $E_{\gamma_1} = E_{e_2}$, then Equation (VI.14) can be written as:

$$E_{\gamma_0} = E_{e_1} + E_{e_2} \tag{VI.17}$$

and Equation (VI.15) becomes

$$\cos\bar{\theta} = 1 + m_0c^2\left(\frac{1}{E_{e_1} + E_{e_2}} - \frac{1}{E_{e_2}}\right) \quad \text{(VI.18)}$$

where now $\bar{\theta}$ is generally an approximation to the true value of θ.

Now since actually $E_{e_2} < E_{\gamma_1}$ (unless a photoelectric absorption occurs) it follows that the true scattering angle, θ, is always less than the approximate angle, $\bar{\theta}$. A closer approximation to the correct value of the scattering angle can be found by correcting E_{e_2} to αE_{e_2} giving a better value for E_{γ_1}, where α is a calculated quantity which depends on E_{e_2}. This approach is originally due to Schönfelder *et al.* (1973), and Herzo *et al.* (1975) have determined a best value of α by Monte Carlo calculations. For values of E_{e_2} ranging from ~10 to 0.5 MeV, α varies smoothly from ~1.2 to 1.6. In summary then, for each event the maximum value for the scattering angle can be found and the approximate value for E_{γ_0}.

The properties of the Compton telescope designed by Herzo *et al.* (1975) have been determined by Monte Carlo simulations and checked with laboratory sources. The threshold energy loss in each scintillator cell was set at 0.2 MeV. As an example we list some typical parameters for a large Compton telescope of geometrical area 10^4 cm^2 and height ~10^2 cm from Herzo *et al.* (1975) for $E_{\gamma_0} \sim 2.2$ MeV:

Angular Resolution = 15° → 17° FWHM;
Energy Resolution = 50% FWHM; and
Sensitivity = 280 cm^2.

By comparison with the properties of the collimated scintillators discussed in Section VI-6.3.1, it does not appear that Compton telescopes will be useful for γ-ray line astronomy particularly because the energy resolution is so poor. The angular resolution and sensitivities given above are also easily attainable with collimated scintillators. The Compton telescope is still attractive, however, since the coincidence requirement reduces dramatically the background counting rate of the telescope. Background is one of the major limitations of the collimated instruments. The energy resolution of a Compton telescope may in fact be considerably improved if the lower hydrogenous scintillator S_2 is larger and made of other material such as NaI(Tl). This development could make the Compton telescope a much more useful instrument. Some preliminary measurements on the atmosphere γ-ray flux have been obtained with this instrument by White *et al.* (1973) and were described in Section VI-6.1.1b (see Figure VI-5). The limiting flux sensitivities attainable with this instrument for several cosmic or solar sources have been estimated by Herzo *et al.* (1975). For example at 2.2 MeV the 2σ flux sensitivity is 8×10^{-5} photons cm^{-2} s^{-1} for an observation time of ~65 000 s (~19 h).

The Compton telescope just described is a wide angle instrument capable of recording γ-rays which can enter the telescope from about π sr (60° zenith angle) of the sky at one position of the instrument. The instrument can also be used with reduced aperture by selecting only events which have a calculated scattering angle $\bar{\theta}$ (Equation (VI.18)) less than some small value, e.g. 15°, then the true scattering angle will be less than this. If in addition the selected events trigger a pyramid arrangement of cells lined up in a particular direction, which defines the axis of the cone of arrival directions of the incident photons, then the detector 'sees' mainly in that direction with an aperture given by the restriction on the scattering angle. This approach reduces the number of cells of the telescope effective for a given source direction and so the total effective area (sensitivity) of the instrument is reduced. Schönfelder *et al.* (1973) have used this approach to construct a Compton telescope which is symmetrical about the vertical and uses a single scattering cell at the top and nine at the bottom. This instrument was flown by Schönfelder and Lichti (1974) and results were obtained on the diffuse γ-ray spectrum and the atmospheric γ-ray spectrum (see Sections V-5.3 and VI-6.1.1b). Besides using the pyramid geometry, the detector of Schönfelder *et al.* (1973) uses plastic scintillators for the scattering cells whereas Herzo *et al.* (1975) have used hydrogenous liquid scintillators which allows the possibility of using electronic pulse shape discrimination techniques to reduce backgrounds. The instrument of Schönfelder *et al.* (1973) had an angular resolution of $\sim 30°$ FWHM and an energy resolution of $\sim 40\%$ in the energy range 2 to 4 MeV and an effective area (sensitivity) of 1.1 cm^2.

The Compton telescope proposed by Alvarez *et al.* (1973) has two interesting features not utilized in the instrument discussed above. A basic design difference is the use of multiwire liquid Xe proportional counters in place of the upper and lower scintillators. According to Alvarez *et al.* (1973), use of the liquid Xe counters permits the localization of a scattering event to an accuracy of ~ 2 mm FWHM and the energy resolution available is comparable to that of NaI(Tl). Since the position of the first and second Compton interactions in the upper and lower counters is accurately known, each event determines a circle in the sky giving the allowed directions for the incident photon as we have discussed above. The important point emphasized by Alvarez *et al.* (1973) is that other events will determine different circles of location in the sky and if the γ-ray source is a point source then all 'circles of location' will intersect at a common position which is the position of the source. By computer simulation it has been shown that this device can determine a source location to a 'circle of confusion' with angular accuracy $\delta_{rms} \simeq 1.6°$ for γ-rays in the energy range 0.3 to 2 MeV, if the main source of the background is the diffuse cosmic gamma radiation. The ultimate accuracy in source location will depend of course on the counting statistics. With a liquid Xe Compton telescope of geometric area 400 cm^2 it has been estimated that a point source of monoenergetic γ-rays can be detected to a 3σ level of 5×10^{-4} photons cm^{-2} s^{-1} for photons in the energy range 0.2 to 3 MeV for an observation time of 10 h (Alvarez *et al.*, 1973). Because of the good angular resolution properties of this Compton telescope and its good energy resolution, it would appear that further development of this approach is desirable.

As a final remark on Compton telescopes, it should be noted that in spite of the use of a coincidence requirement and the well known Compton equations a serious background can result from neutron interactions that simulate true γ-ray coincidence events. This problem, first pointed out by White and Schönfelder (1975), was discussed in Sections VI-6.1.1d and 6.1.5c. These authors have estimated that neutron interactions in C can provide as much as 50% of the apparent γ-ray events in the instruments flown by Schönfelder *et al.* (1973) and Herzo *et al.* (1975).

CHAPTER VII

CONCLUSIONS

A review of the major theoretical and experimental efforts in γ-ray astronomy, which was the intent of this monograph, would not be complete without noting recent significant advances. These were reviewed at the International Gamma Ray Symposium held in Greenbelt, Md. in 1976, June 2–4, after the body of this text was completed. The reader is referred to the proceedings of this conference, which have recently been published.* Although, in the author's opinion, no significant new evidence for cosmic γ-ray lines was reported at this symposium beyond that referred to in this monograph, some interesting new data were presented. These include: further measurements of high energy γ-rays from the galaxy by spark chambers on SAS-2 and COS-B, additional observations of Vela-type γ-ray bursts, balloon-borne investigations using cooled Ge(Li) spectrometers searching for γ-ray lines from supernova remnants.

It should also be mentioned here that the Solar Maximum Mission spacecraft to be launched in late 1979 will carry a γ-ray spectrometer intended primarily for solar observations. The instrument is a scaled-up version of the OSO-7 NaI spectrometer discussed in Section V-5.1.1. It will have the capability of measuring γ-ray lines at a flux level as low as 10^{-3} cm^{-2} s^{-1} with good energy resolution up to ~ 20 MeV; solar γ-rays from π^0 decay can also be detected at a minimum flux level of 10^{-3} photons cm^{-2} s^{-1}. Results from this mission should advance our knowledge of solar processes. Unfortunately, the weight limitations for experiments on the SMM spacecraft do not allow for incorporation of a high resolution Ge(Li) spectrometer.

As far as the general field of γ-ray line astronomy is concerned, it is abundantly clear that in the next several years many experimenters will attempt to detect γ-ray lines using large volume ($\gtrsim 100$ cm^3) Ge(Li) spectrometers from both balloon and space platforms in addition to the HEAO-C experiment discussed in Section VI-6.3.1d. It is important that the satellite missions to detect γ-ray lines be complemented by balloon programs and that balloon and satellite missions be continued using existing instrumentation, since transient events such as nearby ($\lesssim$ Mpc) supernovae and solar flares can produce γ-ray line fluxes well above the instrument thresholds.

The obstacles before us in advancing this field are overwhelming, as we have discussed here several times. It is expected that progress will be slow in achieving line flux sensitivities below 10^{-4} photons cm^{-2} s^{-1} at MeV photon energies using existing techniques.

* These proceedings should be available in preprint form in the summer of 1976 and in final form by the end of 1976. They can be obtained from the conference organizers at the Goddard Space Flight Center, Greenbelt, Maryland – Dr. C.E. Fichtel, Code 662, or Dr. F.W. Stecker, Code 602.

On the other hand, the potential of γ-ray line astronomy warrants continued and new efforts to develop techniques to overcome the problems limiting our advance. In conclusion, then, summing up the problems and prospects of the field of γ-ray astronomy, the author believes that ultimately γ-ray line astronomy will become a very effective tool in probing the Universe because of the unique signature of spectral lines, the penetrability of γ-rays, and their origin in high energy astrophysical processes. It is because of this belief that this book has been written.

NOTES ADDED IN PROOF

The following significant publications present recent developments in the topics discussed in the indicated sections.

Chapter II [Mechanisms for γ-Ray Line and Continuum Production]

Tucker, W.H.: 1976, *Radiation Processes in Astrophysics,* MIT Press, Cambridge, Massachusetts.

A new book that describes several important radiation processes from a semi-classical point of view.

Section II-2.4.3c [Thermonuclear (Exoergic) Reactions]

Fowler, W.A., Caughlan, G.R., and Zimmerman, B.A.: 1975, 'Thermonuclear Reaction Rates, II', in G.R. Burbidge (ed.), *Ann. Rev. of Astron. and Astrophys.* Vol. 13, Annual Reviews, Inc., Palo Alto, California, p. 69.

A recently published revised discussion of the thermonuclear reaction rates of importance in astrophysics.

Section III-3.1.2 [Positron and Neutron Production]

Crannell, C.J., Joyce, G., Ramaty, R., and Werntz, C.: 'Formation of the 0.511 MeV Line in Solar Flares', *Astrophys. J.* (in press).

This paper gives a detailed analysis of the rates of free positron annihilation and positronium formation in a solar flare plasma taking into account the density and temperature dependence of the competing processes. The authors conclude that the asymmetrically broadened line at 0.51 MeV from annihilation through positronium could be detectable late in a solar flare. The relationship of 0.51 line shape to flare temperature and density is also discussed.

Section III-3.1.3 [Excited Nuclear States]

(A) Ramaty, R. and Crannell, C.J.: 1976, 'Solar Gamma Ray Lines as Probes of Accelerated Particle Directionalities in Flares', *Astrophys. J.* **203**, 766.

These authors give a calculation of the relation between the expected peak position of the solar flare excited 6.1 MeV transition (^{16}O) and the angular distribution of solar cosmic rays at the Sun. A shift in the photopeak position as large as 40 keV could occur.

(B) Crannell, C.J. and Crannell, H.: 1976, 'High Energy Gamma Ray Lines from Excited

Nuclei in Solar Flares', *Bull. Am. Phys. Soc.* **21**, No. 4, 587.

This paper considers the inelastic process $^{12}C(p, p')^{12}C^{*(15.1\,MeV)}$ giving a prompt 15.1 MeV gamma ray. By using measured excitation cross sections for the reaction and the measured branching ratio for the decay of the excited state to the ground state these authors conclude that the flux of this line in solar flares could be ~ 1→ 10% of the flux of the prompt 4.4 MeV γ-ray line from the reaction $^{12}C(p, p')^{12}C^{*(4.4\,MeV)}$. The relative intensity of the two γ-ray lines is very sensitive to solar flare proton spectral shape.

Section V-5.2 [Cosmic Observations (Point and Localized Sources)]

Hall, R.C., Meegan, C.A., Walraven, G.D., Djuth, F.T., Shelton, D.W., and Haymes, R.C.: 1975, Conference papers for *14th International Cosmic Ray Conference,* Max-Planck Institute for Extraterrestrial Physics, Munich, Vol. 1, p. 84.

Evidence for γ-ray line emission from the strong radio source Centaurus A (NGC-5128), reported by the Rice University group from a balloon flight on 1974, April 2. Weak lines (3.3σ) have been reported at ~ 1.6 MeV and at ~ 4.5 MeV.

Section V-5.4 [Transient γ-Ray Bursts]

The γ-ray bursts observed through 1973 by various satellites and space probes have been presented in Table V-5. An up-to-date tabulation is being prepared at Los Alamos, but is not yet available; however, some interesting but incomplete additional information is available as follows (personal communication from R.W. Klebesadel):

(1) A complete survey of the Vela data has detected only 3 γ-ray bursts in 1974.

(2) An incomplete survey has given evidence for 2 events in 1975 but only a portion of a year's data has been analyzed.

(3) A new search procedure has detected 5 events in the first 6 months of 1976.

(4) An event was observed on 1976, 12 June by 4 Velas and the Solrad-Hi satellite. This circumstance could permit a well defined location for the event.

Section VI-6.1.2 [Instrument Activation (Local Production)]

Dyer, C.S., Trombka, J.I., and Seltzer, S.M.: 1975, 'Nuclear Data for Assessment of Activation of Scintillator Materials During Space Flight', *Proceedings of the Nuclear Cross Sections and Technology Conference, NBS SP-425,* Vol. 2, Washington, D.C., p. 480.

The authors describe an improved scheme for calculating activation by trapped protons and cosmic rays.

Section VI-6.1.4 [Space Probes]

Dyer, C.S., Trombka, J.I., Schmadebeck, R.L., Eller, E., Bielefeld, M.J., O'Kelley, G.D., Eldridge, J.S., Northcutt, K.J., Metzger, A.E., Reedy, R.C., Schonfeld, E., Seltzer, S.M., Arnold, J.R., and Peterson, L.E.: 1975, 'Radioactivity Observed in Sodium Iodide Gamma Ray Spectrometer Returned on Apollo 17 Mission', *Space Sci. Inst.* **1**, 279.

Direct evidence is provided for thermal and fast neutron background effects during the Apollo 17 mission.

Appendix A [Attenuation Coefficients for γ-Ray Interactions]

Veigel, W.J.: 1973, 'Photon Cross Sections from 0.1 keV to 1 MeV for Elements Z = 1 to Z = 94', *Atomic Data Tables* **5**, 51.

This paper gives a semi-empirical compilation of various photon cross sections and also includes a bibliography of experimental cross section data.

General

Atomic Data and Nuclear Tables published monthly by Academic Press, Inc., 111 Fifth Ave., New York, N.Y. 10003.

For up-to-date nuclear data on photon interactions, reference should be made to this periodical.

APPENDIX A

ATTENUATION COEFFICIENTS FOR γ-RAY INTERACTIONS

A very valuable discussion of the general problem of the interaction of radiation with matter was given several years ago by Evans (1955). This remains as the standard reference for students and research workers; however, a more recent review has been given by Davisson (1966) and should be consulted for basic theory and references up to 1966. Evans (1958, 1968) also gives useful summaries. A recent review on the photoelectric process for photon energies above 10 keV was given by Pratt *et al.* (1973) and should be consulted for the latest theoretical calculations and comparison with experiments.

In any detector application, a quantity of major interest is the total probability of attenuation of γ-rays or the probability of occurrence of a specific γ-ray interaction. We will summarize here the more important aspects for our area of interest (see Evans, 1955 and Davisson, 1966 for details). The linear attenuation coefficients for the three main processes and their relation to the basic cross sections discussed in Chapter IV are:

Photoelectric: $\tau(\text{cm}^{-1}) = {}_a\tau\ N$; where ${}_a\tau$ is the atomic cross section (cm^2/atom) and N is the atomic number density (cm^{-3}).

Compton: $\sigma(\text{cm}^{-1}) = {}_e\sigma\ N\ Z$; where ${}_e\sigma$ is the probability of removal (cm^2/electron) of the photon from a collimated beam, N is the same as above, and Z is the number of electrons per atom. [${}_e\sigma = {}_e\sigma_s + {}_e\sigma_a$, where ${}_e\sigma_s$ is the Compton scattering coefficient and ${}_e\sigma_a$ is the Compton absorption coefficient (see Davisson, 1966).]

Pair production: $\kappa(\text{cm}^{-1}) = {}_a\kappa\ N$; where ${}_a\kappa$ is the total pair production cross section per nucleus (cm^2).

Since these linear attenuation coefficients are for independent processes and each determines the probability of complete removal of the photon from a collimated beam, the total linear attenuation coefficient is written as $\mu_0(\text{cm}^{-1}) = \tau + \sigma + \kappa$. A collimated beam of γ-rays is attenuated in number by a factor $e^{-\mu_0 x}$. Some recent experimental values of the mass attenuation coefficient (expressed in cm^2 g^{-1}) for high energy mononergetic γ-rays in various materials are given in Table A-1 from Ahmed and Cochran (1973). The quantity tabulated is $\Sigma_t \equiv \mu_0/\rho$ where ρ (g cm^{-3}) is the density of the absorber.

Mass attenuation coefficients (cm^2 g^{-1}) for total or individual processes are simply the linear coefficients divided by the mass density ρ (g cm^{-3}). Thus for the photoelectric effect, $\tau/\rho = {}_a\tau \cdot \Lambda/A$ (cm^{2}g^{-1}); for the Compton effect, $\sigma/\rho = {}_e\sigma \cdot Z \cdot \Lambda/A$ (cm^{2}g^{-1}) and $\kappa/\rho = {}_a\kappa \cdot \Lambda/A$ (cm^{2}g^{-1}) for pair production, where Λ is Avogadro's number and A

TABLE A-1
Total γ-ray attenuation coefficient (From K.U. Ahmed and R.G. Cochran: 1973, *Nucl. Tech.* **17**, 66. Used by permission American Nuclear Society.)

Energy (MeV)	Corrected Σ_t (cm^2 g^{-1})				
	Al	Fe	Cu	Zr	Pb
6.02	0.02670	0.03074	0.03101	0.03388	0.04388
7.28	0.02523	0.02999	0.03081	0.03401	–
7.72	0.02465	0.03004	0.03079	0.03425	0.04643
8.49	0.02406	0.02990	0.03091	–	0.04754

is the atomic weight. As Evans (1955) points out, mass attenuation coefficients are most valuable for γ-ray detector work, since they are independent of the actual density and physical state of the absorber, whether it be in a gaseous, liquid or solid form. It is also interesting to note that the mass attenuation coefficient for the Compton process is a very slowly varying function of Z since $Z/A \sim 0.5$ for all elements except for hydrogen where $Z/A = 1$. For detector considerations, the next several pages present curves of the mass attenuation coefficients (cm^2 g^{-1}) versus initial photon energy for the individual and the total of all processes for the following elements and compounds:

1	Hydrogen;	Figure A-1;	p. 280
2	Air;	Figure A-2;	p. 281
3	Pb;	Figure A-2;	p. 281
4	Al;	Figure A-3;	p. 282
5	H_2O;	Figure A-3;	p. 282
6	NaI(Tl);	Figure A-4;	p. 283
7	CsI(Tl);	Figure A-5;	p. 284
8	Ge;	Figure A-6;	p. 285
9	Xe;	Figure A-7;	p. 286
10	Polystyrene;	Figure A-8;	p. 287

These curves were prepared from data in the tabulations of Hubbell (1969) and Storm and Israel (1970) following the general format originally used by Evans (1955). The Storm and Israel (1970) work is the most recent tabulation readily available and incorporates recent revisions to theory particularly in the case of photoelectric cross sections. The incoherent scattering calculations of Storm and Israel (1970) include the effect of bound electrons in Compton scattering. This is denoted as σ/ρ in Figures 1, 4, 5, 6, 7, and 8. The total coherent or Rayleigh scattering from atomic electrons is also indicated in these curves as σ_r/ρ, the total photoelectric interaction as τ/ρ, and pair production as κ/ρ. Since these curves give effectively the total probability for each separate process and their combination, it is necessary to refer to other sources if one is interested, for example, in the energy absorbed in the material by the given process. For energy absorption by the electrons in incoherent scattering, the photoelectric effect, and pair production, see the tabulations of Storm and Israel (1970) and a discussion by Davisson (1966). Additional

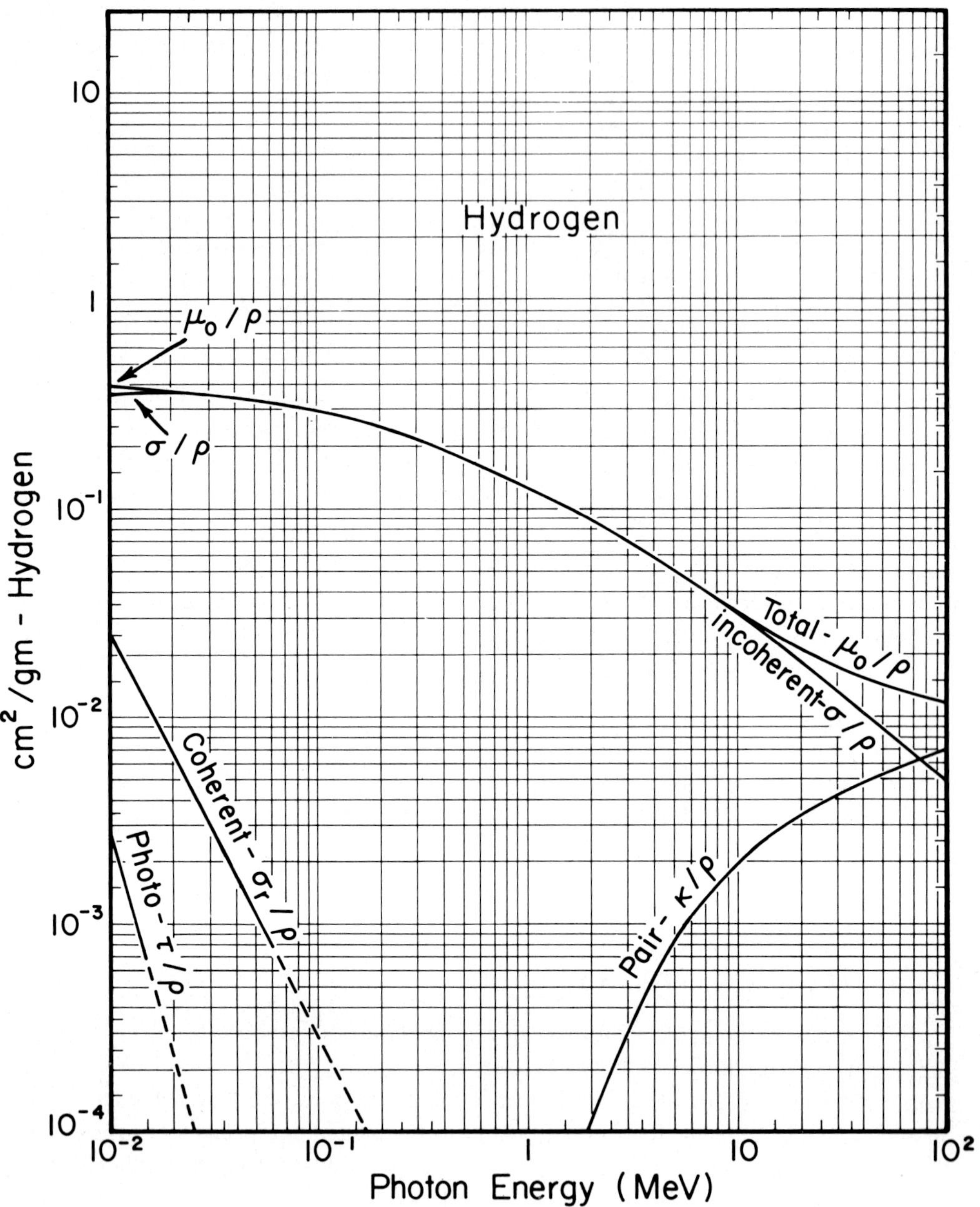

Fig. A-1. The mass attenuation coefficients (cm^2 g^{-1}) are shown for γ-rays in hydrogen for the major γ-ray interaction processes in the energy range $10^{-2}-10^2$ MeV. (Plotted from data in Storm and Israel, 1970.)

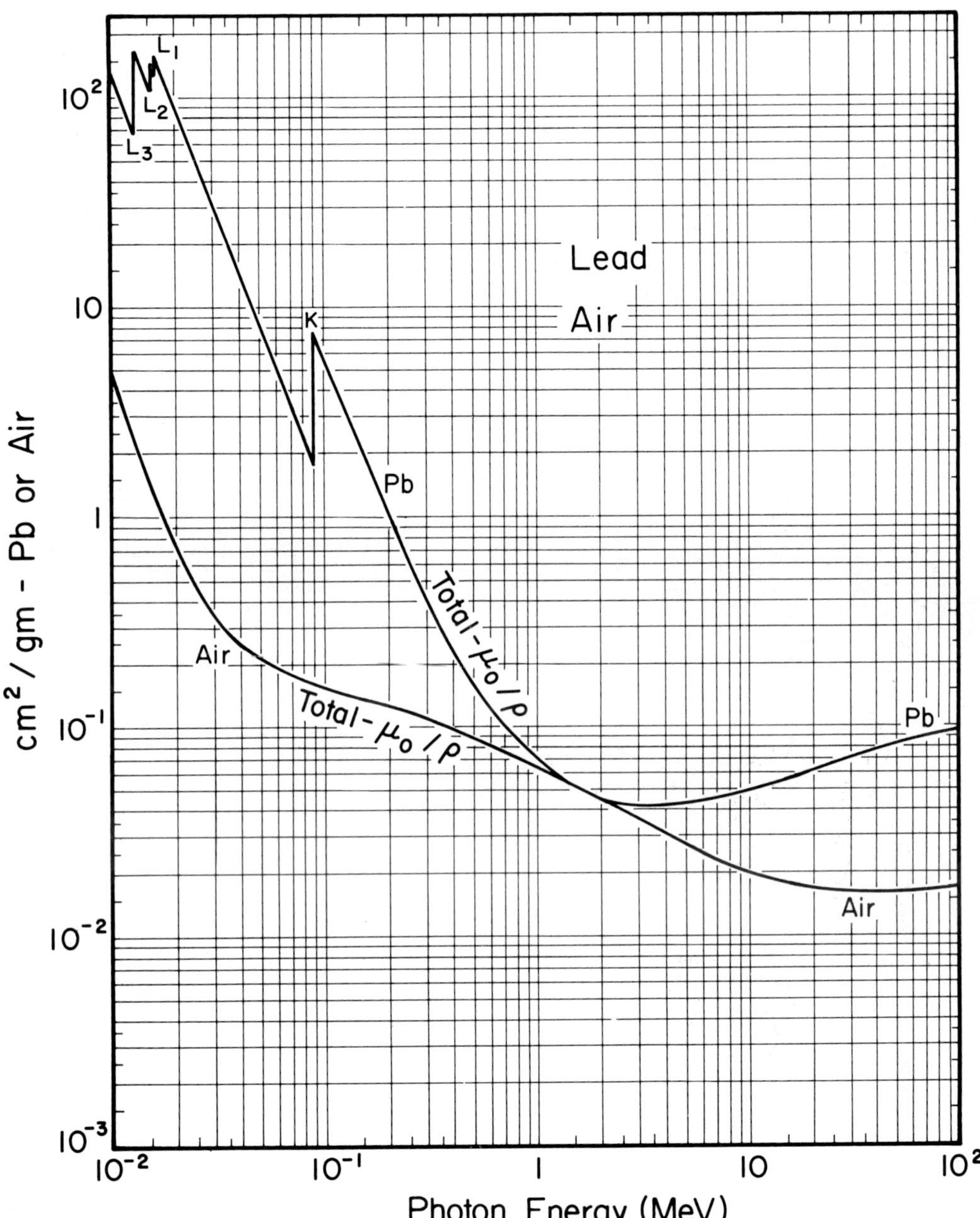

Fig. A-2. The total mass attenuation coefficients ($cm^2 g^{-1}$) are shown for γ-rays in air and lead the energy range $10^{-2} - 10^2$ MeV. (Plotted from data in Hubbell, 1969.)

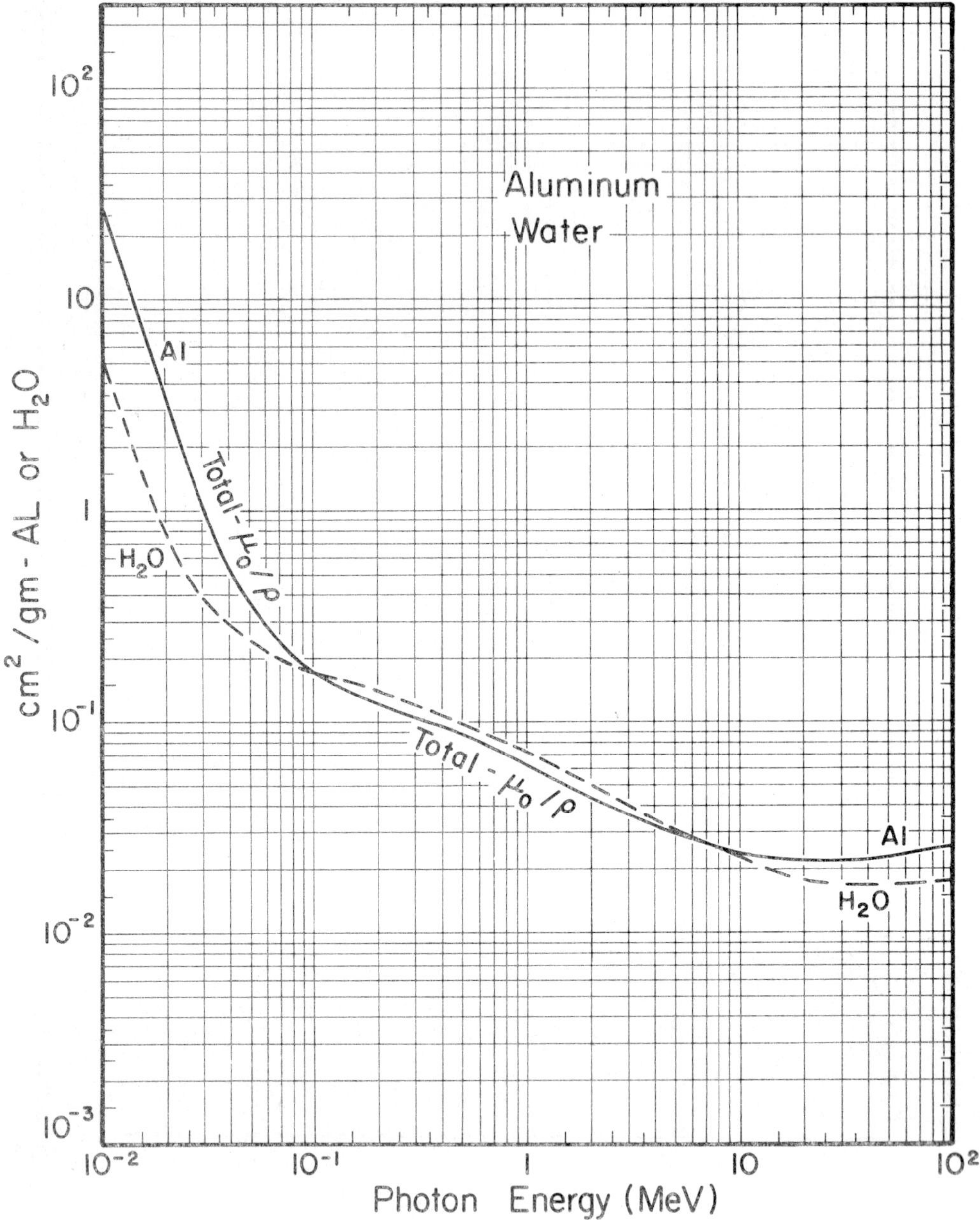

Fig. A-3. The total mass attenuation coefficients ($cm^2 g^{-1}$) are shown for γ-rays in aluminum and water in the energy range 10^{-2}–10^2 MeV. (Plotted from data in Hubbell, 1969.)

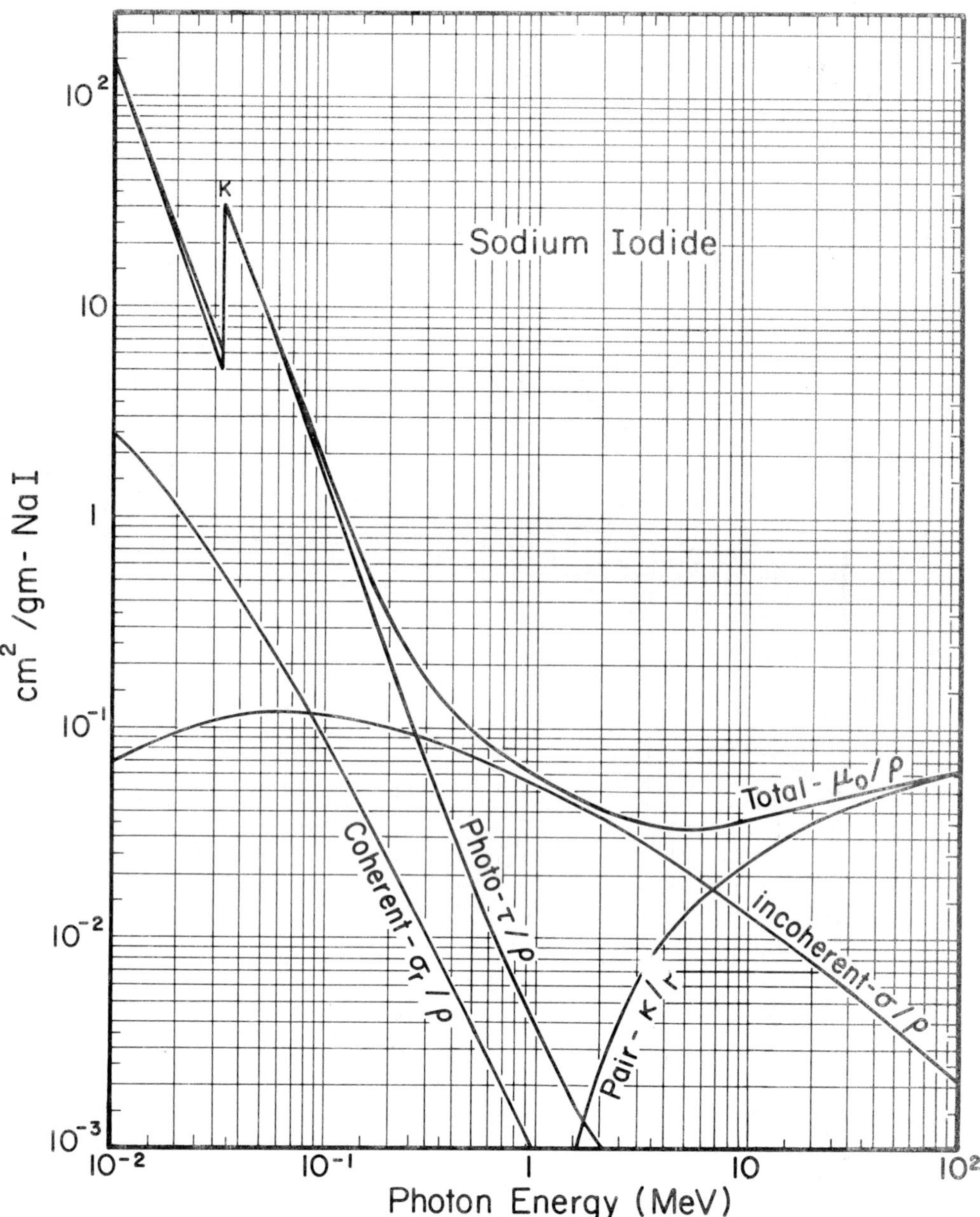

Fig. A-4. The mass attenuation coefficients ($cm^2\ g^{-1}$) are shown for γ-rays in sodium iodide for the major γ-ray interaction processes in the energy range $10^{-2} - 10^2$ MeV. (Plotted from data in Storm and Israel, 1970.)

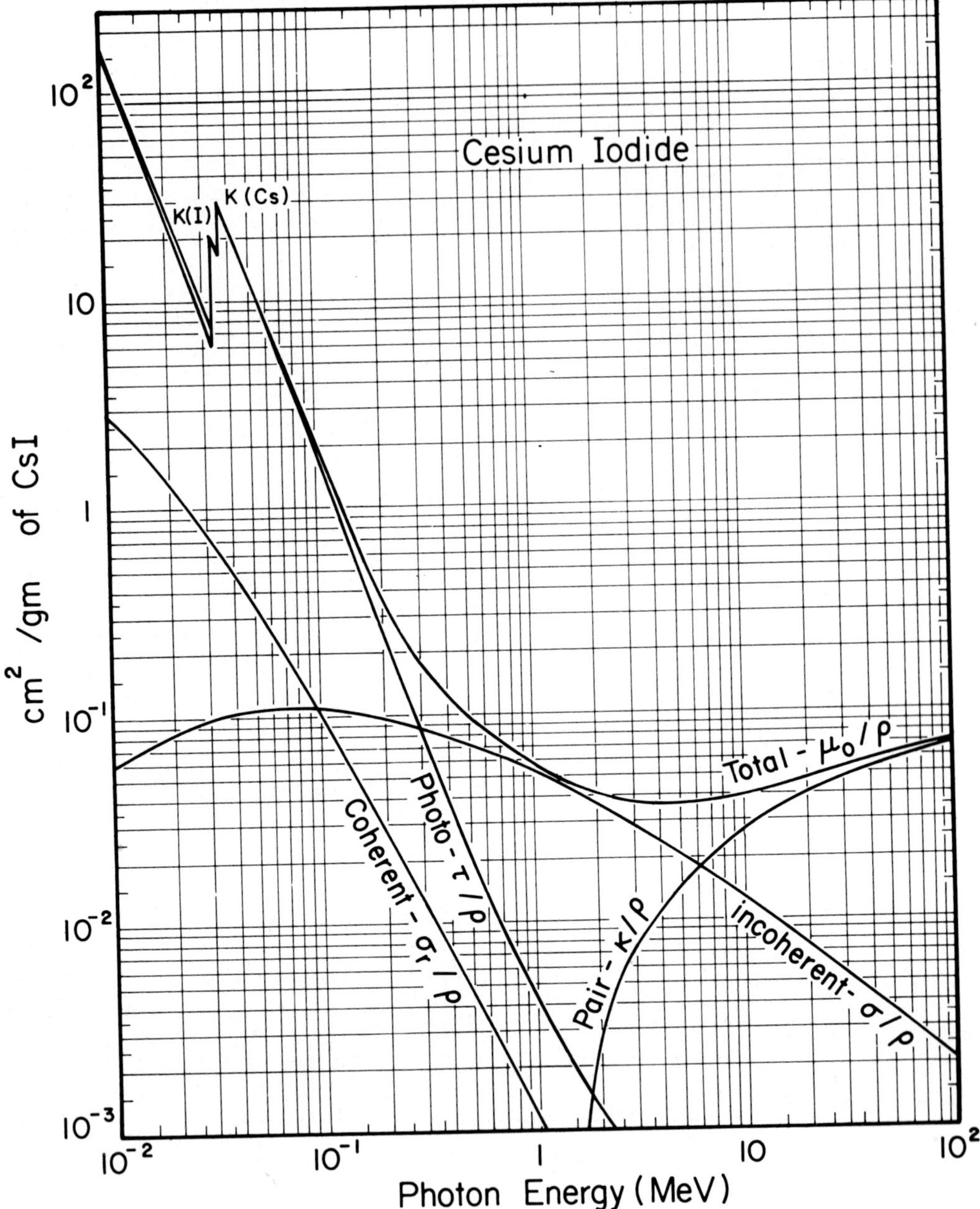

Fig. A-5. The mass attenuation coefficients ($cm^2 g^{-1}$) are shown for γ-rays in cesium iodide for the major γ-ray interaction processes in the energy range $10^{-2} - 10^2$ MeV. (Plotted from data in Storm and Israel, 1970.)

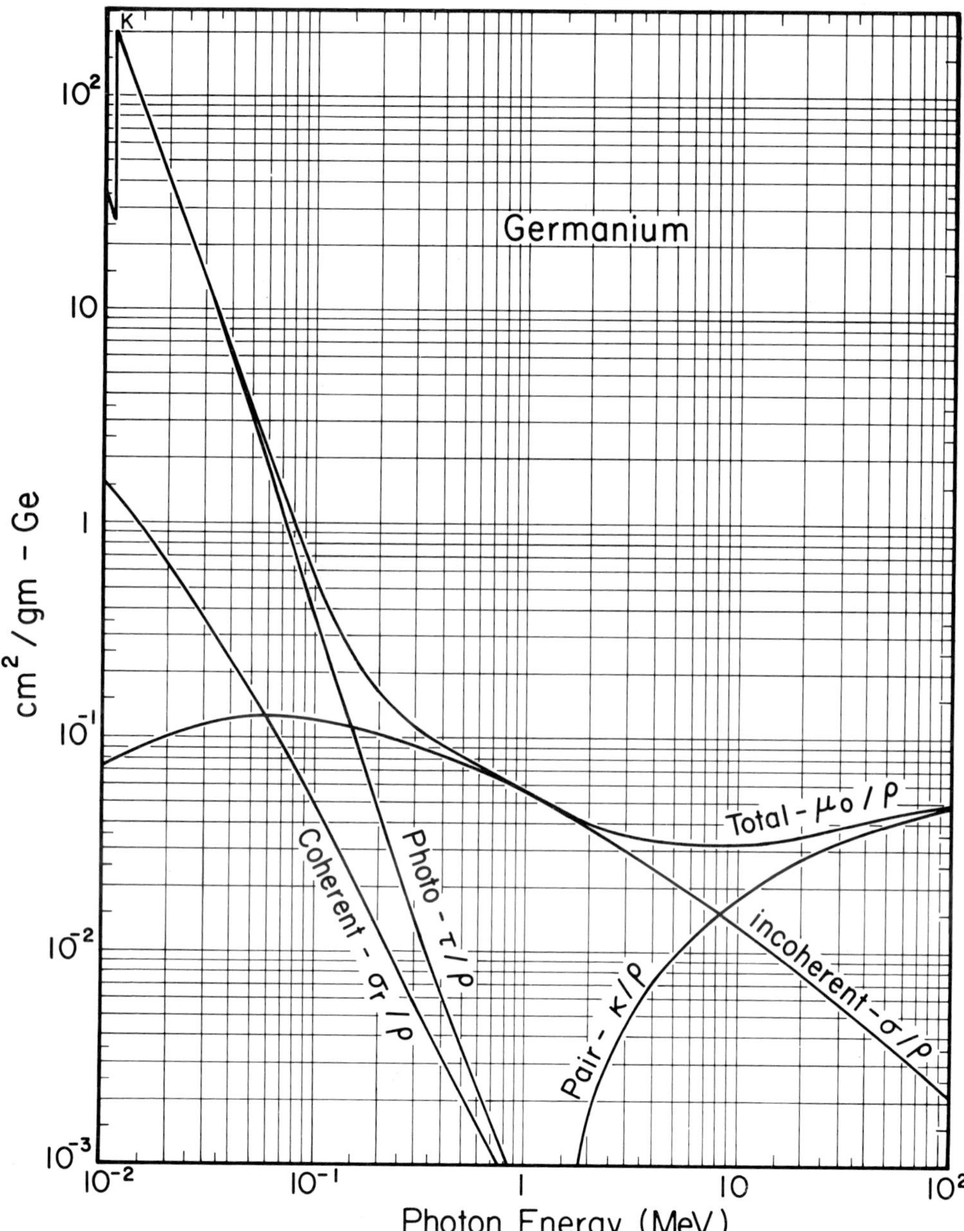

Fig. A-6. The mass attenuation coefficients (cm^2 g^{-1}) are shown for γ-rays in germanium for the major γ-ray interaction processes in the energy range $10^{-2} - 10^2$ MeV. (Plotted from data in Storm and Israel, 1970.)

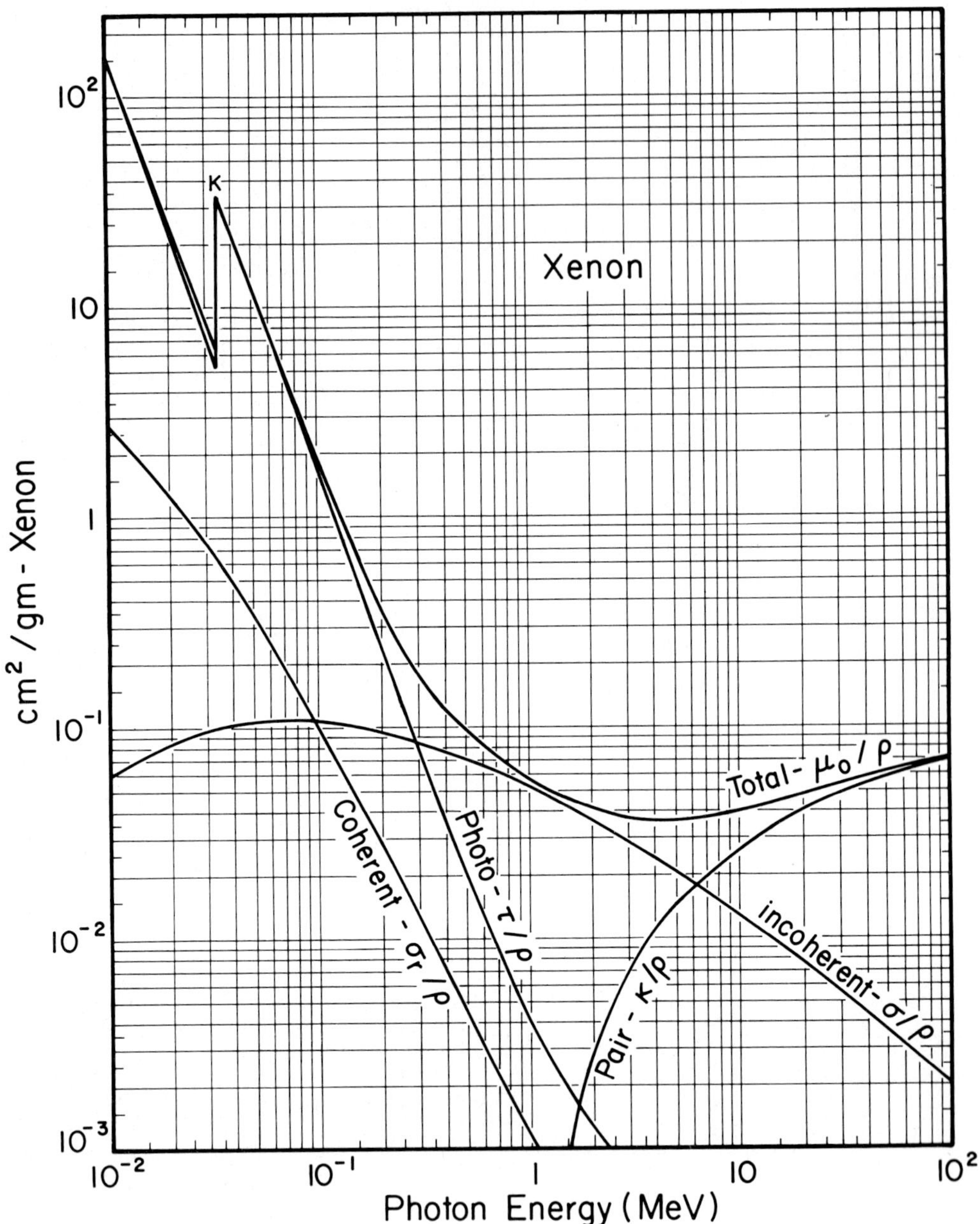

Fig. A-7. Mass attenuation coefficients ($cm^2\ g^{-1}$) for γ-rays in xenon for the major γ-ray interaction processes in the energy range $10^{-2} - 10^2$ MeV. (Plotted from data in Storm and Israel, 1970.)

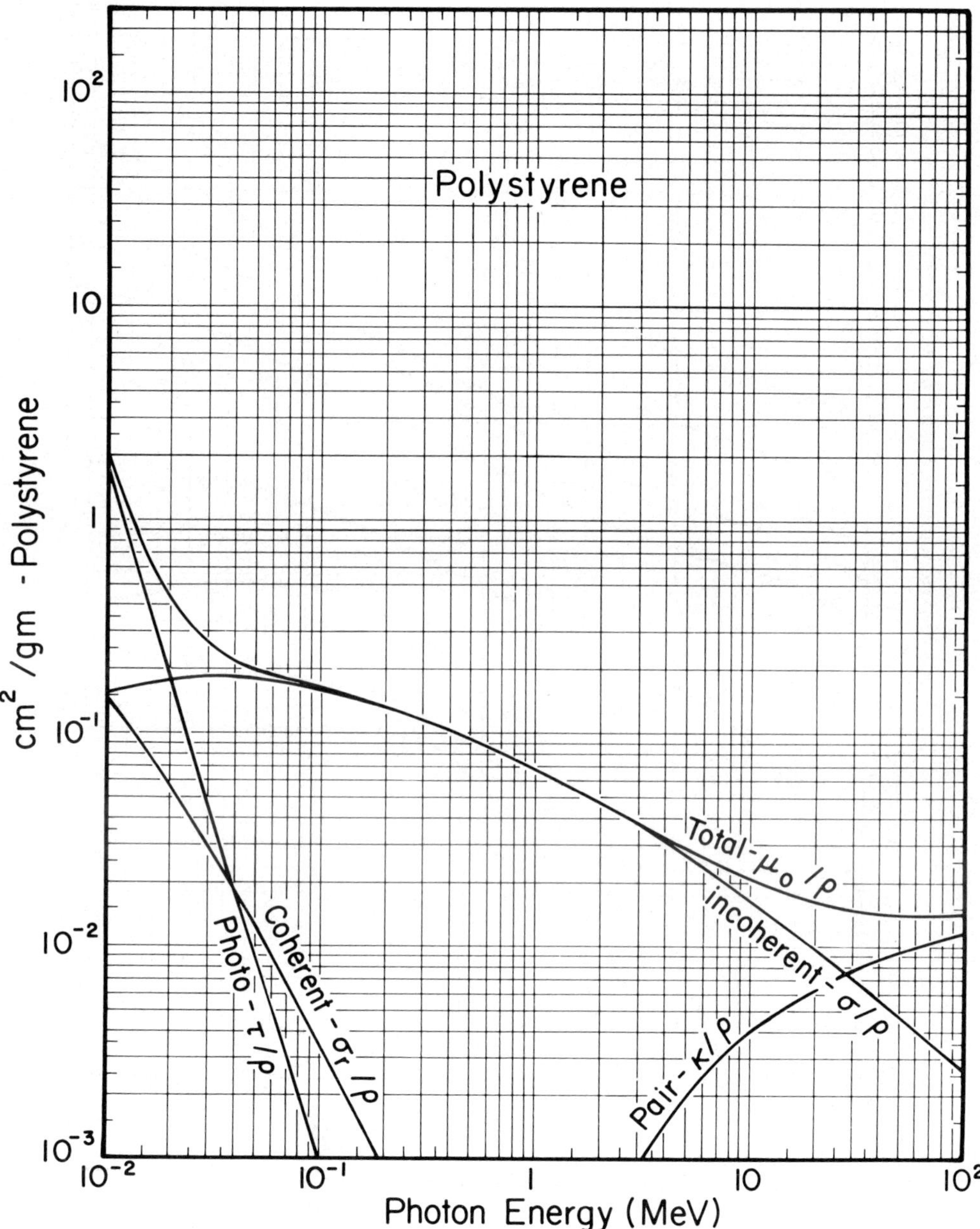

Fig. A-8. The mass attenuation coefficients ($cm^2\ g^{-1}$) are shown for γ-rays in polystyrene for the major γ-ray interaction processes in the energy range $10^{-2} - 10^2$ MeV. The carbon to hydrogen ratio of unity corresponds approximately to that in some organic scintillators such as NE 102 and NE 213. (Plotted from data in Hubbell, 1969 and Storm and Israel, 1970.)

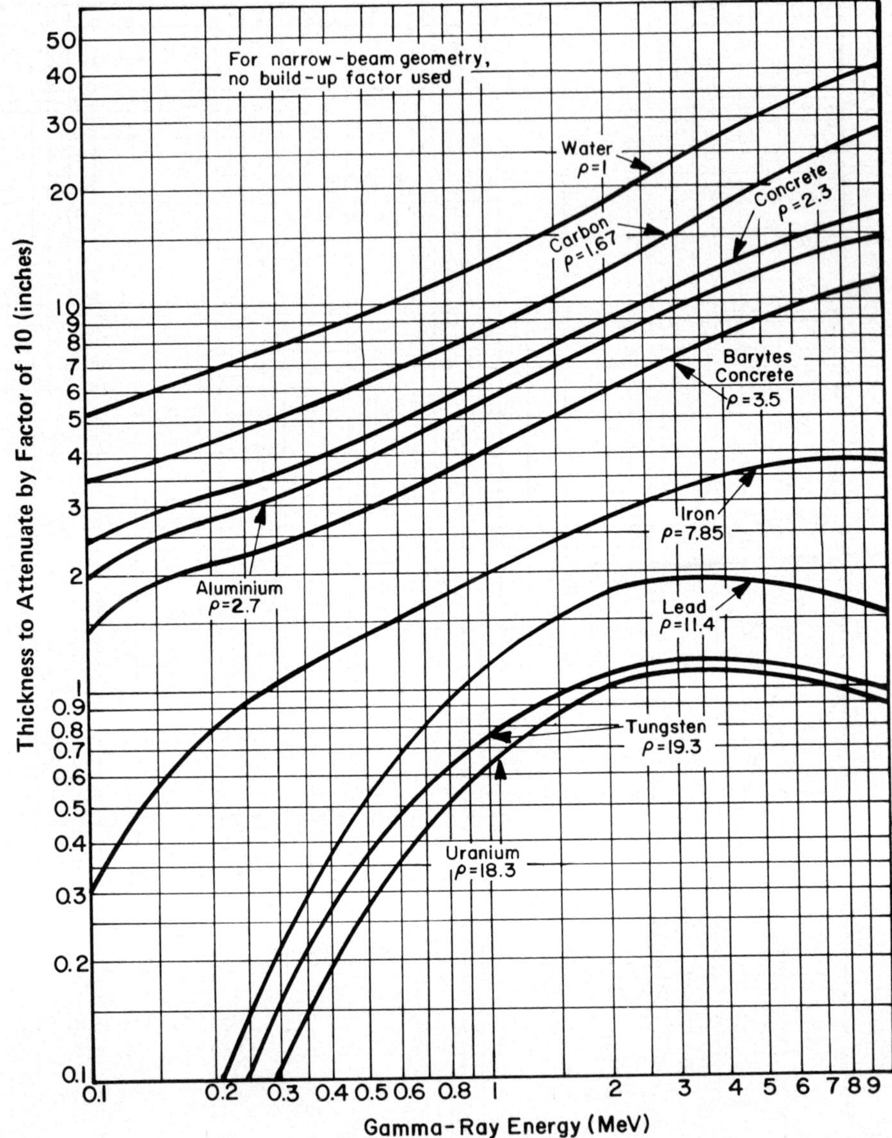

Fig. A-9. The thickness of a given material necessary to attenuate by a factor of 10 a narrow monenergetic beam of γ-rays of energy indicated on the abscissa. The mass density ρ of each material is also indicated relative to water. (From Moteff, 1955.)

data and new tabulations on radiation interactions may be found in the monthly periodical *Atomic Data and Nuclear Data Tables* published by Academic Press, Inc., New York.

Another important class of interactions which are usually not included in the contribution to total mass attenuation coefficients are photonuclear reactions. Hubbell (1969) and Hayward (1970) should be consulted for discussions and further references on these special reactions.

Figure A-9 is a plot of the approximate thickness of different materials needed to attenuate by 1/10 the intensity of an incident monoenergetic narrow photon beam. The quantity plotted is sometimes called the 'tenth value thickness' and is useful in assessing the gross-shielding properties of various standard shielding materials in laboratory situations.

APPENDIX B

CONVERSION FACTORS FOR ENERGY UNITS

Conversion factors between various units equivalent to energy are given for convenience. The relation establishing the conversion factor is indicated at the bottom of the appropriate column. The average energy of the photons in a Planck distribution is given by $\bar{\epsilon}_{BB} = 2.7kT$, while the average kinetic energy for a particle in a Maxwell-Boltzmann distribution is $\bar{E} = \frac{3}{2}kT$.

See Table B-1 on next page.

TABLE B-1[a]
Conversion Factors for Energy Units.

erg	eV	s^{-1}	cm^{-1}	deg K	g	Atomic mass units (u)	cal
1 erg	6.2418×10^{11}	1.50929×10^{26}	5.0345×10^{15}	7.2436×10^{15}	1.112649×10^{-21}	6.7010×10^{2}	2.3901×10^{-8}
1.60210×10^{-12}	1 eV	2.41804×10^{14}	8.0657×10^{3}	1.16049×10^{4}	1.78258×10^{-33}	1.07356×10^{-9}	3.8291×10^{-20}
6.6256×10^{-27}	4.13558×10^{-15}	1 s^{-1}	3.335640×10^{-11}	4.7992×10^{-48}	7.3720×10^{-11}	4.4398×10^{-24}	1.5836×10^{-34}
1.9863×10^{-16}	1.23981×10^{-4}	2.997925×10^{10}	1 cm^{-1}	1.4388	2.2101×10^{-37}	1.33101×10^{-13}	4.7474×10^{-24}
1.3805×10^{-16}	8.6171×10^{-5}	2.0836×10^{10}	6.9503×10^{-1}	1 K	1.5361×10^{-37}	9.2509×10^{-14}	3.2996×10^{-24}
8.987554×10^{20}	5.6098×10^{32}	1.35649×10^{47}	4.5248×10^{36}	6.5102×10^{36}	1 g	6.0225×10^{23}	2.1481×10^{13}
1.49232×10^{-3}	9.31478×10^{8}	2.2524×10^{23}	7.5131×10^{12}	1.0810×10^{13}	1.66043×10^{-24}	1 u	3.5667×10^{-11}
4.1840×10^{7}	2.6116×10^{19}	6.3149×10^{33}	2.1064×10^{23}	3.0307×10^{23}	4.6554×10^{-14}	2.8037×10^{10}	1 cal
		$E = h\nu = hc/\lambda$		kT	$E = mc^2$		

[a] From Lederer *et al.*, 1968.

REFERENCES

Adams, N. and Braddick, H.J.: 1951, *Z. Naturforsch.* **6a,** 592.

Adams, J.A.S. and Lowder, W.M.: 1964, *The Natural Radiation Environment,* University of Chicago Press, Chicago, p. 1033.

Adler, I. and Trombka, J.I.: 1970, in J.G. Roederer and J. Zahringer (eds.), *Physics and Chemistry in Space,* Vol. 3, Springer-Verlag, New York.

Agrinier, B., Forichon, M., Leray, J.P., Parlier, B., Montmerle, T., Boella, G., Maraschi, L., Sacco, B., Scarsi, L., Da Costa, J.M., and Palmieri, R.: 1973, *Proceedings 13th International Conference Cosmic Rays,* Vol. 1, University of Denver, Denver, Colorado, p. 8.

Ahmed, K.U., and Cochran, R.G.: 1973, *Nucl. Tech.* **17,** 66.

Alburger, D.: 1966, in K. Siegbahn (ed.), *Alpha-, Beta-, and Gamma-Ray Spectroscopy,* Vol. 1, North-Holland Publishing Company, Amsterdam, p. 745.

Alfvén, H.: 1965, *Rev. Mod. Phys.* **37,** 652.

Allen, C.W.: 1973, *Astrophysical Quantities,* 3rd Edition, Athlone Press, London.

Aller, L.H.: 1963, *Astrophysics (The Atmospheres of the Sun and Stars),* 2nd Edition, The Ronald Press Company, New York, p. 111.

Alvarez, L.W., Dauber, P.M., Smith, L.H., Buffington, A., Derenzo, S.E., Muller, R.A., Orth, C., and Smoot, G.: 1973, *The Liquid Xenon Compton Telescope: A New Technique for Gamma-Ray Astronomy,* Series 14, Issue 17, Space Sciences Laboratories Berkeley, California.

Anderson, K.A.: 1961, *Phys. Rev.* **123,** 1435.

Anglin, J.D., Dietrich, W.F., and Simpson, J.A.: 1973, in R. Ramaty and R.G. Stone (eds.), *High Energy Phenomena in the Sun, NASA SP-342,* p. 315.

Apparao, M., Krishna, V., Daniel, R.R., Vijayalakshmi, B., and Bhatt, V.L.: 1966, *J. Geophys. Res.* **71,** 1781.

Armstrong, T.W., Chandler, K.C., and Barish, J.: 1973, *J. Geophys. Res.* **78,** 2715.

Arnold, J.R., Metzger, A.E., Anderson, E.C., and Van Dilla, M.A.: 1962, *J. Geophys. Res.* **67,** 4878.

Arons, J.: 1971a, *Astrophys. J.* **164,** 437.

Arons, J.: 1971b, *Astrophys. J.* **164,** 457.

Arons, J. and McCray, R.: 1969, *Astrophys. J. (Letters)* **158,** L91.

Arons, J., McCray, R., and Silk, J.: 1971, *Astrophys. J.* **170,** 431.

Baade, W., Burbidge, G.R., Hoyle, F., Burbidge, E.M., Christy, R.F., and Fowler, W.A.: 1956, *Publ. Astron. Soc. Pacific* **68,** 296.

Baird, G.A., Delaney, T.J., Lawless, B.G., Griffiths, D.J., Shakeshaft, J.R., Drever, R.W.P., Meikle, W.P.S., Jelley, J.V., Charman, W.N., and Spencer, R.E.: 1975, *Astrophys. J. (Letters)* **196,** L11.

Bartholomew, G.A., Doveika, A., Eastwood, K.M., Monaro, S., Groshev, L.V., Demidov, A.M., Pelekhov, V.I., and Sokolovskii, L.L.: 1967, *Nuclear Data* **3A,** 367.

Becquerel, H.: 1896, *Compt. Rend. Acad. Sci. Paris* **122,** 420, 501, 1086.

Beers, Y.: 1957, *Introduction to the Theory of Error,* Addison-Wesley Publishing Company, Inc., Reading, Mass., p. 46.

Bell, R.E.: 1966, in K. Siegbahn (ed.), *Alpha-, Beta-, and Gamma-Ray Spectroscopy,* Vol. 2, North-Holland Publishing Company, Amsterdam, p. 905.

Bennett, K., Penego, P., Rochester, G.K., Sanderson, T.R., and Sood, R.K.: 1972, *Nature* **238,** 31.

Bertini, H.W.: 1963, *Phys. Rev.* **131,** 1801.

Bertini, H.W.: 1965, *Phys. Rev.* **138,** AB2.

Bertini, H.W.: 1966, *Nucl. Phys.* **87,** 138.

Bethe, H.A.: 1934, *Proc. Camb. Philos. Soc.* **30,** 524.

Bethe, H.A. and Ashkin, J.: 1953, in E. Segré (ed.), *Experimental Nuclear Physics,* Vol. 1, John Wiley and Sons, Inc., New York, p. 166.

Bethe, H.A. and Heitler, W.: 1934, *Proc. Roy. Soc.* **A146,** 83.

Bethe, H.A. and Maximon, L.C.: 1954, *Phys. Rev.* **93,** 768.

Beuermann, K.P.: 1971, *J. Geophys. Res.* **76,** 4291.

Biermann, L., Haxel, O., and Schülter, A.: 1951, *Z. Naturforsch.* **6a,** 47.

Blatt, J.M. and Weisskopf, V.F.: 1952, *Theoretical Nuclear Physics,* John Wiley and Sons, Inc., New York.

Bodansky, D., Clayton, D.D., and Fowler, W.A.: 1968a, *Phys. Rev. (Letters)* **20,** 161.

Bodansky, D., Clayton, D.D., and Fowler, W.A.: 1968b, *Astrophys. J. Suppl.* **16,** 299.

Boldt, E.: 1969, in H. Ögelman and J.R. Wayland (eds.), *Lectures in High Energy Astrophysics, NASA SP-199,* p. 49.

Boldt, E. and Serlemitsos, P.: 1969, *Astrophys. J.* **157,** 557.

Boldt, E., Holt, S., Rothschild, R., and Serlemitsos, P.: 1974, *What is Special About Cygnus X-1*?, *GSFC X-661-74-323,* and 1974, in D. Venkatesan (ed.), *Proceedings of International Conference on X-Rays in Space,* University of Calgary, Canada.

Borsellino, A.: 1953, *Phys. Rev.* **89,** 1023.

Brancazio, P. and Cameron, A.G.W. (eds.): 1969, *Supernovae and Their Remnants,* Gordon and Breach Publishing Company, New York.

Brecher, K. and Morrison, P.: 1969, *Phys. Rev. (Letters)* **23,** 802.

Brown, J.C.: 1971, *Solar Phys.* **18,** 489.

Brown, R.L.: 1970a, *Astrophys. J. (Letters)* **159,** L187.

Brown, R.L.: 1970b, *Lettere al Nuovo Cimento* **4,** 941.

Brown, R.T.: 1973, *Astrophys. J.* **179,** 607.

Browning, R., Ramsden, D., and Wright, P.J.: 1972, *Nature* **238,** 138.

Brysk, H. and Zerby, C.D.: 1968, *Phys. Rev.* **171,** 292.

Burbidge, G.: 1966, in L. Gratton (ed.), *High Energy Astrophysics,* Academic Press, New York, p. 145.

Burbidge, G.R., Hoyle, F., Burbidge, E.M., Christy, R.F., and Fowler, W.A.: 1956, *Phys. Rev.* **103,** 1145.

Burbidge, E.M., Burbidge, G.R., Fowler, W.A., and Hoyle, F.: 1957, *Rev. Mod. Phys.* **29,** 547.

Burrus, W.R.: 1960, *IRE Trans. Nucl. Sci.* **NS-7,** 102.

Buselli, G., Clancy, M.C., Davison, P.J.M., Edwards, P.J., McCracken, K.G., and Thomas, R.M.: 1968, *Nature* **219,** 1124.

Butler, J.W.: 1959, *Table of* (p, γ) *Resonances, NRL-5282,* Naval Research Laboratory, Washington, D.C.

Cameron, A.G.W.: 1967, in L.H. Ahrens (ed.), *Origin and Distribution of Elements,* Pergamon Press, London, p. 125.

Cameron, A.G.W.: 1973, *Space Sci. Rev.* **15,** 121.

Canizares, C.R., Neighbours, J.E., and Matilsky, T.: 1974, *Astrophys. J. (Letters)* **192,** L61.

Carpenter, G. and Dyer, C.S.: 1973, *Astrophys. Space Sci.* **14,** 95.

Castelli, J.P., Barron, W.R., and Badillo, V.L.: 1973, *World Data Center A, Report UAG-28,* Part I, p. 183.

Cavallo, G. and Gould, R. J.: 1971, *Nuovo Cimento* **28,** 77.

Chang, C.C., Wall, N.S., and Frankel, Z.: 1974, *Phys. Rev. (Letters)* **33,** 1493.

Cheng, C.: 1972, *Space Sci. Rev.* **13,** 3.

Chiu, H.: 1968, *Stellar Physics* Blaisdell Publishing Company, Waltham, Mass.

Chubb, T.A., Kreplin, R.W. and Friedman, H.: 1966, *J. Geophys. Res.* **71,** 3611.

Chupp, E.L.: 1963, in W.N. Hess (ed.), *AAS-NASA Symposium on the Physics of Solar Flares, NASA SP-50,* p. 445.

Chupp, E.L.: 1971, *Space Sci. Rev.* **12,** 486.

Chupp, E.L., Sarkady, A.A., and Gilman, H.P.: 1967, *Planet. Space Sci.* **15,** 881.

Chupp, E.L., Forrest, D.J., Sarkady, A.A., and Lavakare, P.J.: 1970, *Planet. Space Sci.* **18,** 939.

Chupp, E.L., Forrest, D.J., and Suri, A.N.: 1973a, in R. Ramaty and R.G. Stone (eds.), *High Energy Phenomena on the Sun, NASA SP-342,* p. 285.

Chupp, E.L., Forrest, D.J., Higbie, P.R., Suri, A.N., Tsai, C., and Dunphy, P.P.: 1973b, *Nature* **241,** 333.

Chupp, E.L., Forrest, D.F., and Suri, A.N.: 1974a, in M.J. Rycroft and R.D. Reasenberg (eds.), *Space Research XIV,* Akademie-Verlag, Berlin, p. 463.

Chupp, E.L., Forrest, D.J., Suri, A.N., Adams, R., and Tsai, C.: 1974b, in C.B. Cosmovici (ed.), *Supernovae and Supernova Remnants,* D. Reidel Publishing Company, Dordrecht, Holland, p. 311.

Chupp, E.L., Forrest, D.J., and Suri, A.N.: 1975, in S.R. Kane (ed.), *Solar Gamma-, X-, and EUV Radiation, IAU Symposium No. 68,* D. Reidel Publishing Company, Dordrecht, Holland, p. 341.

Clark, G.: 1971, in F. LaBuhn and R. Lüst (eds.), *New Techniques in Space Astronomy, IAU Symposium No. 41,* D. Reidel Publishing Company, Dordrecht, Holland, p. 3.

Clark, G.W., Garmire, G.P., and Kraushaar, W.L.: 1968, *Astrophys. J. (Letters)* **153,** L203.

Clayton, D.D.: 1968, *Principles of Stellar Evolution and Nucleosynthesis,* McGraw-Hill Book Company, New York.

Clayton, D.D.: 1971, *Nature* **234,** 291.

Clayton, D.D.: 1972, *Nucleosynthesis: Origin and Abundances of the Elements.* A contribution to Encyclopedia of the Twentieth Century (Istituto della Enciclopedia Italiana, Roma) (preprint circulated by D.D. Clayton.)

Clayton, D.D.: 1974, *Astrophys. J.* **188,** 155.

Clayton, D.D. and Craddock, W.L.: 1965, *Astrophys. J.* **142,** 189.

Clayton, D.D. and Fowler, W.A.: 1969, *Comments on Astrophys. and Space Phys.* **1,** No. 4, 147.

Clayton, D.D. and Hoyle, F.W.: 1974, *Astrophys. J. (Letters)* **187,** L101.

Clayton, D.D. and Silk, J.: 1969, *Astrophys. J. (Letters)* **158,** L43.

Clayton, D.D. and Woosley, S.E.: 1969, *Astrophys. J.* **157,** 1381.

Clayton, D.D., Colgate, S.A., and Fishman, G.J.: 1969, *Astrophys. J.* **155,** 75.

Cline, T.L. and Desai, U.D.: 1975, *Astrophys. J. (Letters)* **196,** L43.

Cline, T.L. and Hones, E.W.: 1968, *Can. J. Phys.* **46,** S527.

Cline, T.L., Desai, U.D., Klebesadel, R.W., and Strong, I.B.: 1973, *Astrophys. J. (Letters)* **185,** L1.

Cohen, B.L.: 1971, *Concepts of Nuclear Physics,* McGraw-Hill Book Company, New York.

Colgate, S.A.: 1968, *Can. J. Phys.* **46,** S476.

Colgate, S.A.: 1970, *Acta Physica Academiae Scientiarum Hungaricae,* Vol. 29, Suppl. 1, p. 353.

Colgate, S.A.: 1973, *Astrophys. J. (Letters)* **181,** L53.

Colgate, S.A.: 1974a, *Astrophys. J.* **187,** 333.

Colgate, S.A.: 1974b, *Astrophys. J.* **187,** 321.

Colgate, S.A. and McKee, C.: 1969, *Astrophys. J.* **157,** 623.

Colgate, S.A., Grasberger, W.H., and White, R.H.: 1961, *University of California Radiation Laboratory, Report No. 6741,* Berkley, California.

Compton, A.H. and Allison, S.K.: 1935, *X-Rays in Theory and Experiment,* D. Van Nostrand Company, New York.

Comstock, G.M., Hsieh, K.C., and Simpson, J.A.: 1972, *Astrophys. J.* **173,** 691.

Cortellesa, P., Benedetto, D.P., and Paizis, C.: 1971, *Solar Phys.* **20,** 474.

Cosmovici, C.B. (ed.): 1974, *Supernovae and Supernova Remnants,* D. Reidel Publishing Company, Dordrecht, Holland.

Cowsik, R.: 1971, *Proceedings 12th International Conference Cosmic Rays*, Vol. 1, Hobart, Tasmania, p. 334.

Cowsik, R.: 1973, in F.W. Stecker and J.I. Trombka (eds.), *Gamma-Ray Astrophysics, NASA SP-339,* p. 185.

Crannell, C.J., McClintock, J.E., and Moffett, T.: 1974, *Nature* **252,** 659.

Critchfield, C.P., Ney, E.P., and Oleksa, S.: 1952, *Phys. Rev.* **85,** 461.

Croom, D.L. and Harris, L.D.J.: 1973, *World Data Center A, Report UAG-28,* Part I, p. 210.

Cruddace, R., Silk, J., and Reeves, H.: 1972, *Bull. Am. Astron. Soc.* **4,** 258.

Dahlbacka, G.H., Freier, P.S., and Waddington, C.J.: 1973, *Astrophys. J.* **180,** 371.

Dahlbacka, G.H., Chapline, G.F., and Weaver, T.A.: 1974, *Nature* **250,** 36.

Dalgarno, A. and McCray, R.A.: 1972, in L. Goldberg (ed.), *Ann. Rev. of Astron. and Astrophys.* Vol. 10, Annual Reviews, Inc., Palo Alto, California, p. 375.

Daniel, R.R. and Stephens, S.A.: 1974, *Revs. of Geophys. and Space Phys.* **12,** 233.

Daniel, R.R., Joseph, G., Lavakare, P.J., and Sunderrajan, R.: 1967, *Nature* **213,** 21.

Daniel, R.R., Gokhale, G.S., Joseph, G., and Lavakare, P.J.: 1971, *J. Geophys. Res.* **76,** 3152.

Daniel, R.R., Joseph, G., and Lavakare, P.J.: 1972, *Astrophys. Space Sci.* **18,** 462.

Davies, H., Bethe, H.A., and Maximon, L.C.: 1954, *Phys. Rev.* **93,** 788.

Davisson, C.M.: 1966, in K. Siegbahn (ed.), *Alpha-, Beta-, and Gamma-Ray Spectroscopy,* Vol. 1, North-Holland Publishing Company, Amsterdam, p. 37.

Davisson, C.M. and Evans, R.D.: 1952, *Revs. Mod. Phys.* **24,** 79.

Dennis, B.R., Suri, A.N., and Frost, K.J.: 1973, *Astrophys. J.* **186,** 97.

Dere, K.P., Horan, D.M., and Kreplin, R.W.: 1973, *World Data Center A, Report UAG-28,* Part II, p. 298.

Deutsch, M.: 1951, *Phys. Rev.* **82,** 455.

Dilworth, C., Maraschi, L., and Perola, G.C.: 1968, *Nuovo Cimento* **56B,** 334.

Dirac, P.A.M.: 1930, *Proc. Camb. Philos. Soc.* **26,** 361.

Dolan, J.F. and Fazio, G.G.: 1965, *Rev. of Geophys.* **3,** 319.

Donahue, T.M.: 1951, *Phys. Rev.* **84,** 972.

Dovshenko, O.I. and Pomanskii, A.A.: 1964, *Sov. Phys. JETP* **18,** 187.

Dunphy, P.P., Forrest, D.J., Chupp, E.L., and Dyer, C.S.: 1975, Conference papers for *14th International Cosmic Ray Conference,* Max-Planck Institute for Extraterrestrial Physics, Munich, Vol. 9, p. 3116.

Duthie, J.G.: 1968, *Can. J. Phys.* **46,** S401.

Dyer, C.S.: 1973, in F.W. Stecker and J.I. Trombka (eds.), *Gamma-Ray Astrophysics, NASA SP-339,* p. 83.

Dyer, C.S. and Morfill, G.E.: 1971, *Astrophys. Space Sci.* **14,** 243.

Dyer, C.S., Engel, A.R., and Quenby, J.J.: 1972, *Astrophys. Space Sci.* **19,** 359.

Dyer, C.S., Dunphy, P.P., Forrest, D.J., and Chupp, E.L.: 1975, Conference papers for *14th International Cosmic Ray Conference,* Max-Planck Institute for Extraterrestrial Physics, Munich, Vol. 9, p. 3122.

Ehmert, A.: 1948, *Z. Naturforsch.* **3a,** 264.

Ehrmann, C.H., Fichtel, C.E., Kniffen, D.A., and Ross, R.W.: 1967, *Nucl. Inst. and Methods* **56,** 109.

Eisenbud, M.: 1973, *Environmental Radioactivity,* Academic Press, New York.

Elliot, H.L.: 1964, *Planet. Space Sci.* **12,** 657.

Elliot, H.: 1969, in C. DeJager and Z. Svestka (eds.), *Solar Flares and Space Research,* North-Holland Publishing Company, Amsterdam, p. 356.

Elliot, H.: 1973, in R. Ramaty and R.G. Stone (eds.), *High Energy Phenomena on the Sun, NASA SP-342,* p. 12.

Evans, R.D.: 1955, *The Atomic Nucleus,* McGraw-Hill Book Company, New York.

Evans, R.D.: 1958, in S. Flugge (ed.), *Handbuch der Physik,* Vol. XXXIV, Springer-Verlag, Berlin, p. 218.

Evans, R.D.: 1968, in F.H. Attix and W.C. Roesch (eds.), *Radiation Dosimetry 1,* 2nd Edition, Academic Press, New York, p. 93.

Eyles, C.J., Linney, A.D., and Rochester, G.K.: 1972, *Solar Phys.* **24,** 483.

Fano, U.: 1953a, *Nucleonics* **11,** No. 8, 8.

Fano, U.: 1953b, *Nucleonics* **11,** No. 9, 55.

Fazio, G.G.: 1967, in L. Goldberg (ed.), *Ann. Rev. of Astron. and Astrophys.* Vol. 5, Annual Reviews, Inc., Palo Alto, California, p. 481.

Fazio, G.G.: 1970, *Nature* **225,** 905.

Fazio, G.G. and Stecker, F.W.: 1970, *Nature* **226,** 135.

Fazio, G.G., Helmken, H.F., Cavrak, S.J., Jr., and Hern, D.R.: 1968, *Can. J. Phys.* **46,** 427.

Feenberg, E. and Primakoff, H.: 1948, *Phys. Rev.* **73,** 449.

Felten, J.E. and Morrison, P.: 1963, *Phys. Rev. (Letters)* **10,** 453.

Felten, J.E. and Morrison, P.: 1966, *Astrophys. J.* **146,** 686.

Fichtel, C.E.: 1974, *Primary Gamma Rays, Report GSFC X-662-74-57,* Goddard Space Flight Center, Greenbelt, Md.

Fichtel, C.E., Cline, T.L., Ehrmann, C.H., Kniffen, D.A., and Ross, R.W.: 1968, *Can. J. Phys.* **46,** S419.

Fichtel, C.E., Kniffen, D.A., and Ögelman, H.B.: 1969, *Astrophys. J.* **158,** 193.

Fichtel, C.E., Hartman, R.C., Kniffen, D.A., and Sommer, M.: 1972 *Astrophys. J.* **171,** 31.

Fichtel, C.E., Kniffen, D.A., and Hartman, R.C.: 1973, *Astrophys. J. (Letters)* **186,** L99.

Fichtel, C.E., Hartman, R.C., Kniffen, D.A., Thompson, D.J., Bignami, G.F., Ögelman, H., Özel, M.E., and Tümer, T.: 1974, *Report X-662-74-304,* Goddard Space Flight Center, Greenbelt, Md.

Fichtel, C.E., Hartman, R.C., Kniffen, D.A., Thompson, D.J., Bignami, G.F., Ögelman, H., Özel, M.E., and Tümer, T.: 1975, *Astrophys. J.* **198,** 163.

Fireman, E.L.: 1963, in W.N. Hess (ed.), *AAS-NASA Symposium on the Physics of Solar Flares, NASA SP-50,* p. 279.

Fireman, E.L., DeFelice, J., and Tilles, D.: 1961, *Phys. Rev.* **123,** 1935.

Fishman, G.J.: 1972a, *Astrophys. J.* **171,** 163.

Fishman, G.J.: 1972b, *Proton Induced Radioactivity in* NaI(Tl) *Scintillation Detectors, Summary Report No. SE-SSL-1497,* Teledyne Brown Engineering, Huntsville, Ala.

Fishman, G.J.: 1973, in F.W. Stecker and J.I. Tromkba (eds.), *Gamma-Ray Astrophys, NASA SP.339,* p. 61.

Fishman, G.J.: 1974, *Neutron and Proton Activation Measurements from Skylab, AIAA/AGU Conference on Scientific Experiments of Skylab, AIAA Paper No. 74-1227,* Huntsville, Alabama, American Institute of Aeronautics and Astronautics, New York.

Fishman, G.J. and Clayton, D.D.: 1972, *Astrophys. J.* **178,** 337.

Flamm, E., Lingenfelter, R.E., MacDonald, G.J.F., and Libby, W.F.: 1962, *Science* **138,** 48.

Forrest, D.J., Higbie, P.R., Orwig, L.E., and Chupp E.L.: 1972, *Nucl. Instr. and Methods* **101,** 567.

Forrest, D.J., Chupp, E.L., and Suri, A.N.: 1975a, *Proceedings International Conference on X-Rays in Space,* University of Calgary, Calgary, Alberta, Canada, p. 341.

Forrest, D.J., Chupp, E.L., and Gleske, I.U.: 1975b, Internal Report, Department of Physics, University of New Hampshire, Durham, N.H.

Fowler, W.A. and Hoyle, F.: 1964, *Astrophys. J. Supplement* **9,** 201.

Fowler, W.A., Caughlan, G.R., and Zimmerman, B.A.: 1967, *Ann. Rev. of Astron. and Astrophys.* Vol. 5, p. 525.

Fowler, W.A., Reeves, H., and Silk, J.: 1970, *Astrophys. J.* **162,** 49.

Frost, K.J., Rothe, E.D., and Peterson, L.E.: 1966, *J. Geophys. Res.* **71,** 4079.

Frye, G.M., Jr. and Smith, L.H.: 1966, *Phys. Rev. (Letters)* **17,** 733.

Frye, G.M., Jr., Reines, F., and Armstrong, A.H.: 1966, *J. Geophys. Res.* **71,** 3119.

Frye, G.M., Jr., Staib, J.A., Zych, A.D., Hopper, V.D., Rawlinson, W.R., and Thomas, J.A.: 1969, *Nature* **223,** 1320.

Frye, G.M., Jr., Albats, P.A., Zych, A.D., Staib, J.A., Hopper, V.D., Rawlinson, W.R., and Thomas, J.A.: 1971, *Nature* **231,** 372.

Fukada, Y., Hayakawa, S., Kashara, I., Makino, F., and Tanaka, Y.: 1975, *Nature* **254,** 398.

Funk, H. and Rowe, M.W.: 1967, *Earth Planet. Sci. (Letters)* **2,** 215.

Garmire, G. and Kraushaar, W.L.: 1965, *Space Sci. Rev.* **4,** 123.

Giaconni, R., Gursky, H., Paolini, F.R., and Rossi, B.B.: 1962, *Phys. Rev. (Letters)* **9,** 439.

Giaconni, R., Murray, S., Gursky, H., Kellogg, E., Schreier, E., and Tananbaum, H.: 1972, *Astrophys. J.* **178,** 281.

Ginzburg, V.L.: 1969, *Elementary Processes for Cosmic Ray Astrophysics,* Gordon and Breach Publishing Company, New York.

Ginzburg, V.L. and Syrovatskii, S.I.: 1964a, *Soviet Phys. – Usp.* **84,** 201. [1965, *Soviet Phys. – Usp.* (English Translation) 7, 696.]

Ginzburg, V.L. and Syrovatskii, S.I.: 1964b, *The Origin of Cosmic Rays,* Macmillan Company, New York.

Gold, T.: 1968, *Nature* **218,** 731.

Goldberg, L., Mohler, O.C., and Müller, E.A.: 1958, *Astrophys. J.* **127,** 302.

Golenetskii, S.V., Mazets, E.P., Il'inskii, V.N., Aptekar, R.L., Bredov, M.M., Gur'yan, Yu. A., and Panov, V.A.: 1971, *Astrophys. Letters* **9,** 69.

Gorenstein, P. and Gursky, H.: 1970, *Space Sci. Rev.* **10,** 770.

Grannan, R.T., Koga, R., Millard, W.A., Preszler, A.M., Simnett, G.M., and White, R.S.: 1972, *A Large Area Detector for Neutrons Between 2 and 100 MeV, IGPP-UCR-72-2,* University of California, Riverside, Riverside, Calif.

Green, J. and Lee, J.: 1964, *Positronium Chemistry,* Academic Press, New York.

Greene, J.: 1959, *Astrophys. J.* **130,** 693.

Greisen, K.: 1966a, in R.E. Marshak (ed.), *Perspectives in Modern Physics,* John Wiley and Sons, New York, p. 355.

Greisen, K.: 1966b, *Phys. Rev. (Letters)* **16,** 748.

Greisen, K.: 1971, *The Physics of Cosmic X-Ray, γ-Ray, and Particle Sources,* Gordon and Breach Publishing Company, New York.

Grindlay, J.E.: 1970, *Astrophys. J.* **162,** 187.

Grindlay, J.E.: 1972, *Astrophys. J. (Letters)* **174,** L9.

Grindlay, J.E., Wright, E.L., and McCrosky, R.E.: 1974, *Astrophys. J. (Letters)* **192,** L113.

Gruber, D.E.: 1974, Doctoral Thesis, University of California San Diego, La Jolla, Calif.

Guthrie, M.P.: 1969, *EVAP-2 and EVAP-3: Modifications of a Code to Calculate Evaporation from Excited Compound Nuclei, ORNL-4379,* Oak Ridge National Laboratory, Oak Ridge, Tenn.

Guthrie, P. and Tademaru, E.: 1973, *Nature* **241,** 77.

Harrison, E.R.: 1967, *Phys. Rev. (Letters)* **18,** 1011.

Harwit, M.: 1973, *Astrophysical Concepts,* John Wiley and Sons, New York.

Hayakawa, S.: 1969, *Cosmic Ray Physics,* Wiley-Interscience Division of John Wiley and Sons, New York.

Hayakawa, S.: 1970, in L. Gratton (ed.), *Non-Solar X- and Gamma-Ray Astronomy, IAU Symposium No. 37,* D. Reidel Publishing Company, Dordrecht, Holland, p. 372.

Hayakawa, S. and Tanaka, Y.: 1970, in L. Gratton (ed.), *Non-Solar X- and Gamma-Ray Astronomy, IAU Symposium No. 37,* D. Reidel Publishing Company, Dordrecht, Holland, p. 374.

Hayakawa, S., Okuda, H., Tanaka, Y., and Yamamoto, Y.: 1964, *Suppl. Prog. Theoret. Phys.* **30,** 153.

Haymes, R.C., Ellis, D.V., Fishman, G.J., Kurfess, J.D., and Tucker, W.H.: 1968, *Astrophys. J. (Letters)* **151,** L9.

Haymes, R.C., Glenn, S.W., Fishman, G.J., and Harnden, F.R., Jr.: 1969, *J. Geophys. Res.* **74,** 5792.

Haymes, R.C., Walraven, G.D., Meegan, C.A., Hall, R.D., Djuth, F.T., and Shelton, D.H.: 1975, *Astrophys. J.* **201,** 593.

Hayward, R.W.: 1967, in E.U. Condon and H. Odishaw (eds.), *Handbook of Physics,* 2nd Edition, McGraw-Hill Book Company, New York, pp. 9–106.

Hayward, E.: 1970, *Photonuclear Reactions, NBS118,* National Bureau of Standards, Washington, D.C.

Heath, R.L.: 1964, *Scintillation Spectrometry – Gamma Ray Spectrum Catalogue,* Vol. 2, 2nd ed., AEC Research and Dev. Report TID-4500, Phillips Petroleum Company, Atomic Energy Division, Idaho Operations Office, USAEC.

Heitler, W.: 1954, *The Quantum Theory of Radiation,* Oxford University Press, London.

Hellwege, K.H.: 1961, *Landolt-Börnstein Numerical Data and Functional Relationships in Science and Technology; Group I,* Vol. 1, Springer-Verlag, Berlin.

Herterich, K., Pinkau, K., Rothermel, H., and Sommer, M.: 1973, *13th International Conference Comic Rays,* Vol. 1, University of Denver, Denver, Colo., p. 21.

Herzo, D., Koga, R., Millard, W.A., Moon, S., Ryan, J., Wilson, R., Zych, A.D., and White, R.S.: 1975, *Nucl. Instr. and Methods* **123,** 583.

Hess, W.N., Canfield, E.H., and Lingenfelter, R.E.: 1961, *J. Geophys. Res.* **66,** 665.

Higbie, P.R., Chupp, E.L., Forrest, D.J., and Gleske, I.U.: 1972, *IEEE Trans. Nucl. Sci.* **NS-19,** No. 1, 606.

Higbie, P.R., Forrest, D.J., Gleske, I.U., Chupp, E.L., and Burtis, D.: 1973, *Nucl. Instr. and Methods* **108,** 167.

Hirasima, Y., Okudaira, K. and Yamagami, T.: 1969, *Acta Phys. Hungaria (Suppl. 2)* **29,** 683.

Holt, S.S.: 1967, *J. Geophys. Res.* **72,** 3507.

Hopper, V.D., Mace, O.B., Thomas, J.A., Albats, P., Frye, G.M., Jr., Thomson, G.B., and Staib, J.A.: 1973, *Astrophys. J. (Letters)* **186,** L55.

Hoyle, F. and Fowler, W.A.: 1960, *Astrophys. J.* **132,** 565.

Hoyle, F., Fowler, W.A., Burbidge, G.R., and Burbidge, E.M.: 1964, *Astrophys. J.* **139,** 909.

Hubbell, J.H.: 1969, *Photon Cross Sections, Attenuation Coefficients, and Energy Absorption Coefficients from 10 keV to 100 GeV, NBS-29,* National Bureau of Standards, Washington, D.C.

Hudson, H. and Tsikoudi, V.: 1973, *Nature (Phys. Sci.)* **245**, 88.

Hulsizer, R. and Rossi, B.B.: 1948, *Phys. Rev.* **73**, 1402.

Hutchinson, G.W., Pearce, A.J., Ramsden, D., and Wills, R.D.: 1970, in L. Gratton (ed.), *Non-Solar X- and Gamma-Ray Astronomy, IAU Symposium No. 37,* D. Reidel Publishing Company, Dordrecht, Holland, p. 300.

Imhof, W.L., Nakano, G.H., Johnson, R.G., and Reagan J.B.: 1973, in F.W. Stecker and J.I. Trombka (eds.), *Gamma Ray Astrophysics, NASA SP-339,* p. 77.

Imhof, W.L., Nakano, G.H., Johnson, R.G., Kilner, J.R., Reagan, J.B., Klebesadel, R.W., and Strong, I.B.: 1974, *Astrophys. J. (Letters)* **191**, L7.

Imhof, W.L., Nakano, G.H., Johnson, R.G., Kilner, J.R., Reagan, J.B., Klebesadel, R,W. and Strong, I.B.: 1975, *Astrophys. J.* **198**, 717.

Jackson, J.D.: 1962, *Classical Electrodynamics,* John Wiley and Sons, Inc., New York.

Jacobson, A.S.: 1968, Doctoral Thesis, University of California San Diego, La Jolla, Calif.

Jacobson, A.S.: 1975, Personal communication.

Jacobson, A.S., Bishop, R.J., Culp, G.W., Jung, L., Mahoney, W.A., and Willett, J.B.: 1975, *Nucl. Instr. and Methods,* **127**, 115.

Jelley, J.V.: 1966, *Nature* **211**, 472.

Johansson, S.A.E. and Kleinheinz, P.: 1966, in K. Siegbahn (ed.), *Alpha-, Beta-, and Gamma-Ray Spectroscopy,* Vol. 1, North-Holland Publishing Company, Amsterdam, p. 805.

Johnson, W.N. III, and Haymes, R.C.: 1973, *Astrophys. J.* **184**, 103.

Johnson, W.N. III, Harnden, F.R., Jr., and Haymes, R.C.: 1972, *Astrophys. J. (Letters)* **172**, L1.

Jones, F.C.: 1961, *J. Geophys. Res.* **66**, 2029.

Jones, F.C.: 1965, *Phys. Rev.* **137**, B1306.

Jones, F.C.: 1971, *Cosmic Gamma Rays from Suprathermal Photon Bremsstrahlung, GSFC X-641-71-372,* Goddard Space Flight Center, Greenbelt, Maryland.

Joseph, G.: 1970, Doctoral Thesis, University of Bombay, TATA Institute for Fundamental Research, Bombay, India.

Kanbach, G., Reppin, C., and Schönfelder, V.: 1974, *J. Geophys. Res. (Space Physics)* **79**, 5159.

Kanbach, G., Reppin, C., Forrest, D.J., and Chupp, E.L.: 1975, Conference papers of *14th International Cosmic Ray Conference,* Max-Planck Institute for Extraterrestrial Physics, Munich, Germany, Vol. 5, p. 1644.

Kasturirangan, K., Rao, U.R., and Bhavsar, P.D.: 1972, *Planet. Space Sci.* **20**, 1961.

Kerr, F.J.: 1971, in S.P. Maran, J.C. Brandt, and T.P. Stecher (eds.), *The Gum Nebula and Related Problems, GSFC Report X-683-71-375,* Goddard Space Flight Center, Greenbelt, Md., p. 1.

Kinzer, R.L., Share, G.H., and Seeman, N.: 1974, *J. Geophys. Res. (Space Physics)* **79**, 4567.

Klebesadel, R.W.: 1974, in I.B. Strong (ed.), *Proceedings of Conference on Transient Cosmic Gamma- and X-Ray Sources, Los Alamos Report LA-5505-C,* Los Alamos, N.M., p. 1.

Klebesadel, R.W., Strong, I.B., and Olson, R.A.: 1973, *Astrophys. J. (Letters)* **182**, L85.

Klein, O. and Nishina, Y.: 1929, *Z. Physik* **52**, 853.

Kniffen, D.A. and Fichtel, C.E.: 1970, *Astrophys. J. (Letters)* **161**, L157.

Kniffen, D.A., Hartman, R.C., Thompson, D.J., and Fichtel, C.E.: 1973, *Astrophys. J. (Letters)* **186**, L105.

Koch, D., Gursky, H., Tananbaum, H., and Kellogg, E.: 1974, in I.B. Strong (ed.), *Proceedings of Conference on Transient Cosmic Gamma- and X-Ray Sources, Los Alamos Report LA-5505-C,* Los Alamos, N.M., p. 16.

Koga, R., Simnett, G.M., and White, R.S.: 1974, in B.G. Taylor (ed.), *Proceedings of the 9th ESLAB Symposium, ESRO SP-106,* Frascati, Italy, p. 31.

Kondo, I. and Nagase, F.: 1969, in C. DeJager and Z. Švestka (eds.), *Solar Flares and Space Research,* North-Holland Publishing Company, Amsterdam, p. 314.

Konstantinov, B.P., Golenetskii, S.V., Mazets, E.P., Il'inskii, V.N., Aptekar, R.L., Bredov, M.M., Gur'yan, Yu, A., and Panov, V.A.: 1971, *Investigation of Variations of Annihilation γ-Radiation by the Artificial Earth Satellite 'COSMOS-135' in Connection with the Possible Antimatter Nature of Meteor Streams, UDC 523.15*, Consultants Bureau. Translation of *Kosmicheskie Issledovaniga*, 1970, Vol. 8, p. 931, Plenum Press, New York.

Korchak, A.A.: 1967, *Sov. Astron. – AJ* **11**, 258.

Kowal, C.T.: 1972, *IAU Circular No. 2405,* Central Bureau for Astronomical Telegrams, Smithsonian Astrophysical Observatory, Cambridge, Mass.

Kozlovsky, B. and Ramaty, R.: 1974, *Astron. and Astrophys.* **34,** 477.

Kraushaar, W.L., Clark, G.W., Garmire, G.P., Helmken, H., Higbie, P., and Agogino, M.: 1965, *Astrophys. J.* **141,** 845.

Kraushaar, W.L., Clark, G.W., Garmire, G.P., Borken, R., Higbie, P., Leong, V., and Thorsas, T.: 1972, *Astrophys. J.* **177,** 341.

Kreger, W.E. and Mather, R.L.: 1967, in S.M. Shafroth (ed.), *Scintillation Spectroscopy of Gamma Radiation,* Vol. 1, Gordon and Breach Publishers, London, p. 33.

Kuo, F., Frye, G.M., Jr., and Zych, A.D.: 1973, *Astrophys. J. (Letters)* **186,** L51.

Kurfess, J.D.: 1971, *Astrophys. J. (Letters)* **168,** L39.

Kuzhevskii, B.M.: 1969, *Sov. Astron. – AJ* **12,** 595.

Lange, I. and Forbush, S.E.: 1942, *Terrest. Magnetism Atmos. Elec.* **47,** 185 (cf. also Forbush, S.E.: 1946, *Phys. Rev.* **70,** 771.)

Laros, J.G.: 1973, Doctoral Thesis, University of California San Diego, La Jolla, Calif.

Laros, J.G., Matteson, J.L., and Pelling, R.M.: 1973, *Astrophys. J.* **179,** 375.

Leavitt, C.P., Robb, D.S., and Young, F.: 1972, *Bull. Am. Phys. Soc.* **17,** 687.

Lederer, C.M., Hollander, J.M., and Perlman, I.: 1968, *Table of Isotopes,* 6th Edition, John Wiley and Sons, New York. (Corrected printing.)

Leventhal, M.: 1973a, *Astrophys. J. (Letters)* **183,** L147.

Leventhal, M.: 1973b, *Nature (Phys. Sci.)* **246,** 136.

Levy, D.J. and Goldsmith, D.W.: 1972, *Astrophys. J.* **177,** 643.

Lewin, W.H.G., Ricker, G.R., and McClintock, J.E.: 1971, *Astrophys. J. (Letters)* **169,** L17.

L'Heureux, J.: 1974, *Astrophys. J. (Letters)* **187,** L53.

Light, E.S., Merker, M.E., Verschell, H.J., Mendell, R.B., and Korff, S.A.: 1973, *J. Geophys. Res.* **78,** 2741.

Lin, W.C., Venkatesan, D., and Van Allen, J.A.: 1963, *J. Geophys. Res.* **68,** 4885.

Lindskog, J., Sundstrom, T., and Sparrman, P.: 1966, in K. Siegbahn (ed.), *Alpha-, Beta-, and Gamma-Ray Spectroscopy,* Vol. 2, North-Holland Publishing Company, Amsterdam, Holland, p. 1599.

Ling, J.C.: 1974, Doctoral Thesis, University of California San Diego, La Jolla, Calif.

Ling, J.C.: 1975, *J. Geophys. Res. (Space Physics),* **80,** 3241.

Lingenfelter, R.E.: 1963, *J. Geophys. Res.* **68,** 5633.

Lingenfelter, R.E.: 1969, *Sol. Phys.* **8,** 341.

Lingenfelter, R.E. and Ramaty, R.: 1967, in B.S.P. Shen (ed.), *High-Energy Nuclear Reactions in Astrophysics,* W.A. Benjamin, Inc., New York, p. 99.

Lockwood, J.A.: 1973, *Space Sci. Rev.* **14,** 663.

Lockwood, J.A., Ifedili, S.O., and Jenkins, R.W.: 1973, *Sol. Phys.* **30,** 183.

Lovell, B.: 1971, *Q. J. Roy. Astron. Soc.* **12,** 98.

Lüst, R., and Pinkau, K.: 1967, in J.G. Emming (ed.), *Electromagnetic Radiation in Space,* Springer-Verlag, New York, p. 231.

Luyten, W.J.: 1949, *Astrophys. J.* **109,** 532.

Maglich, B. (ed.): 1974, *Adventures in Experimental Physics,* Vol. 4, World Science Education, Princeton, N.J.

Maran, S.P., Brandt, J.C., and Stecher, T.P.: 1971, *The Gum Nebula and Related Problems, NASA/GSFC Report X-683-71-735,* Goddard Space Flight Center, Greenbelt, Md.

Marinelli, L.D., Miller, C.E., May, H.A. and Rose, J.E.: 1962, in C.A. Tobias and J.H. Lawrence (eds.), *Advances in Biological and Medical Physics,* Volume VIII, Academic Press, New York, p. 131.

Marion, J.B., Arnette, T.L., and Owens, H.C.: 1959, *Tables for the Transformation Between the Laboratory and Center-of-Mass Coordinate Systems and for the Calculations of the Energies of Reaction Products, AEC, ORNL 2574,* Oak Ridge National Laboratory, Oak Ridge, Tenn.

Marmier, P. and Sheldon, E.: 1969, *Physics of Nuclei and Particles,* Vol. I, Academic Press, New York.

Marmier, P. and Sheldon, E.: 1970, *Physics of Nuclei and Particles,* Vol. II, Academic Press, New York.

Massey, H.S.W. and Mohr, C.B.: 1954, *Proc. Phys. Soc. London* **A67,** 695.

Matteson, J.L.: 1974, in I.B. Strong (ed.), *Proceedings Conference on Transient Cosmic Gamma- and X-Ray Sources, LA-5505-C,* Los Alamos Scientific Laboratory, Los Alamos, N.M., p. 237.

Maximon, L.C. and Bethe, H.A.: 1952, *Phys. Rev.* **87,** 156.

May, T.C. and Waddington, C.J.: 1969, *Astrophys. J.* **156,** 437.

Mazets, E.P., Golenetskii, S.V., Il'inskii, V.N.: 1973, *A Cosmic Gamma-Ray Burst Detected on Kosmos-461 Satellite,* preprint 454, Academy of Science of the U.S.S.R., A.F. Ioffe Physico-Technical Institute, Leningrad.

Mazets, E.P., Golenetskii, S.V., Il'inskii, V.N., Guryan, Yu, A., and Kharitonova, T.V.: 1975, *Astrophys, Space Sci.* **33,** 347.

McVittie, G.C.: 1965, *General Relativity and Cosmology,* The University of Illinois Press, Urbana, I11.

Mendell, R.B., Verschell, H.J., Merker, M., Light, E.S., and Korff, S.A.: 1973, *J. Geophys. Res.* **78,** 2763.

Meneguzzi, M. and Reeves, H.: 1973, *Proceedings 13th International Conference Cosmic Rays,* Vol. 1, Denver, Colo., p. 478, (reprint).

Meneguzzi, M. and Reeves, H.: 1975, *Astron. and Astrophys.* **40,** 91.

Merker, M., Light, E.S., Verschell, H.J., Mendell, R.B., and Korff, S.A.: 1973, *J. Geophys. Res.* **78,** 2727.

Metzger, A.E.: 1973, in F.W. Stecker, and J.I. Trombka (eds.), *Gamma Ray Astrophys, NASA SP-339,* p. 97.

Metzger, F.E. and Deutsch, M.: 1950, *Phys. Rev.* **78,** 551.

Metzger, A.E., Anderson, E.C., Van Dilla, M.A. and Arnold, J.R.: 1964, *Nature-Letters to the Editor* **204,** 766.

Metzger, A.E., Parker, R.H., Gilman, D., Peterson, L.E., and Trombka, J.I.: 1974, *Astrophys. J. (Letters)* **194,** L19.

Minkowski, R.: 1964, in L. Goldberg (ed.), *Ann. Rev. of Astron. and Astrophys.,* Vol. 2, Annual Reviews, Inc., Palo Alto, Calif., p. 247.

Morfill, G. and Pieper, G.F.: 1974, in I.B. Strong (ed.), *Proceedings Conference on Transient Cosmic Gamma- and X-Ray Sources, LA-5505-C,* Los Alamos Scientific Laboratory, Los Alamos, N.M., p. 206.

Morgan, D.L., Jr. and Hughes, V.W.: 1970, *Phys. Rev.* **D2,** 1389.

Morrison, P.: 1958, *Nuovo Cimento* **7,** 858.

Morrison, P.: 1966, in R.E. Marshak (ed.), *Perspectives in Modern Physics,* John Wiley and Sons, New York, p. 343.

Morrison, P.: 1967, in L. Goldberg (ed.), *Ann. Rev. Astron. and Astrophys.,* Vol. 5, Annual Reviews, Inc., Palo Alto, Calif. p. 325.

Moteff, J.: 1955, *Nucleonics* **13,** No. 7, McGraw-Hill, Inc. New York, p. 24.

Motz, J.W. and Missoni, G.: 1961, *Phys. Rev.* **124,** 1458.

Motz, H. and Bäckström, G.: 1966, in K. Siegbahn (ed.), *Alpha-, Beta-, and Gamma-Ray Spectroscopy,* Vol. 1, North-Holland Publishing Company, Amsterdam, Holland, p. 769.

Nakagawa, S., Tsukada, M., Okudaira, K., Hirasima, Y., Yoshimori, M., Yamagami, T., Murakami, H., and Iwama, S.: 1971, *Proceedings 12th International Conference Cosmic Rays,* Vol. 1, Hobart, Tasmania, p. 77.

Nakano, G.H., Imhof, W.L., Reagan, J.B., and Johnson, R.G.: 1973, in F.W. Stecker and J.I. Trombka (eds.), *Gamma-Ray Astrophysics, NASA SP-339,* p. 71.

Nelms, A.T.: 1953, *Graphs of the Compton Energy-Angle Relationship and the Klein-Nishina Formula from 10 keV to 500 MeV, NBS-542,* National Bureau of Standards, Washington, D.C.

Novik, R., Weisskopf, M.C., Berthelsdorf, R., Linke, R., and Wolff, R.S.: 1972, *Astrophys. J. (Letters)* **174,** L1.

Oke, J.B. and Searle, L.: 1974, in G.R. Burbidge (ed.), *Ann. Rev. Astron. and Astrophys.,* Vol. 12, Annual Reviews, Inc., Palo Alto, Calif., p. 315.

Omnès, R.: 1969, *Ann. Phys. (Paris)* **4,** 515; See also *Proceedings of the 6th Texas Meeting on Relativistic Astrophysics,* New York, 1972.

Ore, A. and Powell, J.L.: 1949, *Phys. Rev.* **75,** 1696.

Orwig, L.E.: 1971, Doctoral Thesis, University of New Hampshire, Durham, N.H.

Orwig, L.E., Chupp, E.L., and Forrest, D.J.: 1971, *Nature (Phys. Sci.)* **231,** 171.

Pacheco, J.A. DeFreitas: 1973, *Astrophys. Letters* **13,** 97.

Pal, Y.: 1973, in H. Bradt and R. Giaconni (eds.), *X- and Gamma-Ray Astronomy, IAU Symposium No. 55,* D. Reidel Publishing Company, Dordrecht, Holland, p. 279.

Parlier, B., Agrinier, B., Forichon, M., Leray, J.P., Boella, G., Maraschi, L., Buccheri, R., Robba, N.R., and Scarsi, L.: 1973, *Nature (Phys. Sci.)* **242,** 117.

Perlow, G.J. and Kissinger, C.W.: 1951, *Phys. Rev.* **81,** 552.

Peterson, L.E.: 1963, *J. Geophys. Res.* **68,** 979.

Peterson, L.E.: 1965, *J. Geophys. Res.* **70,** 1762.

Peterson, L.E.: 1973, in H. Bradt and R. Giaconni (eds.), *X- and Gamma-Ray Astronomy, IAU Symposium No. 55,* D. Reidel Publishing Company, Dordrecht, Holland, p. 51.

Peterson, L.E.: 1975, in G.R. Burbidge (ed.), *Ann. Rev. Astron. and Astrophys.,* Vol. 13, Annual Reviews, Inc., Palo Alto, Calif., p. 423.

Peterson, L.E., Jerde, R.L., and Jacobson, A.S.: 1967, *AIAA J.* **5,** 1921.

Peterson, L.E., Pelling, R.M., and Matteson, J.L.: 1971, *Techniques in Balloon X-Ray Astronomy.* Presented at the *SPARMO Symposium in Conjunction with XIV COSPAR Plenary Meeting,* June 17 – July 2, Seattle, Washington. Report from University of California San Diego, La Jolla, Calif.

Peterson, L.E., Trombka, J.I., Metzger, A.E., Arnold, J.R., Matteson, J.L., and Reedy, R.C.: 1973a, in F.W. Stecker and J.I. Trombka (eds.), *Gamma-Ray Astrophysics, NASA SP-339,* p. 41.

Peterson, L.E., Schwartz, D.A., and Ling, J.C.: 1973b, *J. Geophys. Res.* **78,** 7942.

Petschek, A.G.: 1967, *Science* **156,** 239.

Pizzichini, G. Palumbo, G.G.C., and Spizzichino, A.: 1975, *Astrophys. J. (Letters)* **195,** L1.

Pollack, J.B. and Fazio, G.G.: 1963, *Phys. Rev.* **131,** 2684.

Pollack, J.B., Guthrie, P.D., and Shen, B.S.P.: 1971, *Astrophys. J. (Letters)* **169,** L113.

Post, R.F.: 1956, *Rev. of Mod. Phys.* **28,** 338.

Pratt, R.H., Levee, R.D., Pexton, R.L., and Aron, W.: 1964, *Phys. Rev.* **134,** A898.

Pratt, R.H., Ron, A. and Tseng, H.K.: 1973, *Rev. Mod. Phys.* **45,** 273.

Preszler, A.M., Simnett, G.M., and White, R.S.: 1974, *J. Geophys. Res. (Space Phys.)* **79,** 17.

Puskin, J.S.: 1970, *Low Energy Gamma Rays in the Atmosphere,* Smithsonian Astrophysical Observatory, Special Report 318, Cambridge, Mass.

Ramaty, R. and Boldt, E.A.: 1971, in S.P. Maran, J.C. Brandt, and T.P. Stecher (eds.), *The Gum Nebula and Related Problems, GSFC X-683-71-375,* Goddard Space Flight Center, Greenbelt, Maryland, p. 97.

Ramaty, R. and Lingenfelter, R.E.: 1973, in R. Ramaty and G. Stone (eds.), *High Energy Phenomena on the Sun, NASA SP-342,* p. 301.

Ramaty, R. and Stone, R.G. (eds.): 1973, *High Energy Phenomena on the Sun, NASA SP-342.*

Ramaty, R. and Lingenfelter, R.E.: 1975, in S.R. Kane (ed.), *Solar Gamma-, X-, and EUV Radiation, IAU Symposium No. 68,* D. Reidel Publishing Company, Dordrecht, Holland, p. 363.

Ramaty, R., Stecker, F.W., and Misra, D.: 1970, *J. Geophys. Res.* **75,** 1141.

Ramaty, R., Borner, G., and Cohen, J.M.: 1973, *Astrophys. J.* **181,** 891.

Ramaty, R., Kozlovsky, B., and Lingenfelter, R.E.: 1975, *Space Sci. Rev.* **18,** 341.

Reedy, R.C., Arnold, J.R., and Trombka, J.I.: 1973, *J. Geophys. Res.* **78,** 5847.

Rees, M.J.: 1966, *Nature* **211,** 468.

Reeves, H., Fowler, W.A., and Hoyle, F.: 1970, *Nature* **226,** 727.

Reina, C., Treves, A., and Tarenghi, M.: 1974, *Astron. and Astrophys.* **32,** 317.

Reppin, C., Chupp, E.L., Forrest, D.J., and Suri, A.N.: 1973, *Proceedings 13th International Conference Cosmic Rays,* University of Denver, Denver, Colo., p. 1577.

Rest, F.G., Reiffel, L., and Stone, C.A.: 1951, *Phys. Rev.* **81,** 894.

Rocchia, R.: 1966, *Rayonnement Gamma Dans L'espace et Dans L'atmosphere,* Ph.D. Thesis, University of Paris, France.

Rocchia, R., Labyrie, J., Ducros, G., and Boclet, D.: 1965, *Proceedings International Conference Cosmic Rays,* London, p. 423.

Roentgen, W.C.: 1895, *Sitzungsber. der Würtzburger Physik – Medic. Gesellschaft.*

Rossi, B.B.: 1952, *High Energy Particles,* Prentice-Hall Inc., New York.

Roughton, N.A., Fritts, M.J., Peterson, R.J., Zaidins, C.S., and Hansen, C.J.: 1974, *Astrophys. J.* **188,** 595.

Roy, R.R. and Reed, R.D.: 1968, *Interactions of Photons and Leptons with Matter,* Academic Press, New York.

Roy, R.R., Goes, M.L., and Berger, J.: 1955, *Compt. Rend. Acad. Sci. Paris* **241**, 1936.
Rudstam, G.: 1966, *Z. Naturforsch.* **21a**, 1027.
Rutherford, E.: 1899, *Phil. Mag.* **47**, 109.
Rutherford, E. and da C. Andrade, E.N.: 1914, *Phil. Mag.* **27**, 854.
Rutherford, E. and da. C. Andrade, E.N.: 1914, *Phil. Mag.* **28**, 263.
Rygg, T.A. and Fishman, G.J.: 1973, *Proceedings 13th International Conference Cosmic Rays*, Vol.2, University of Denver, Denver, Colo. p. 472.
Samimi, J., Share, G.H., and Kinzer, R.L.: 1974, in B.G. Taylor (ed.), *Proceedings of 9th ESLAB Symposium, ESRO SP-106*, Frascati, Italy, p. 211.
Sandhu, H.S., Webb, E.H., Mohanty, R.C., and Roy, R.R.: 1962, *Phys. Rev.* **125**, 1017.
Sargent, W.L.W., Searle, L., and Kowal, C.T.: 1974, in C.B. Cosmovici (ed.), *Supernovae and Supernova Remnants,* D. Reidel Publishing Company, Dordrecht, Holland, 33.
Savedoff, M.P.: 1959, *Nuovo Cimento* **13**, 12.
Schaeffer, O.A. and Zähringer, J.: 1962, *Phys. Rev. (Letters)* **8**, 389.
Scheel, J. and Röhrs, H.: 1972, *Z. Physik.* **256**, 226.
Schnopper, H.W., Bradt, H.V., Rappaport, S., Boughan, E., Burnett, B., Doxsey, R., Mayer, W., and Watt, S.: 1970, *Astrophys. J. (Letters)* **161**, L161.
Schönfelder, V. and Lichti, G.: 1974, *Astrophys. J. (Letters)* **191**, L1.
Schönfelder, V. and Lichti, G.: 1975, *J. Geophys. Res. (Space Physics)* **80**, 3681.
Schönfelder, V., Hirner, A., and Schneider, K.: 1973, *Nucl. Instr. and Methods* **107**, 385.
Schwartz, D.A.: 1970, *Astrophys. J.* **162**, 439.
Schwartz, D.A.: 1974, *Comments on the Diffuse X-Ray Background – 2 – 82 keV, Proceedings of International Conference on X-Rays in Space,* University of Calgary, Calgary, Alberta, Canada, p. 1096.
Schwartz, D.A. and Gursky, H.: 1973, in F.W. Stecker and J.I. Trombka (eds.), *Gamma-Ray Astrophysics, NASA SP-339,* 15.
Schwartz, D.A., Hudson, H.S., and Peterson, L.E.: 1970, *Astrophys. J.* **162**, 431.
Schwinger, J.: 1949, *Phys. Rev.* **75**, 1912.
Sciama, D.W.: 1971, *Modern Cosmology,* Cambridge University Press, Cambridge, Great Britain.
Seeger, P.A., Fowler, W.A., and Clayton, D.D.: 1965, *Astrophys. J. Supplement* **11**, 121.
Seltzer, S.M.: 1975, *Nucl. Instr. and Methods* **127**, 293.
Sérsic, J.L., Pastoriza, G., and Carranza, M.: 1972, *Astrophys. Space Sci.* **19**, 469.
Severnyi, A.B.: 1957, *Sov. Astron. – AJ* **1**, 324.
Severnyi, A.B.: 1958, *Sov. Astron. – AJ* **2**, 310.
Severnyi, A.B.: 1964, in L. Goldberg (ed.), *Ann Rev. of Astron. and Astrophys.*, Vol. 2, Annual Reviews, Inc., Palo Alto, Calif., p. 363.
Severnyi, A.B. and Shabanskii, V.P.: 1961, *Sov. Astron – AJ* **4**, 583.
Shapiro, S.L.: 1973, *Astrophys. J.* **180**, 531.
Share, G.H.: 1975, *Discovery of Two Cosmic X-Ray Bursts in 1970*, Presented at the *Symposium on Fast Transients in X- and Gamma Rays*, Varna, Bulgaria.
Share, G.H., Kinzer, R.L., and Seeman, N.: 1974a, *Astrophys. J.* **187**, 45.
Share, G.H., Kinzer, R.L., and Seeman, N.: 1974b, *Astrophys. J.* **187**, 511.
Share, G.H., Meekins, J.F., and Kreplin, R.W.: 1974c, in B.G. Taylor (ed.), *The Context and Status of Gamma-Ray Astronomy, Proceedings of 9th ESLAB Symposium, ESRO SP-106*, Frascati, Italy, p. 25.
Shima, Y. and Alsmiller, R.G., Jr.: 1970, *Nucl. Sci. and Eng.* **41**, 47.
Shklovsky, I.S.: 1968, *Supernovae,* John Wiley and Sons, New York.
Shvartsman, V.F.: 1970, *Astrofizika* **6**, 123.
Shvartsman, V.F.: 1971, *Sov. Astron. – AJ* **15**, 377.
Silberberg, R. and Tsao, C.H.: 1973a, *Proceedings 13th International Conference Cosmic Rays,* Vol. I, University of Denver, Denver, Colo., p. 528.
Silberberg, R. and Tsao, C.H.: 1973b, *Astrophys. J. Supplement* **25**, 315.
Silberberg, R. and Tsao, C.H.: 1973c, *Cross Sections of Proton-Nucleus Interactions at High Energies,* Naval Research Laboratory *Report No. 7593,* Washington, D.C.

Silk, J.: 1973, in L. Goldberg (ed.), *Ann. Rev. of Astron. and Astrophys.*, Vol. 11, Annual Reviews, Inc., Palo Alto, Calif., p. 269.

Sood, R.K.: 1972, *Sol. Phys.* **23**, 183.

Sood, R.K., Bennett, K., Clayton, P.G., and Rochester, G.K.: 1974, *8th ESLAB Symposium*, Frascati, Italy, to be published.

Staib, J., Frye, G., and Zych, A.: 1973, *Proceedings 13th International Conference Cosmic Rays*, University of Denver, Denver, Colo., p. 916.

Stecker, F.W.: 1969, *Astrophys. Space Sci.* **3**, 579.

Stecker, F.W.: 1970, *Astrophys. Space Sci.* **6**, 377.

Stecker, F.W.: 1971, *Cosmic Gamma-Rays, NASA SP-249* and Monobook Company, Baltimore, Md.

Stecker, F.W.: 1973a, *Astrophys. J.* **185**, 499.

Stecker, F.W.: 1973b, *Nature (Phys. Sci.)* **242**, 59.

Stecker, F.W.: 1973c, in F.W. Stecker and J.I. Trombka (eds.), *Gamma-Ray Astrophysics, NASA SP-339*, p. 211.

Stecker, F.W.: 1975, in J.L. Osborne and A.W. Wolfendale (eds.), *Origin of Cosmic Rays*, D. Reidel Publishing Company, Dordrecht, Holland, p. 267.

Stecker, F.W. and Frost, K.J.: 1973, *Nature (Phys. Sci.)* **245**, 70.

Stecker, F.W. and Trombka, J.I (eds.): 1973, *Gamma-Ray Astrophysics, NASA SP-339.*

Stecker, F.W., Morgan, D.L., Jr., and Bredekamp, J.: 1971a, *Phys. Rev. (Letters)* **27**, 1469.

Stecker, F.W., Morgan, D.L., Jr., and Bredekamp, J.: 1971b, *Cosmic Matter-Antimatter Annihilation and the γ-Ray Background Spectrum, GSFC X-647-71-237*, Goddard Space Flight Center, Greenbelt, Md.

Stecker, F.W., Puget, J.L., Strong, A.W., Bredekamp, J.: 1974, *Astrophys. J. (Letters)* **188**, L59.

Stephens, S.A.: 1970, *Proc. Indian Acad. Sci.* **72**, 214.

Strong, I.B. (ed.): 1974, *Proceedings Conference on Transient Cosmic Gamma- and X-Ray Sources, LA-5505-C*, Los Alamos Scientific Laboratory, Los Alamos, N.M.

Strong, I.B., Klebesadel, R.W., and Olson, R.: 1974, *Astrophys. J. (Letters)* **188**, L1.

Sturrock, P.: 1971, *Astrophys. J.* **164**, 529.

Suri, A.N., Dunphy, P.P., Chupp, E.L., and Forrest, D.J.: 1974, *Bull. Am. Phys. Soc.* **19**, 531 and an internal UNH report.

Suri, A.N., Chupp, E.L., Forrest, D.J., and Reppin, C.: 1975, *Solar Phys.* **43**, 415.

Thompson, D.J.: 1974, *J. Geophys. Res.* **79**, 1309.

Thompson, D.J., Bignami, G.F., Fichtel, C.E., and Kniffen, D.A.: 1974, *Goddard Space Flight Center Report, GSFC X-662-74-58*, Greenbelt, Md.

Trombka, J.I.: 1970, *Nature* **226**, 827.

Trombka, J.I., Senftle, F., and Schmadebeck, R.: 1970, *Nucl. Instr. and Methods* **87**, 37.

Trombka, J.I., Metzger, A.E., Arnold, J.R., Matteson, J.L., Reedy, R.C., and Peterson, L.E.: 1973, *Astrophys. J.* **181**, 737.

Trombka, J.I., Eller, E.L., Schmadebeck, R.L., Adler, I., Metzger, A.E., Gilman, D., Gorenstein, P., and Bjorkholm, P.: 1974, *Astrophys. J. (Letters)* **194**, L27.

Truran, J.W., Arnett, W.D., and Cameron, A.G.W.: 1967, *Can. J. Phys.* **45**, 2315.

Ullrich, H., Boschitz, E.T., Engelhardt, H.D. and Lewis, C.W.: 1974, *Phys. Rev. (Letters)* **33**, 433.

Ulmer, M.P., Baity, W.A., Wheaton, W.A., and Peterson, L.E.: 1974, *Astrophys. J.* **193**, 535.

Unsöld, A.: 1969, *The New Cosmos*, translated by W.H. McCrea, Springer-Verlag, New York.

Valdez, J.V., Freier, P.S., and Waddington, C.J.: 1970, *Acta Physica Academiae Scientiarum Hungaricae*, Vol. 29, Suppl. 1, p. 79.

Van Beek, H.F.: 1973, Doctoral Thesis, University of Utrecht, Holland.

Van Lieshout, R., Wapstra, A.H., Ricci, R.A., and Girgis, R.K.: 1966, in K. Siegbahn (ed.), *Alpha-, Beta- and Gamma-Ray Spectroscopy*, Vol. I, North-Holland Publishing Company, Amsterdam, Holland, p. 510.

Vedrenne, G., Albernhe, F., Martin, I., and Talon, R.: 1971, *Astron. and Astrophys.* **15**, 50.

Vette, J.I.: 1962, *J. Geophys. Res.* **67**, 1731.

Vette, J.I., Gruber, D., Matteson, J.L., and Peterson L.E.: 1970, *Astrophys. J. (Letters)* **160**, L161.

Villard, P.: 1900, *Compt. Rend. Acad. Sci. Paris* **130**, 1010.

Von Buttlar, H.: 1968, *Nuclear Physics – An Introduction*, Academic Press, New York.

Waddington, C.J.: 1969, *The Fragmentation of Cosmic Ray Nuclei in Interstellar Hydrogen,* University of Minnesota, preprint.

Wang, H.T.: 1975, Doctoral Thesis, University of Maryland, College Park, Md.

Wang, H.T. and Ramaty, R.: 1974, *Solar Phys.* **36**, 129.

Webber, W.R.: 1967, *Handbuch der Physik* **46**, Springer-Verlag, Berlin, p. 181.

Weekes, T.C.: 1969, *High-Energy Astrophysics*, Chapman and Hall, London.

Wheaton, W.A., Ulmer, M.P., Baity, W.A., Datlowe, D.W., Elcan, M.J., Peterson, L.E., Klebesadel, R.W., Strong, I.B., Cline, T.L., and Desai, U.D.: 1973, *Astrophys. J. (Letters)* **185,** L57.

White, R.S.: 1973, *Rev. Geophys. and Space Phys.* **11,** 595.

White, R.S. and Schönfelder, V.: 1975, *Astrophys. and Space Sci.,* **38,** 19.

White, R.S., Herzo, D., Koga, R. and Simnett, G.: 1973, *Atmospheric Gamma Rays from 1.5 to 15 MeV,* Paper presented at *13th International Conference on Cosmic Rays,* University of Denver, Denver, Colo., unpublished.

Williams, D.J.: 1972, in E.R. Dyer and J.G. Roederer (eds.), *Part III of Solar Terrestrial Physics, 1970,* D. Reidel Publishing Company, Dordrecht, Holland, p. 66.

Womack, E.A., and Overbeck, J.W.: 1970, *J. Geophys. Res* **75,** 1811.

Yang, C.N.: 1950, *Phys. Rev.* **77,** 242.

Zeldovich, Ya. B. and Novikov, I.D.: 1971, *Relativistic Astrophysics.* Vol. 1, University of Chicago Press, Chicago, Ill.

Zobel, W., Maienschein, F.C., Todd, J.H., and Chapman, G.T.: 1968, *Nucl. Sci. and Eng.* **32,** 392.

Zwicky, F.: 1965, in L.H. Aller and D.B. McLaughlin (eds.), *Stellar Structure,* Vol. 8, The University of Chicago Press, Chicago, Ill., p. 367.

Zwicky, F.: 1969, in P.J. Brancazio and A.G.W. Cameron (eds.), *Supernovae and Their Remnants,* Gordon and Breach Publishers, New York, p. 1.

Zych, A.D., Herzo, D., Koga, R., Millard, W.A., Moon, S., Ryan, J., Wilson, R., White, R.S., and Dayton, B.: 1975, *IEEE Trans. on Nucl. Sci.* **NS-22,** 605.

INDEX OF SUBJECTS

GEOPHYSICS AND ASTROPHYSICS MONOGRAPHS

AN INTERNATIONAL SERIES OF FUNDAMENTAL TEXTBOOKS

1. R. Grant Athay, *Radiation Transport in Spectral Lines.* 1972, XIII + 263 pp.
2. J. Coulomb, *Sea Floor Spreading and Continental Drift.* 1972, X + 184 pp.
3. G. T. Csanady, *Turbulent Diffusion in the Environment*, 1973, XII + 248 pp.
4. F. E. Roach and Janet L. Gordon, *The Light of the Night Sky.* 1973, XII + 125 pp.
5. H. Alfvén and G. Arrhenius, *Structure and Evolutionary History of the Solar System*, 1975, XVI + 276 pp.
6. J. Iribarne and W. Godson, *Atmospheric Thermodynamics*, 1973, X + 222 pp.
7. Z. Kopal, *The Moon in the Post-Apollo Era*, 1974, VIII + 233 pp.
8. Z. Švestka, *Solar Flares*, 1976, XV + 376 pp.
9. A. Vallance Jones, *Aurora*, 1974, XVI + 301 pp.
10. C.-J. Allègre and G. Michard, *Introduction to Geochemistry*, 1974, XI + 142 pp.
11. J. Kleczek, *The Universe*, 1976, VII + 259 pp.
12. E. Tandberg-Hanssen, *Solar Prominences*, 1974, XVI + 155 pp.

Forthcoming:

A. Giraud and M. Petit, *Physics of the Earth's Ionosphere.*
J. Audouze and S. Vauclair, *Nuclear Astrophysics.*
W. D. Heinz, *Double Stars.*
M. M. Millán, *Dispersive Correlation Spectroscopy for In-situ and Remote Pollution Monitoring.*